Entrepreneurship in Agricultural Development

The Editors

Dr. Dipak De is Professor, Department of Extension Education, Institute of Agricultural Sciences, Banaras Hindu University, Varanasi, Uttar Pradesh, India. He completed his Post Graduation and PhD from Indian Agricultural Research Institute, Pre PhD from Indian Institute of Technology, New Delhi. He was awarded IARI Junior and Senior Research Fellowships and IIT Senior Research Fellowship. His contribution towards academic growth includes a) Axiomatic Theory in adoption of farm technology b) Member of the Information Systems and their characteristics c) Axiomatic theory of communication behaviour and d) Axiomatic Theory of Entrepreneurial behaviour of farmers. Prof.De has published more than 75 research papers, written 10 book chapters, Authored 6 books and reviewed 2 books.

Dr. Basavaprabhu Jirli did his graduation and post graduation from University of Agricultural Sciences, Bangalore, obtained PhD from National Dairy Research Institute, Karnal. Earlier served as Documentation Scientist at Karnataka Rajya Vigyana Parishat, Banagalore, Research Associate, ICAR-National Research Centre for Rapeseed Mustard, Bharatpur and Assistant Professor in Central Agricultural University. Presently working as Associate Professor in the Department of Extension Education, Institute of Agricultural Sciences, Banaras Hindu University. He has been engaged in teaching and research for more than 14 years. Have more than 30 research publications, five edited volumes of proceedings and three edited and two authored books in his credit.

Dr. N.R. Gangadharappa is working as Professor and Head in the Department of Agricultural Extension, University of Agricultural Sciences, Bangalore. He has focused on Agriculture extension research in areas of Women empowerment, Natural resource management, Empowerment of backward class and resource poor farmers, performance evaluation of various extension and outreach activities. Gangadharappa is a contributing author of 76 research publications in refereed journals, 21 full length research publications in National and International seminars workshops and 20 other publications.

Entrepreneurship in Agricultural Development

– Editors –

Dipak De
Basavaprabhu Jirli
Gangadharappa N.R.

2017

Daya Publishing House®

A Division of

Astral International Pvt. Ltd.

New Delhi – 110 002

Cataloging in Publication Data--DK
Courtesy: D.K. Agencies (P) Ltd. <docinfo@dkagencies.com>

Entrepreneurship in agricultural development / editors, Dipak De, Basavaprabhu Jirli, Gangadharappa N.R.
    pages cm
    Includes bibliographical references and index.
    ISBN 978-93-86071-25-5 (International Edition)

    1. Entrepreneurship--India. 2. Agriculture--Economic aspects--India.   I. De, Dipak, editor. II. Jirli, Basavaprabhu, editor. III. Gangadharappa, N. R., editor.

    HB615.E58 2016   DDC 338.040954   23

Published by   : **Daya Publishing House®**
                                 A Division of
                                 Astral International Pvt. Ltd.
                                 – ISO 9001:2015 Certified Company –
                                 4736/23, Ansari Road, Darya Ganj
                                 New Delhi-110 002
                                 Ph. 011-43549197, 23278134
                                 E-mail: info@astralint.com
                                 Website: www.astralint.com

Preface

"A farmer does not become an entrepreneur only by adopting a new agricultural technology but he becomes an entrepreneur only when he comes to be an operator of a farm business"

Rural development is more than ever before linked to entrepreneurship development. Institutions and individuals promoting rural development now see entrepreneurship as a strategic development intervention that could accelerate the agricultural and rural development process. Furthermore, institutions and individuals seem to agree on the urgent need to promote agricultural enterprises: development agencies see agricultural entrepreneurship as an enormous employment potential; politicians see it as the key strategy to prevent rural unrest; farmers see it as an instrument for improving farm earnings; and women see it as an employment possibility near their homes which provides autonomy, independence and a reduced need for social support. To all these groups, however, entrepreneurship stands as a vehicle to improve the quality of life for individuals, families and communities and to sustain a healthy economy and environment. Entrepreneurship education is often seen as a special kind of training to become an entrepreneur or to enlarge job prospects of the individual. In a broader view entrepreneurial attitudes and skills shall be promoted, in a narrow sense a specific training on how to create a business is meant.

An entrepreneur is one who combines the land of one, labor of another and the capital of yet another, and, produces a product. By selling the product in the market, he pays interest on capital, rent on land and wages to laborers. What remains is his or her profit. Around 80 percent of entrepreneurs fail with their business enterprise. To be successful your product needs to be new to the consumer, and something you think will sell.

This book provides more information regarding entrepreneurship development that students want or need. It contains theory and practical information. It is oriented towards students who not ony enjoy science and technology but also study. It will provide students to meet out wide range of challanges in the class room or the work place.

The book has been organised in eight sections namely conceptual issues in entrepreneurship development, managerial aspects of enterprise development, legal issues in enterprise development, information and communication technologies (ICT) for enterprise development, intellectual property rights (IPR) and enterprise development, prospective areas of enterprise promotion and success stories of enterprise development. Under these sections different aspects are contributed by learned authors. We hope that a publication of such magnitude will act as reference material for students and researchers on entrepreneurship development. We believe that this book will try to find effective guidance to satisfy the requirements of the students.

We are thankful to all the authors for contributing valuable information in the form of book chapters. We are indebted to PFC, TIFAC, DST, New Delhi for according permission to reprint the topic related to IPR.

Dipak De
Basavaprabhu Jirli
Gangadharappa N.R.

Contents

Management Issues in Entrepreneurship Development

Information and Communication Technologies (ICT)
in Entrepreneurship Development

Success Stories of Entrepreneurship

Conceptual Issues in Entrepreneurship Development

Chapter 1

Entrepreneurship: A Conceptual Framework

Dipak De, M.S. Rao and Basavaprabhu Jirli

Department of Extension Education, Institute of Agricultural Sciences
Banaras Hindu University, Varanasi – 221 005, U.P.

Introduction

All round development of agriculture is possible only with effective exploitation of entrepreneurial behaviour skills as well as material resources. But, our country is scarce of material resources but abundant of human resources. So, we can identify individuals in all segments of the population, who have the requisite entrepreneurial behaviour skills. The entrepreneur is an economic man, who tries to maxinlise his profits by identification and adoption of innovations. However, the entrepreneurs are not simply innovators but they are the persons with a will to act to assure risk and bring about a change through organisation of human efforts. It plays a key role in economic development of the country. Importance of development of entrepreneurship as an ingredient of economic development has been recognised long time back (Hmachalam, 1990). It was as early as in 1950, that the need for entrepreneurial behaviour development was first felt and since then substantial amount of research has been done in this sphere. This chapter has been written in the following subheads namely: Concept of entrepreneur, concept of entrepreneurship, concept of entrepreneurial behaviour and factors affecting entrepreneurial behaviour.

Concepts of Entrepreneur

An overview of literature pertaining to concept of entrepreneur is absolutely essential. The word entrepreneur is derived from **French** word **'Entreprendre'** and the **German** word **"Unternehmen"**, both mean **"To undertake"**. For a long time there was no equivalent for the term 'entrepreneur' in the English language.

Three words were commonly used to connote the sense the French term carried: adventurer, undertaker and projector; these were used interchangeably and lacked the precision and characteristics of a scientific expression (Gopakumar, 1995). Hence the term 'entrepreneur' did not find any prominence in the history of economic thought. The earliest attempt to invest the concept with some economic content could be traced to the works of an 18th century French writer, Bernard F. de Belidor, who defines entrepreneurship as buying labour and materials at uncertain prices and selling the resultant output at contracted prices (Hoselitz, 1960). Entrepreneurship as a concept gathered prominence in economic literature mainly through the writings of Richard Cantillon (1680-1734), who gave the concept some analytical treatment and assigned the entrepreneur an economic role by emphasizing on 'risk' as a prominent entrepreneurial function (Gopakumar, 1995). According to Encyclopaedia Britanica entrepreneur means individual response for the operation of a business, including the choice of a product, the mobilization of necessary capital, decisions on product prices and quantities, the employment of labour and expanding or reducing the productive facilities.

According to Weber (1930), entrepreneurs are a product of the particular social conditions in which they live, and it is the society which shapes the personality of individuals as entrepreneurs. Entrepreneur represents an individual or a group of individuals, who conceive, initiate and maintain for a sufficiently long period of time a social institution which produces economic goods (Cole, 1949) or to put it differently, who perceive a business opportunity and create an organization to pursue it (Bygrave and Hofer, 1991).

Cole (1959) described entrepreneur as decision maker and indicated the following functions of an entrepreneur.

1. The determination of those objectives of the enterprise and the change of those objectives as conditions required or made advantages.
2. The development of an organisation including efficient relationship with subordinates and all employees.
3. The securing of adequate financial resources the relation with existing and potential investors.
4. The requisition of efficient technological equipment and the revision of it as new machinery appeared.
5. The development of market for the products and devising of new products to meet or anticipate consumer demand.
6. The maintenance of good relationship with public authorities and with society at large.

According to Schumpter (1961) an entrepreneur was a dynamic agent of change, or the catalyst who transformed increasingly physical, natural and human resources into corresponding production possibilities.

David (1962) opined that the prime motive of the entrepreneurs was to accumulate and without motive there should be no accumulation to facilitate capital formation and economic development. Entrepreneurs universally acknowledged as a new men, provide the industrial push to a society and project it in to the path of economic growth and modernization.

According to Hagen (1962) entrepreneur as a creative problem solver interested in things in the practical and technological realm most entrepreneurial activities do not involve innovative techniques to any considerable degree but rather involve coping with the method of doing business and of combining inputs quite similar to those combinations already in existence.

According to Harbison and Hyers (1964) capital cannot itself produce anything. It must be harnessed for producing goods and the capital goods into harnessed for producing consumer goods and this investment process requires the services of some agent or intermediary who initiates, organises, makes decision, takes risk, innovates and some time also manages.

According to Cole (1968) the term entrepreneur represents an individual or a group of individuals who conceive, initiate and maintain for a sufficiently long period of time a social institution which produces economic goods or to put it differently, who perceive a business opportunity and create an organisation to pursue it.

According to Joshi and Kapur (1973) farm entrepreneur as the person or a group of persons who organises and operate the business and is responsible for the results. *i.e.* losses and gains from the business. and is pioneer in organising and developing the farmers.

Rao (1975) opined that any person actively engaged in inventing or developing or expanding or effectively maintaining an organisation was an entrepreneur. According to him if a person is starting a new organisation, developing it or expanding it, the very act of understanding these activities qualifies him to be called as entrepreneur.

According to Leeds and Stainton (1978) entrepreneur as a person who initiates production, takes decision bears risks and involve in organising and coordinating the other factors.

Pareek and Nadkarni (1978) defined entrepreneur as one who initiates, establishes an economic activity or enterprise. Entrepreneurship thus refers to the general trend setting up new enterprise in a society. Operationally entrepreneurship development could mean development of entrepreneurs and promotion of increased flow of individuals to entrepreneurial ranks.

Haredero (1979) described agricultural entrepreneur as a person who introduces changes which directly or indirectly lead to higher agricultural inputs.

Kirzner (1979) stated the entrepreneurs as performing various functional roles as risk taker, decision maker, organizer or coordinator, innovator, employer of factors of production, gap seeker and input completer, arbitrageur. According to him entrepreneur being alert to economic opportunities uses information advantages for his own profits.

Massie (1982) stated that entrepreneur means a risk taker one who promotes a business activity.

Battacharya (1983) observed that the South-East Asian Countries in their efforts for developing agriculture have provided some area's of their countryside with better infrastructure in relation to other area's irrigation, communication and cooperative institutions. Some fanners have responded to the new enterprise commonly associated with the use of fertilizer, pesticides better than other fanners in these area's of those countries. The fanner group of fanners were known as agricultural entrepreneurs.

According to Drucker (1985) entrepreneur is one who always searches for change, responds to it, and exploits it as an opportunity. Entrepreneurs innovate and innovation is a specific instrument of entrepreneurship.

De (1986) stated that a farmer does not become an entrepreneur only by adopting a new agricultural technology but he becomes an entrepreneur only when he comes to be an operator of a farm business. A business involves rational decisions on investment after assessing risk, other alternatives and possibilities or profit and loss. An entrepreneur is a dynamic agent of change or the catalyst who increasingly transforms the physical, natural, and human resources into corresponding production possibilities.

Singh (1986) stated that those were the concepts of individuals who grasped an idea, developed it and.perused it's success doggedly with unflagging spirit. These individuals were entrepreneurs or the man who organises the business unit and increases it's productive capacity. According to Mokry (1988) entrepreneur as "a symbol of individualism and mysterious link between idea and product, has gained visibility as attention on began to focus on innovation competitiveness and productivity".

According to Shailendra (1990) a rural entrepreneur is some one who is prepared to take risk for self betterment, but is also willing to give of himself for the community by staying and creating local wealth.

Bisht and Sharma (1991) stated that economic development is the outcome of production or industrialisation, and production or industrialisation leads us to man-known in the business world as an "Entrepreneur" the change producing force in economic life. Entrepreneurs are the persons who initiate, organise, manage and control the affairs of a business unit that combine the factors of production to supply goods and services, whether business pertains to agriculture, industry, trade or profession.

Singh (1992) stated that important business motivation for women is the need to provide security and to the family. Women entrepreneurs are often motivated by a desire to have flexibility in their work and family. The major sources of motivation for women could thus be need to achieve, desire to be independent, need for Job satisfaction, economic necessity, desire to make use of one's talent or skill and the desire to be one's own boss.

Jyothi and Prasad (1993) stated that an entrepreneur is that person who put's together resources and takes great strain to start a new business venture.

According to Sundari (1995) an entrepreneur is a person who organises, runs and is responsible for a business enterprise to make a profit. In the process, he/she has different roles to play and differentiation to take.

According to Busenitz (1996) entrepreneurs are more alert to new opportunities and use information differently. It further, indicates that entrepreneurs make a habit of scanning their environment for information that may lead to new business opportunities.

Jhamtani (1996) opined that entrepreneur is a person who organizes, manages and assumes the risk of a business. Entrepreneurs are self employed and income generating persons. But all self employed and income generating persons may not necessarily be self employee.

According to Stoner (1996) entrepreneur is the ability to take the factors of production land, labour and capital -and use them to produce new goods or services or the originator of a new business venture and a new organisation for that venture.

Polo (1996) specifies the strategic role of the entrepreneur in the current economic and socio cultural context. A complex and articulated entrepreneurial function has come to playa crucial role in modern enterprise, particularly in public companies with many share holders.

According to Agarwal (1997) there has been rapid globalisation of the Indian economy, so much so that growth and prosperity in the years ahead. Call for a totally new breed of managers. We now' need manager entrepreneurs, not manager administrators. It has become essential for professional management training institutes to impart new skills to manager who will have to be more people/customer oriented rather than boss oriented. They will have to identify themselves the quality rather than quantity. According to Richart and Maurer (1997) entrepreneurs are defined as those individuals who have started or purchased a small business and who are still leading the business they started or purchased.

According to Scott (1997) entrepreneur is one who takes risks to initiate business activity.

Khanka (1998) opined that who has an urge to do (or) create something new, organize production, undertake risk and handle the economic uncertainty involved in running enterprise is called entrepreneur. The set of such attributes the entrepreneur possesses is called entrepreneurship. Entrepreneurship refers to enterprising or achieving attitude.

Manimala (1998) stated that the entrepreneurial process involves organising scarce resources from the environment. Personal networks are an important means for the entrepreneur to secure such resources. Entrepreneurs make use of their personal and professional networks. The analyses of the anecdotes are grouped as follows:

1. Search for new ideas
2. Expertise development

3. Mobilising funds
4. Organising for the initial production
5. Marketing through net works.
6. Acquiring/developing people
7. Building the corporate image.
8. Management of risk
9. Management of growth

Ramana and Papaiah (1998) stated that an entrepreneur is an economic leader who possess the ability to recognize opportunities for the successful introduction of new commodities, new techniques and new source of supply, and to assemble the necessary plant and equipment, management and labour force, and organise them into a running concern. Whatever be the economic and political set up of a country, entrepreneurship is essential for economic development. According to them entrepreneur as a recipient of pore profit. and is bearing the cost of uncertainty. and identifies uncertainty with a situation where the probabilities of alternative outcomes cannot be determined either by a priori reasoning or by statistical inference. A priori reasoning is simply irrelevant to economic situation involving a unique event or the entrepreneur is the prime mover in economic development his.function, to innovate or carry out new combinations or entrepreneurs are individuals motivated by a will for power. Their special characteristic being an inherent capacity to select correct answers, energy, will and mind to overcome fixed talents of thoughts, and a capacity to with stand social opposition. Two main roles for the entrepreneur is which involves making available inputs that improve the efficiency of existing production methods or facilitate the introduction of new ones. The role of the entrepreneur is to improve the flow of information in the market. The second role gap filling is closely a kin to the arbitrage function.

According to Ripsas (1998) entrepreneur is a bearer of uncertainty who is compensated for by the residual income called profits. According to him primary function of the entrepreneur is to decide what to do and how to do it without being certain about possible future benefits but according to schumpeter entrepreneur is not the risk bearer but the driving force in economic development from with in the economy: According to him entrepreneurship as the carrying out of innovations. For an entrepreneur the motivation is intrinsic and not driven by the desire to make profit.

Sara (1998) concluded that farm household productivity was a major factor in enabling the survival of small scale agricultural production. Many farm businesses combine agricultural production with other income generating activities and such multiplicity of activities has always been an important and distinctive feature of the farming sector. That's why the farmer must be treated as a part of rural entrepreneurial class.

Prakasam (1998) concluded that the entrepreneur, being an economic man aims to optimise his profits through innovative means. He has a will to act, is ready to assume risk and to organize human resources for production of foods and services which in turn generate and multiply economic and social activities that induce change in the area and in the society.

Ramana (1999) entrepreneurs are defined as those people who work for themselves are called "entrepreneurs". The.word entrepreneur is derived from French word "Entreprendre" meaning to "Undertake". The entrepreneur is thus a person who organizes and manages in an activity/organisation, undertaking the risks for fulfilling some of his needs. His job involves the quality of boldness, courage, dynamism and risks taking in sufficient measure. The r' entrepreneur in this context is defined as one who could start a new activity or a new enterprise which is a deviation from his traditional family occupation or profession.

Many individuals are born with "entrepreneurial DNA" *i.e.*, they exhibit a predisposition to entrepreneurship. In fact, early studies found a strong correlation between entrepreneurial predisposition or propensity and firm start-up decisions (Learned, 1992).

Professor of Psychology Alan Jacobowitz, believes that entrepreneurs are born, not made (Cohen, 1980). Through interviews with over 500 entrepreneurs over a three-year period, Jacobowitz observed that entrepreneurs commonly share certain personality characteristics. These include: restlessness, independence, a tendency to be a loner, and extreme self-confidence.

There is a distinct difference between possessing entrepreneurial tendencies and acting on them. Many latent entrepreneurs reach a level of comfort in their jobs and careers that may never result in an outward expression of entrepreneurship, while others grow bored of routine and seek out new challenges. Many entrepreneurs are born out of an "event." This event could take many forms including: losing one's job, threat of bankruptcy, loss of a significant other, frustration on the job, and discovering a marketplace gap (Knudson. *et al.,* 2004).

Concept of Entrepreneurship

An overview of literature pertaining to concept of; entrepreneurship is absolutely essential. Cole (1949) stated that entrepreneurship comprises any purposeful activity that initiates maintains or develops a profit-oriented business in interaction with the internal situation of the business or with the economic, political and social circumstances surrounding the business.

According to Diamond (1957) entrepreneurship is equivalent to "enterprise" which involves the willingness to assume risks in undertaking an economic activity particularly new one it may involve an innovation but not necessary so. It always involves risk taking and decision-making, although neither risk nor decision making may be of great significance.

Hostelits (1957) opined that entrepreneurship was associated with personality pattern in which achievement motivation was strong. But the presence of strong achievement motivation in a group of individuals did not necessarily produce an abundance of entrepreneurs unless certain other conditions of social structure and culture strongly favour achievement oriented individuals.

Hagen (1968) stated that entrepreneurship is not only the conceiving of the idea behind a venture but also designing and maintaining the organisation for carrying it out.

According to Schumpeter (1970) entrepreneurship is

☆ A function of group level pattern.

☆ A function of managerial skill and leadership.

☆ An organisational building function.

☆ A function of high achievement.

☆ Input-completing and gap filling.

☆ A function of status withdrawal, function.

☆ A function of social, political and economic structure.

Nandi (1973) found that for successful entrepreneurship, a high need for independence and a high need for influencing other is required.

Bhatt (1974) described at broad level the entrepreneurship in social organisation as a highly complex process. It is 'the result of interaction of various characteristics like natural endowments, historical tradition, educational-and cultural standards, social stratification, religious and moral values, family organizations development at any given period of time. According to Gaikwad (1978) entrepreneurship connotes innovativeness, an urge to take risk in face of uncertainties, and intuition *i.e.* a capacity of seeing things in a way which afterwards proves to be true.

According to Pareek and Nadkami (1978) entrepreneurship is not simply adoption of a new activity or a practice. It is transformation of a person his acquiring a new identity.

According to Rao and Mehta (1978) entrepreneurship can be described as a creative and innovative response to the environment. Such response can take place in any field of social endeavour-business, industry, agriculture, education, social work, and the like. Being new things or doing things that are already being done in a new way is therefore, a simple definition of entrepreneurship.

Rao and Mehta (1978) opined that entrepreneurship is a package of personality characteristics of entrepreneurs.

De (1981) concluded that entrepreneurship is a package of personality characteristics of entrepreneurs. The characteristics conventionally associated with entrepreneurship -leadership, innovativeness, risk taking and so on -are so associated precisely because, in a profitable farming culture, they are essential features of effective farm business.

Carland (1984) stated that nature or intensity of entrepreneurship would depend on the intensity of the complete set of behavioural dispositions related to being an entrepreneur.

Rani (1986) concluded that women took up entrepreneurship to fulfill economic needs and to satisfy some of their personality needs like power and achievement and to gain a novel experience.

Singh (1986) stated that doing new things or doing things that are already being done in a new way is entrepreneurship.

Vmze (1987) revealed that emergence of entrepreneurship depends on certain personality traits. The traits and attributes like involvement, decision making, access to and control over resources, calculated risk taking, innovativeness, need for achievement, placing family and friends second to business *etc*. Increase probability of an entrepreneur emerging out successful and therefore, should be possessed by them to be able to perceive an opportunity and translate that into productive enterprise.

Stevenson and Jarillo (1990) observed that entrepreneurship literature can be identified to address by and large three main questions.

1. What happens to the business and the economy when entrepreneurs act? *i.e.* the outcome of entrepreneurial activities.

2. Why do they act as entrepreneur? *i.e.* the cause of entrepreneurial action of individuals.

3. How do they act as entrepreneurs? *i.e.* the process of entrepreneurship. Entrepreneurship broadly involves identification of market opportunities and innovative creation of combinations of resources for altering the aggregate economy.

Bisht and Sharma (1991) stated that increased production and small industry strategy is obviously the outcome of human activity. Human activity in this context is known as entrepreneurship.

Covin and Slevin (1991) stated that entrepreneurship is a firm behaviour. Entrepreneurship is a phenomenon of emergence, it evolves overtime (Gartner *et al.*, 1992; Learned, 1992).

According to Gupta and Srinivasan (1992) entrepreneurship is described as the function of handling economic activity, undertaking risk, creating something new and organising and coordinating resources. According to Bull and Willard (1993) entrepreneurship is the carrying out of new combinations causing discontinuity.

Mall (1993) opined that entrepreneurship is an important factor in economic development. It is now widely recognised. While economic development depends largely on the rate of applied technical advancement, innovation and the level of technical progress in the economy, indicating applied technical advancement depend on the supply of entrepreneurs in society.

According to Thomas and Padmakar (1993) entrepreneurship in an individual is to productively integrated resources and enhances economic growth. Need for achievement (n -Ach) as the factor that instigates people to be entrepreneurial and venture into innovative and productive activities is enhancing economic growth. Entrepreneurship is that factor which urges an individual to take advantage of favourable situations by understanding innovative practices with a concern for excellence and assessment of self and the environment. Entrepreneurship is a way of life, a thought process, to bring any sustainable change; effort has to be broader based.

According to Ali (1995) entrepreneurship is considered to be a process with entrepreneurs making the links between opportunities and resources.

According to Pareek and Rao (1995) entrepreneurship is not only a career but also a way of life. A person taking up this career undergoes a transformation in his or her life style.

Patel (1995) stated that entrepreneurship would lead to generation of more income, reduce unemployment/underemployment, minimise incidence of poverty, reduce regional imbalance and promote export trade. This is the area which can provide self employment to illiterate, literate, semiskilled men and women. Entrepreneurs in rural areas, may they be men or women, can now very well, through adoption of new agricultural strategy/technology, not only improve the productivity and production of rural resources (*viz.*, land, labour, livestock, vegetation, forestry, water, fisheries, fruits, flowers, vegetables, milk, mushrooms sericulture *etc.*) but also enhance income substantially, beyond one can expect, l-Hi-tech projects in the area of aquaculture, floriculture, micro-irrigation, bio-fertilizer, embryo-transfer technology, *etc.* have established their unique place in the wake of liberalisation and globalisation of Indian economy.

According to Stevenson and Jarillo (1995) entrepreneurship is a process by which individuals either on their own or inside organizations pursue opportunities without regard to the resources they currently control.

According to Jhaintani (1996) entrepreneurship refers to identifying/innovating ideas, product and services, mobilizing resources, organising production/services and finally marketing those covering the risk with constant strive for growth and excellence.

Bankston and Zhou (1996) examined the development of the fishing industry as a factor for ethnic entrepreneurship among the Vietnamese in using 1980 and 1990 census data. Findings demonstrated that the fishing industry become a major occupational concentration if Vietnamese, 1980 and 90, and self employment in this new ethnic group increased sharply during the decade. This extractive industry also became the largest single source of self employment for Vietnamese, since 33.4 per cent of all self employed Vietnamese were in fishing, and 66 per cent of those were self employed.

Carrier (1996) stated that entrepreneurship was a method of stimulating innovation and using the creative energy of employees by giving them the resources and independence they needed to innovative within the firm or the introduction and implementation of a significant innovation by one or more employees working within an organisation.

Lumpkin and Dess (1996) stated that entrepreneurship constituted new entry.

Prasad (1996) stated that the objective of the entrepreneurship development agency (EDA) was to promote general awareness among the rural masses about setting up a small business or it might be to attract a specific target group like women for an entrepreneurship development programming (EDP).

According to Ray and Rama Chandran (1996) entrepreneurship is an individual's response to a situation *i.e.* the environment around him, and the creation of an organisation is essential for the carrying through of that response, then the entrepreneur, the environment and the organisation must be regarded as crucial elements in any frame work relating to entrepreneurship.

According to Stoner (1996) entrepreneurship is a discontinuous process of combining resources to produce new goods or services. The benefits of entrepreneurship is (1) economic growth (2) productivity (3) create new technologies, products andst1rvices and it changes and rejuvenate market competition.

According to Banerjee and Talukdar (1997) entrepreneurship means a quality destined to increase production spontaneously.

According to Patel and Sanoria (1997) entrepreneurship is a form of human behaviour and this in turn is a manner of acting of an individual in a given situation to achieve certain results. Entrepreneurship is the propensity of mind to take calculated risks with confidence to achieve a predetermined enterprise objective. In instance, it is risk taking ability of the individual broadly coupled with rational decision making to increase production in agriculture business, industry *etc*.

Krishnan and Kumar (1998) defines entrepreneurship as buying labour and materials at uncertain prices and selling the resultant output at contracted price.

Wick (1998) entrepreneurship refers to the intentional creation or transformation of an organization for the purpose of creating or adding value through organization of resources values added by entrepreneurial behaviour result in increases in personal wealth, increases in cash flow for a region, new or better jobs and new or better products or services.

Khanka (1995) states that making the entrepreneurship is a complex phenomenon. This is a crystallisation of social milieu from which entrepreneurs come, family imbibes~' make up of their minds, personal attitudes, caste system, educational background, family occupation and so on. In fact, several factors go in to the making of an entrepreneur and thereby entrepreneurial society, these are grouped under three major headings:

1. Stimulation
2. Support
3. Sustaining

Manimekalai (1998) opined that entrepreneurship development among the rural labour force would strengthen the village economy, promote regional development, bring job diversification, and relieve the dependence on agriculture. The small scale agricultural production in Trichy district was becoming increasingly unprofitable due to the continuing drought and large scale agricultural production has been transforming it self into an agro business in Trichy district. All these suggest that, there has been a need for self employed industrial activity or non farm activity as an alternative to agricultural employment as agriculture fails to support the dependent population.

According to Ran1ana (1999) entrepreneurship is a purposeful activity indulged in initiating, promoting and maintaining economic activities for the production and distribution of wealth. The individual as an entrepreneur is a critical factor in economic development and an integral part of socio-economic transformation. Therefore, the basic concept of entrepreneurship connotes effectiveness, an urge to take risks in the face of uncertainties and intuition.

According to Ma and Tan (2006) entrepreneurship is the process in which pioneers, innovators or champions of innovation, immersed in and guided by the creativity-oriented perspective, engage in the practice of creation and innovation driven activities, which lead to a certain level of performance as indicated by the realized creation and innovation. The specific form of entrepreneurship depends on the patterns of interaction among the pioneer, perspective, and practice, whose effects jointly determine entrepreneurial performance.

Concept of Entrepreneurial Behaviour

Parsons and Shills (1951) proposed a theory of action which assumes that the actor strives to achieve goals. The theory of action is conceptual scheme for the analysis of behaviour of living organisms. There are four points to be noted in the conceptualization of behaviour.

1. Behaviour is oriented to the attainments of ends or goals or other anticipated states of affair.
2. It takes place in a situation.
3. It is normatively regulated.
4. It involves expenditure of energy or effort or motivation. This means that any behaviour of a living organism might be anticipated in terms of the anticipated state of affairs towards which it is directed.

According to Legans (1961) behaviour is response to stimuli refers to what an individual knows (knowledge), what he can do (skill), mental and physical, what he thinks (attitude) and what he actually does (action).

Calder and Ross (1976) stated that behaviour was a function of many personal factors -motives, habits, attitudes, and so on - and of many environmental factors, such as norms law, rewards, and punishment.

According to Fisher (1982) behaviour is a function of continuous process of multi directional interaction between the person and the situations, including other persons that he or she encounters.

Robbins (1996) reported that behaviour was a function of its consequences. and found that people would most likely engage in desired behaviour if they were rewarded for doing so, these rewards were most effective if they immediately followed the desired response, and behaviour that was not rewarded, or was punished, was less likely to be repeated.

Fishbein (1965) reported that entrepreneurial behaviour was a function of the surrounding social structure both past and present, which could be readily influencing by manipulability economic and social incentives.

Mcclelland (1965) stated that casual sequences of entrepreneurial behaviour were as follows:

Ideological values $\longrightarrow$ Family socialisation $\longrightarrow$ Need for Acluevement $\longrightarrow$ E.B.

Javillionar and Peters (1973) listed three dimensions of, entrepreneurial behaviour *viz.*

1. Risk taking ability.
2. Novel or energetic instrumental activity, and
3. Individual responsibility as indicators of entrepreneurial perfoffi1ance

According to Rao (1985) entrepreneurial behaviour is the result of an interaction of individual, situational, psychological, social and experiential factors.

Singh (1986) stated that Entrepreneurial behaviour (EB) was a function of an individual's personality characteristics and environmental factors. This could be represented as:

EB = f (PE)

where,

        P: Personality characteristics

        E: Environmental factors.

Manjula (1995) stated that entrepreneurial behaviour was the change in knowledge, skill and attitude of entrepreneurs towards the selected enterprise.

Patel and Sanoria (1997) stated that the activity of an individual to decide for adopting certain enterprise to make profit was regarded as entrepreneurial behaviour.

Paul (1998) revealed that majority of respondents in all the groups of small (65 per cent), medium (50 per cent), and big farmers (55 per cent) had medium entrepreneurial behaviour whereas low entrepreneurial behaviour constitutes more among small farmers (22.5 per cent) and medium farmers (32.5 per cent). The high entrepreneurial behaviour (32.5 per cent) of big farmers might be due to their sound financial condition to take risk and adopt new technologies and also more innovativeness and leadership abilities.

In the present book entrepreneurial behaviour has been defined as package of personality characteristics and environmental factors related to dynamic agent of change for transforming physical, natural and human resources into corresponding production possibilities.

Dimensions of Entrepreneurial Behaviour

Murry (1938) Conceived achievement motivation as a desire ordendency to do things rapidly and or as well as possible. For specified the desire as to accomplish things something difficult to master, manipulate or organise physical objects, human beings or ideas to overcome obstacles and attain a high standard, to excel one's self to rival and surpass others.

Mcclelland (1953) stated that success in competition with some standard of excellence is need achievement. They characterized achievement motivation as the propensity to strive for success in any and all situations in which a standard of excellence was though to apply.

Edwards (1954) while dealing with the theory of decision making, explain that the crucial factor about economic man is that he is rational. This means two things, can weakly order that states into which he can get and he also can make his choice so as to maximise something. Foe further states that the fundamental content of the maximization is that economic man always choose the best alternative from among those open to him, as he sees it.

Cantillon (1955) emphasized risk taking as the distinguishing attribute of the entrepreneur.

Atkinson (1956) conceived achievement motive as latent disposition to strive for particular goal, state or aim.

Edwards (1957) defined achievement motivation to do one's best, to be successful, to accomplish tasks requiring skill and efforts, to be a recognised authority and the like.

Hoselitz (1957) opined that entrepreneurship was associated with personality pattern in which the achievement motivation was strong. But the presence of strong achievement motivation in a t group of individuals did not necessarily produce an abundance of I entrepreneurs unless certain other conditions of social structure and culture strongly favour achievement oriented individuals to, enter economic pursuits.

English and English (1958) defined knowledge as a body of; understand information possessed by an individual or by culture. Knowledge is generally understood as intimate acquaintance of an individual with facts.

According to Atkinson and Litwin (1960) persons with high need to achieve, by and large tend to take moderate risks *i.e.* they do not avoid risk and also that they do not like situations and commitments where the probability of desired outcome is very low. This behaviour emanates from a conflict between "hope of success" and "fear of failure" resulting a positive net balance of hope of success. Entrepreneurial behaviour flourishes where absolute certainty of both success and failure is absent.

Fraser (1961) conducted an investigation on small line entrepreneurs in a rural village in Orissa and found that people with high need achievement showed more entrepreneurial spirit and less involvement in traditional cultivation of the soil than those with low n-Ach.

According to Mcclelland (1961) achievement motivation is the individuals need or desire to perform or to do better not so P1uch for the sake of social recognition or prestige but to attain an inner feeling of personal accomplishment. and proposed a model of achievement motivation, usually referred as A- T -D model. This model in brief, can be explained as that a rise in the level of n-Ach (A) in a society will be associated with a rise in the rate of development (D) with a time lag (T) which will be always positive. He reported the following characteristics of person with high n-Ach.

1. Likes to take personal responsibility.
2. Likes to moderate risk.
3. Wants to know the results of his efforts.
4. Tends to persist in the face of adversity.
5. Tend to be innovative.
6. Is oriented towards the future.
7. Tend to be mobile and is not completely context.

Neill and Rogers (1963) considered achievement motivation as the value instilled in an individual through the sociological process in which the individual feels a need or desire to excel in reaching certain goals only for the satisfaction of reaching goal and not for rewards of the goals or ends involved.

Neill and Rogers (1963) postulated that farmer's achievement motivation leads to the individual excellence in farming. They found significant correlation between need for achievement of Ohio state farmers (U. S.A.) and some of the indi9ators of excellence in farming such as production man work units, man days, of labour on the farm and number of acres in the farm.

Kogan and wallach (1964) stated the decision making involves the weighing of alternatives in terms of their desirability's and their likelihood issues concerning the avoidance and acceptance of risk in arriving at decisions hence are likely to be important ingredients in things process.

Rogers and Neill (1966) reported correlations between n- Ach scores and agricultural innovativeness in farming communities in both.India and Columbia. The correlations were significantly positive in six out of eight Indian farming communities, yielding or a total correlation for all individuals of 0.32 (N = 702). Among 302 farmers in six different villages in Columbia the average correlation was 0.18 indicating that farmers who think about doing better (high n -Ach) do in fact respond more to greater opportunities be adopting new farm practices.

Wharton (1966) observed that subsistence farmers responded as quickly t6 economic stimuli as the most commercial farmers in the modern world and it was precisely those with high achievement who were responsible to economic stimuli and who found ways of making a better living at farming from adopting new practices.

Mcclelland (1969) stated the characteristics/dimensions of entrepreneurs as follows:

1. Need achievement
2. Desire for responsibility

3. Preference for moderate risk
4. Perception of probability of risk
5. Stimulation by feed back
6. Energetic activity
7. Future oriented
8. Skill in organising
9. Attitude towards money

Rogers and Svenning (1969) defined achievement motivation as a spontaneously expressed desire to do something well for its own sake rather than to gain power or love or recognition. They concluded that farm production was positively related to achievement motivation.

Rogers and Svenning (1969) defined innovativeness as the degree to which an individual adopts new idea's relatively earlier than others in his social system. They further stated that innovativeness was one of the most important indicators of farm excellence. The adoption of technological innovations indicated one way in which a peasant could improve.

Sharma (1970) concluded that self-confidence was considered as one of the important characteristic of entrepreneurs.

Singh (1970) pointed out that mean n-Ach scores of the business entrepreneurs were statistically high than the agricultural entrepreneurs.

Basavanna (1971) stated that self confidence was a consistent behaviour pattern which revealed that one had faith in one's abilities.

Basu and Moulik (1971) stated that support system for developing entrepreneurial capabilities assumed considerable importance. Further, they stated that a business enterprise had internal and external managerial functions and it was the external managerial role that assumed crucial importance, particularly in relation to the procurement of finance, raw material and marketing of products.

Hornaday and Aboud (1971) reported that need for achievement, support; independence and leadership were the most significant entrepreneurial characteristics.

Kilby (1971) indicated that one of the integral parts of entrepreneurial behaviour was introduction of new products or new technology.

Rogers and Shoemaker (1971) reported that innovations were more likely to have a commercial rather than a subsistence economic orientation, they had more favourable attitude towards risk, high level of achievement motivation, greater exposure to change agent, mass media, sought more information, and had a greater knowledge of innovations.

Rogers and Shoemaker (1971) concluded that knowledge of farmers with reference to adoption of innovation divided knowledge into three types. Awareness -Knowledge, how-to-knowledge (information necessary to use knowledge) and principles knowledge (Functioning principles under lying innovation).

Singh and Singh (1971) stated that successful farm entrepreneurs made better use of resources. Hence it was expected that ultilization of available resources by farmers was likely to be related to their farming performance.

Singh and Singh (1971) reported that business entrepreneurs had high score on need for achievement and risk taking when compared with agricultural entrepreneurs.

Singh and Singh (1971) reported that progress entrepreneurs had moderate risk taking. In a study on risk taking ability of farmers they found that progressive agricultural entrepreneurs exhibited moderate risk taking scores further observed that positive but not significant trend of relationship was found between anxiety, and risk taking scores in case of successful agricultural entrepreneurs, whereas significant reverse relationship was found in case of unsuccessful agricultural entrepreneurs.

Sinha and Mehta (1972) indicated that the middle land holders showed greater motivation to achieve and greater readiness to change than both smaller and bigger land holders. They also found that younger farmers were better disposed to achieve and change irrespective of size of land holding. The results identified a curvilinear relationship between size of holding and n-Ach and change proneness.

Javillionar and Peters (1973) reported three dimensions of entrepreneurial behaviour viz; risk taking ability, novel or energetic instrumental activity and individual entrepreneurial performance.

Joshi and Kapur (1973) emphasized the managing a farm as a continuous process of decision making. Successful farm management required the ability and capacity on the part of farmers not only to make decisions, but also to make the correct decisions. According to them, a decision could be taken to take no action and line with status -quo, to postpone a solution while obtaining further information, or to undertake a different course of action. Successful farm managers preferred to try alternate course of action, of-course within reasonable risk 1imits. It might be that the most effective decisions were made by people who are neither extremely high nor extremely low in the willingness to take risks.

Singh and Pal (1974) observed significant association between n-Ach and socio economic status of the farmers. They also found that achievement motivation was significantly correlated with education but not land size. The result of the study also indicated that the respondents belonging to progressive and non progressive villages significantly differed in their achievement motivation.

Battacharjee and Akhouri (1975) indicated that propensity to take risk was found significantly associated characteristic of entrepreneurs. Often it is compared to gambling in India, since farmers have to face so many risks and uncertainties in their day to day agricultural operations. Some of the uncertainties that effect the farm production can be weather variations-failure, excess, untimely, attack of pests-and disease, adulterated, failure of electricity *etc.*, market conditions -low prices and government policies, social responses and value's.

Hundal and Singh (1975) opined that achievement motivation was particularly associated with farm success. Other characteristics linked with success were intelligence, aspiration to advance, need for power, tender minded temperament, and radial outlook. All these factors except need for power were with in the wider concept of achievement motivation.

Hundal and Singh (1975) observed that farmers showed n-Ach, n-power and n-affiliation in the descending order.

Singh and Kumar (1975) while studying wheat farmers in three districts of Uttar Pradesh (India) concluded that achievement motivation of farmers was weakly related to adoption of innovations in wheat cultivation (only two percent of variance was explained by achievement motivation out of 25 per cent of total variance explained by all the variable included in the study.

Sisodia (1976) reported higher motivational intensity with in increase in size of land holding.

East West Centre (1977) the task force set-up by East-West Centre (1977) Honolulu had identified self -confidence as one of the 19 most frequent characteristics of entrepreneurs. The task force also grouped confidence, independence individuality, optimism, leadership, dynamism under self-confidence.

Rousmasset (1977) stated that risk and uncertainty arouse because production was not an instantaneous process. Time lapses between when entrepreneurs decide to go in a line of production and when the product finally reaches the market. The entrepreneurs have to assess cost and price and physical performance before he embarks on his enterprise, and these may change with time.

Kanungo and Bhatnagar (1978) stated that people with high n-Ach were known to seek and assume high degree of personal responsibility, set challenging but realistic goals, work with concrete feed back, research their environment and choose partner with expertise in their work.

According to Pareek and Nadakami (1978) entrepreneurship can be conceived as an innovative action.

Pareek and Nadakami (1978) reported that the knowledge of the world as well as business were among the special qualities of the entrepreneurs. The entrepreneurs need to have knowledge about several areas of activity relevant to his domain of enterprise. Such knowledge helps him plan his strategy and use his skills effectively knowledge about environment, industry and technology is considered important.

Rao and Mehta (1978) observed that the need for independence and the sense of determination were the two chief characteristics that drive the entrepreneurs to start their own business and prefer not to be controlled by others. Further they observed "entrepreneurs are inclined to approach their tasks with a hope of success they attempt any task in the hope that they will succeed rather than with a fear of failure. Such hope of success enhances their confidence.

Singh (1978) in their study on modernization of farmers found that economic motives was most important followed by prestige and recognition, self actualization, innovativeness and affiliation in the decreasing order of importance. It was also observed that achievement motivation was least motive as ranked by the farmers.

Singh (1978) identified that cautious decision making was associated with progressive farm behaviour.

Ramakrishnan (1979) observed that risk-taking was one of the characteristics of the entrepreneurs.

Schwartz (1979) observed that achievement motivation was a potential factor affecting entrepreneurship significantly.

Singh (1979) stated that small entrepreneurs were handicapped with material resources while medium size entrepreneurs were more resourceful, the farmers had low linkage with support agencies while the latter had strong linkage.

Singh (1979) concluded that achievement motivation of farmers was associated with their farm output.

Singh (1979) observed high scores on attitude towards farming, preference for activity, pride on work, upward striving, cautious decision making were associated with progressive farm behaviour, low scores on these variables associated with static or declining farm success.

Singh (1979) studied the differential outlook of small and medium size entrepreneurs and pointed the following observations among small and medium entrepreneurs. Small entrepreneurs had narrow vision of entrepreneurial world, while medium size entrepreneurs indicated wider vision and experience. Small entrepreneurs were aggressive in experimenting, whereas medium size entrepreneurs had insight in putting alternatives.

Small entrepreneurs are handicapped with material resources while medium entrepreneurs were more resourceful. Small entrepreneurs had low linkage with support agencies, whereas medium size entrepreneurs had strong linkage with support agencies. Small entrepreneurs felt socially alienated, while medium size entrepreneurs had social status. Small entrepreneurs were localite in nature, whereas medium size entrepreneurs were cosmopolite in nature. '

Kourilsky (1980) concluded that delineated persistence, academic ability and creativity were critical characteristics.

Prasad (1983) stated that decision making ranked second in contributing to achievement motivation of rice growing farmers in Southern states. He also reported that the same variable has taken third place in discriminating between high and low achievement motivators.

Rao (1985) concluded that innovativeness and self confidence were in second position in explaining variance of entrepreneurial behaviour, while predicting farming performance.

Rao (1985) stated that risk taking was positively and significantly related to farming performance.

Rao (1985) while studying the farming performance of farmers in Andhra Pradesh found significant and positive correlation between farming performance and the entrepreneurial characteristics such as education farm size, cosmopoliteness, caste, training, social participation, knowledge, innovativeness, information seeking, decision making, self confidence, n-Ach, Utilization of assistance, result orientedness and risk orientation.

According to Tripathi (1985) the environmental form conduct is the aggregate and dynamic interaction of those external factors which interact with the entrepreneur, the resultant organization and also among themselves to have an impact on the functioning of the venture. It encompasses the political, economic, legal, social, cultural, demographic, competitive, technological, physical, natural, ecological and all other environmental components.

De (1986) revealed that socio-economic status, education knowledge of HYV of wheat practices, sources of information utilized, innovative orientation and progressive value's had positive and significant association with entrepreneur characteristics farmers.

Tushman and Anderson (1986) stated that product innovator might involve shifting to new technologies.

Swamy (1988) concluded that the farmers were mort innovative and enterprising migrated early by identifying the existence of income generating opportunities outside the village and provided information for their success.

Harper and Vyakaranam (1988) stated that a shift from a family management to enterprise management might be easier than a shift from paid employment to self employment.

Rao (1989) found significant relationship between innovativeness and income of small, medium and big farmers while studying entrepreneurial behaviour characteristics of vegetable growers of Andhra Pradesh.

Jyothi (1990) stated that entrepreneurs of Guntur district, about 73 per cent of the sample entrepreneurs were either graduates or post graduates.

Downing (1991) concluded that majority of women's (69 per cent) independently took their own decisions. Sons (31 per cent) took decisions with the help of spouses, friends or other family members. This pattern of behaviour may be common with men entrepreneurs as well. If the family members fail to give relevant advice then a cooperative effort of the staff and other sources are often sought. However, most of the time of final decision is made by the entrepreneurs.

According to Shaver and Scott (1991) the need for achievement (n-Ach) is responsible for economic development. Greater the development of n-Ach, during early socialization of people, the more likely that economic development will be achieved. A society with a generally high level of n-Ach will produce more rapid economic growth. Achievement motivation could be inculcated through training in self reliance, rewarding, hard work and persistence in goal achievement, and creating interest in excellence.

Ricky (1992) stated that managers were carrying out entrepreneurial roles through which they initiated changes to take advantage of opportunities. Although it was common to search for the entrepreneurial personality.

Jyothi and Prasad (1993) observed that rural women had little knowledge regarding projects or fields of business. They seemed to have limited knowledge of the traditional occupations.

Mall (1993) opined that people particularly in backwards areas suffered from low achievement motivation due to socio cultural and institutional factors.

Kanungo and Mendonca (1994) stated that entrepreneurs played both the roles of managership and leadership. Managerial role was exhibited by entrepreneur in their capacity as head of the enterprise. They also played leadership role when they were driven by their own vision to innovate or bring in a change in the manager events took place. Leadership behaviour had been usually visualized in the context of change. The change might be driven by environmental pressures or values of the leader, but was often related to a vision of a more desirable future state shared by leader and follower.

Koontz (1994) found that entrepreneurs took personal risks in initiating change and they expected to be rewarded for it. They needed some degree of freedom to pursue their idea's, this in turn required that sufficient authority be delegated.

Zahir (1994) opined that a large majority of the entrepreneurs (82.5 per cent) were blow the age of 40 years. Over half of these (42.5 per cent) were in the age group of 20-30 years. These figures support the assumption that the younger generation is relatively more adventurous, prepared to take more risk, dynamic and innovative the characteristics required of a successful entrepreneur. Moreover 30 of the entrepreneur's understudy (75 per cent) were graduates or above of these, 17 were technically/professionally qualified. New generation of entrepreneurs thus possess better educational qualifications and consider it an important factor for the success of any business venture in the changing environment. Finally, only 2 entrepreneurs (5 per cent) belonged to families engaged in agriculture and allied activities, which shows that entrepreneurial development programmes/schemes do not seem to be very effective in decreasing the dependency of Indian people on agriculture.

Bell and Pavitt (1995) stated that innovation was treated as development and initial commercialization of new technology.

Jha and Shiyani (1995) concluded that different socio economic factors, education status, infra structural facilities influenced adoption of dairy innovation positively and significantly.

Kanungo and Conger, (1995) opined that entrepreneur engaged in visioning with a practical bent of mind. A leader's vision was often idealized; where as an entrepreneur's vision was more pragmatic and deeply rooted in the environmental realities representing constraints and opportunities. Hence, he was highly resourceful and innovative, always seeking opportunities to materialize his vision.

Keshari and Kota (1995) opined that the entrepreneurial attributes in the presence of economic factors did not have any significant impact on a firm's propensity to export. Among the economic factors, propensity to export was positively influenced by size and negatively affected by capital intensity.

Kumar (1995) stated that entrepreneurs were seen as performing various functional roles as risk taker, decision maker, organizer or coordinator, innovator, employer of factors of

Production, arbitrageur and allocator of resources to alternative uses.

Manjula (1995) concluded that management orientation was positively and significantly related to entrepreneurial behaviour of participant and non participant women's under DWACRA programme.

According to Pareek and Rao (1995) different types of entrepreneurs may have different characteristics. For example rural entrepreneurs are more likely to be community oriented, less acquainted with business related matters and more dependent. Those who set up business with their own capital may be some what different from those who get loans to set up their enterprises. There will also be a difference between the first generation, (new) entrepreneurs, and those who come from families with entrepreneurial experience.

According to Akhouri (1996) the enterprise may be small or big but it demands management abilities in its owner/manager. The various facts of management such as production, marketing, financial management *etc.* are crucial for entrepreneurs.

Jhamtani (1996) stated that two major factors had played significant role in developing entrepreneurship. One of them the development of human factor the entrepreneurship himself, another major factor was the development of environment –where entrepreneurial activities could flourish and grow. The human factors referred to the attitude, desire and motivation of an individual, his capability to perceive the environment changes and opportunities as well as his ability to solve the problems which he was likely to face.

Sinha (1996) stated that human factors were central to entrepreneurial effectiveness. Human factors may comprise (a) background and demographic characteristics (b) their beliefs, value's orientation and manipulative skills and (c) style of leadership. These factors together playa critical role in the birth

Sinha (1996) concluded that personality characteristics like manipulative skill, achievement motivation, belief in functional value's and positive work value played a greater role in the healthy growth of the enterprise.

Stoner *et al.* (1996) stated that certain psychological and sociological factors were characteristics of entrepreneurs. The psychological factors were need achievement, locus control, tolerance of risk, tolerance of ambiguity and type -A behaviour, lack of viable concept, lack of market familiarity, lack of technical skills, lack of seed capital, lack of business know-how, non motivation, social stigma, time pressers, legal constraints, monopoly, and patent inhibitions were barriers to entrepreneurship. Market contact, local incubator companies, capable local manpower, technical education and support, supplier assistance and credit, local venture capitalists, venture savvy bankers, capable local advisors, education and successful role models were environmental helps to entrepreneurship.

Timmons (1996) stated that successful entrepreneurs had a high level of self confidence. The entrepreneurial role had long, been recognized as a prime source of innovation and creativity.

According to Ameerjan and Naika (1997) achievement motivation is defined as a striving for success in agriculture with an emphasis on doing better than what a farmer has achieved before either in terms of some objective standard or the standard of some other farmers performance.

Banerjee and Talukdar (1997) revealed that in spite of raising nurseries/cut flowers potted plants on a commercial basis, only 10 per cent of entrepreneurs were "high" on knowledge about entrepreneurial operations. Majority of respondents fell in medium category (51.66 per cent) where technical knowledge about enterprises operations was concerned (followed by 38.34 per cent) in low and only 10 per cent in high category. Most of the entrepreneurs had a great extent of risk taking calculation (38.34 per cent) in high and 45 per cent in medium category. Most of the entrepreneurs had high decision making ability and high managerial ability. He further stated that achievement motivation is positively and significantly related with entrepreneurship and the respondents had a high need for achievement (38.70 per cent).

According to Gaikwad (1997) rural women also playa role to help males in decision making. In rural society decisions are mainly taken by males and tl1ey have to take many decisions. They may be about crops to taken. Cultivation practices to be performed, livestock to be maintained and so many. The women participation in decision enhances the decision making power of males and also gives them moral support. The agriculture economy is dependent on such decisions. The involvement of farm women in decision making along with males is therefore, of primary importance.

Gaikwad *et al.* (1997) concluded that there was meager participation of farm women in decision making process, particularly related to agriculture. It would have to be increased. For this there was a need to divert time to farm women from kitchen -child care to other fields. The knowledge of the women needed to be improved by training and education and also exposing them to different social, economic, and political activities. This would not only improve their skill, but also help them to expose their idea's, views and entrepreneurial abilities and increase their participation in the decision making process.

According to Gogi and Talukdar (1997) human motives and values playa very important role in determining agricultural productivity. This achievement motivation (n-Ach) is one such factor which characterize an individual as progressive or development oriented and which has got direct bearing on the individual at as well as society as a whole.

Gogi and Talukdar (1997) concluded that majority of the scientists fell under the category of low group followed by medium and high group respectively. It reflects that most of the scientists (54.5 per cent) were having low level of achievement motivation. Only 30.50 per cent and 15 per cent of scientists were medium and high achievement motivation oriented respectively.

Kumar (1997) concluded that majority of the farmers (54 per cent) had high level of knowledge followed by low and medium (40 per cent, 6 per cent) respectively. The selected independent variables *i.e.* education, local participation, farm size, farm power, farm implements, irrigation potentiality and communication factors were positively and significantly related with the knowledge level of potato growers. The independent variable's *i.e.* caste, age and occupation were not significantly related with knowledge level of potato growers.

Mohanty and Ray (1997) concluded that involvement of farm women in the decision making process suggests that women assumed some role of responsibility only in those decisions which were economically less important and with which they were physically attached.

Palmer and Ahiamadu (1997) stated that task oriented and people oriented leadership were really helping as far as the sustainability and prosperity of the public enterprises in developing countries were concerned.

Parichha and Das (1997) opined that the wife alone took independent decision regarding storage and preservation of seeds and food grains 30.44 per cent followed by harvesting of crops. Marketing of farm produce was a share responsibility of husband and wife 49.96 per cent. Harvesting of crops was also joint responsibility, 40.87 per cent. Decision regarding cropping pattern was a whole family decision 36.53 per cent followed by husband alone took independent decision 32.17 per cent. Since decision making is an important area, and her participation in fant1ing activities were more.

Rao and Mishra (1997) revealed that all the three categories of respondents were having medium level of knowledge about pesticide use in cash crop cultivation. In case of male heads 7.5 per cent were having low level of knowledge whereas in case of off springs it was 17.5 per cent. About one fourth of the total number of respondents in each category were having high level of knowledge. The mean knowledge score was highest in case of off springs, followed by male heads and female heads. The lowest mean knowledge score among female heads might be due to their low level of education, less participation in field activities and less contact with extension agencies.

Singh and Vent1a (1997) reported that knowledge was one of the important components of human behaviour and as such played an important part in the covert as well as overt behaviour of an individual.

Yasunobu (1997) found that Peladang Jaya (successful farmers) were mostly engaged in mixed cropping paddy cultivation, except for 20 per cent who were involved in monocropping paddy, cultivation, 14 people were mainly engaged in monocropping. Others combined the cultivation of paddy rice with other field crops mostly a combination with industrial and/or with short term cash crops, and fruit orchards. Following three major characteristics can be pointed out the leading farmers.

1. They made utmost effort in increase their income besides, farming they held various side Jobs in small industry or were engaged in large scale business.

2. They were interested in the new technologies, including machinery, chemicals and new varieties; they planned the crops, which other farmers in the area did not use. They easily adopted new projects.

3. They had acquired a knowledge and/or experience of farm management practices. All most all the respondents participated in courses and seminars.

Lekha *et al.* (1998) found that illiteracy, ignorance, lack of economic holdings and social sanctions were the root causes of entrepreneurial slackness among women. These factors influenced women's personal and social attitudes and suppressed their entrepreneurial predispositions such as self confidence, risk taking and achievement motivation.

According to Kanungo (1998) innovativeness refers to creation of new products, markets product, market combinations, method of production and organization, and the like that enable the enterprise gain competitive advantage in the market. According to Khan (1998) the process of technological innovation revolves around the entrepreneur. It is often suggested that the process innovation is synonymous with the idea of entrepreneurship.

Paul (1998) found that 65 per cent of small farmers had low risk taking ability. 45 per cent of both medium and big farmers had medium and high risk taking ability.

Paul (1998) concluded that majority of small farmers (57.5 per cent) had low leadership ability compare to medium farmers (40 per cent) who had medium leadership ability and big faffilers (51.5 per cent) who had high leadership ability).

Paul (1998) revealed that majority of respondents in alt the groups of small (52.5 per cent), medium (45 per cent) and big farmers(60 per cent) had medium achievement motivation whereas, high achievement motivation constitutes more among medium farmers (35 per cent) than that of big farmers (22.5 per cent). Majority (80 per cent) of small farmers, 37.5 per cent of medium farmers had low innovativeness whereas 40 per cent of big farmers had high innovativeness. Decision making was a turning point in the adoption of new technology, with respect to farm decision making small farmers (47.5 per cent) had low farm decision making whereas, 65 per cent medium and 50 per cent of big farmers had medium decision making.

Rajak (1998) found out that majority of the existing entrepreneurs belonged to tailors and shop-keepers followed by black-smiths basket makers, weavers and potters. The economic status was very low especially black-smiths, basket makers and weavers. Near about 75 per cent of them came under the land less and the marginal farmers category. The main objective for them was to earn some money for their existence. Innovation and risk taking attitude was less in them, excepting a few of the tailors. The study showed that nearly 60 per cent of them were skilled and others were semi skilled. Except tailoring by and large all other occupations were traditional and hereditary and using traditional implements. There was general poverty among them and they could not afford to buy modem implements.

Sullivan *et al.* (1998) stated that four activities of innovation were (1) market -technology linkage (2) organizing for creative problem solving (3) monitoring and evaluation (4) commitment to the innovation process.

Suresh *et al.* (1998) stated that leader played the role of agents for change responsible for effectively bringing about changes or transformation in the existing system and its members. Leaders on the other hand, used transformational influence to bring about changes in others attitudes, values and behaviours. This was done essentially through effectively articulating future goals or visions and empowering others to work toward the achievement of the goals.

Vijaya and Kamalanabhan (1998) found out that power, self actualisation and achievement motivation were significantly higher in entrepreneurs compare to economic and affiliation motivation.

Vijaya and Kamalanabhan (1998) stated that the single most important causative factor in the rise of entrepreneurship was achievement motivation. It was described as a social psychological drive among the people that lead to economic development of a country. The proposition that a high need for achievement was positively related to entrepreneurship.

Factors Affecting Entrepreneurial Behaviour

An overview of literature pertaining to factors affecting entrepreneurial behaviour of farmers is absolutely essential. Atkinson and Miller (1956) observed that the first born had higher achievement motivation than those later in birth. He further concluded that younger sons might have higher achievement motivation.

Lionberger (1960) opined that there was positive association between extent, of farmers' social participation and his level of adoption of innovations.

Rani (1966) found out that educational background did not influence women in seeking entrepreneurship.

Patel (1968) observed that trained farmers gained significantly higher adoption scores than non-trained farmers. In this study training refers whether an individual farmer had attended the farmers training programme and if so for how many days.

Reddy and Kivlin (1968) stated that there was positive association between extent of farmer's social participation and his level of adoption of innovations.

Christopher (1969) found out that formal education was significantly associated with entrepreneurship.

Christopher (1969) in his study on small entrepreneurs in Hyderabad and Secunderabad reported that a typical small entrepreneurs were either first born or the eldest male child in their father's family.

Christopher (1969) reported that membership in organizations was not found association in case of small industry entrepreneurs.

Rogers and Svenning (1969) reported that farm size was associated significantly with modernization variables *viz.*, achievement motivation, innovativeness, cosmopoliteness *etc*. Therefore, it was expected that the farm size of respondents was likely to be related with their overt entrepreneurial behaviour.

Patel (1970) while describing the indicators for entrepreneurship reported that 70 per cent of the entrepreneurs were in the age group of26-40 and found out that education as a significant indicator as it ranged from fifth standard to Ph.D.

Derossai (1971) found out that majority of the entrepreneurs had higher education. Kilby (1971) stated that perception of market factors contributed to development of entrepreneurship and indicated that one of the integral parts of the entrepreneurial behaviour was introduction of new product or new technology. Rogers and Shoemaker (1971) generalized that earlier adopters had more social participation that late adopters.

Rogers and Shoemaker (1971) stated that earlier adopters had more years of education than late adopters. Pence it is expected that the formal education of respondents was likely to be related with their overt entrepreneurial behaviour.

Rogers and Shoemaker (1971) observed that individuals personality as well as the norms of his system might affect where he sought information, "What messages he receives and how he interprets the information he receives" further they observed generally individuals tended to expose themselves to those ideas which were in accordance with their interests, needs or existing attitudes (selective exposure).

Nandy (1973) observed that majority of the entrepreneurs had higher education.

Sharma (1973) surveyed the social characteristics of 555 business entrepreneurs in 6 different places in the country and reported that more than 50 per cent entrepreneurs were vaisyas, a business community. However, in the case of farming, the castes other than Brahmins and vaisyas were generally found in rural India. But it was not uncommon to find few farm entrepreneurs in these castes also.

Akhouri (1975) found out that entrepreneurs were very tenacious in perusing their entrepreneurial goal inspite of many hurdles spread over a considerable period of time.

Rao (1975) stated that caste and community were potential factors in identifying entrepreneurship.

Rao (1975) identified increase in utilization of information sources as a variable for measuring entrepreneurial effectiveness.

Kittur (1976) stated that values largely influenced the individual's behavioural patterns. Value was tlle relative importance people attribute to different objects, phenomena, and circumstances. People oriented their thinking, feeling and actions towards different things in life, based on values they hold. As such they became important organized themes in the behaviour of individuals.

Leibenstein (1976) stated that in less developed world, the special entrepreneur searched for and discovered new economic information and translated the same in to new markets, the techniques and goods.

Lim (1977) opined that education and training had definitive role in enhancing entrepreneurship. Majority of rural entrepreneurs had a low education level. It appeared that formal education had only a minimum influence among rural micro entrepreneurs in Kerala. Inspite of the minimal role of formal education noticed in our study, education could help in broadening knowledge base, by identifying opportunities, and by pointing out ways to overcome barriers imposed by one's environment.

Uplankar (1977) noticed that majority of self employed industrial entrepreneurs were educated upto matriculation and one third had technical education.

Pareek and Nadkarni (1978) identified four sets of factors which influence entrepreneurship. They are: the individual, the socio-cultural factors, the support system and the environment. These factors are considered to be interacting and influencing entrepreneurship. While the individual, the environment and the support system directly influenced entrepreneurship, socio-cultural factors contribute through the individual.

Mishra (1979) reported that farm educational exposure, as intervening variable had significantly associated with adoption of wheat technology as well as return over the expenditure.

Singh (1979) revealed that successful tribal entrepreneurs were cosmopolite in nature. He also observed that small entrepreneurs were localite in nature whereas medium size entrepreneurs were cosmopolite in nature. Small entrepreneurs had narrow vision of entrepreneurial world, while medium size entrepreneurs indicated wider vision and experience.

According to Nandapurkar (1980) ability to coordinate farming activities is defined as the degree to which an individual co-ordinates action in a time dimension.

Sharma (1980) opined that the entrepreneurs were very mobile. He further concluded that more than 50 per cent entrepreneurs were vaisyas, a business community.

Suhasini (1981) reported that majority of the respondents were belonged to medium socio-economic status. Majority of the women entrepreneurs belonged to medium extension contact and mass media exposure.

Prasad (1983) observed that exposure to information sources ranked third in contributing to achievement motivation and ranked eight in discriminating between high and low.

Rao (1985) Concluded that experience was found slightly positive but not significant with farming performance. Caste is significant positive relationship with their farming performance. Majority of the farmers belonged to high order caste, high on self- confidence, medium on risk orientation, had moderate decision making, and high result orientedness. They were eldest sons to their parents. However, they were found low on remaining entrepreneurial behaviural factors.

Singh and Gupta (1985) reported that about 68.9 per cent of potential women entrepreneurs were graduates or post graduates and around 6.7 per cent had even professional qualifications.

De (1986) found out that socio-economic status, education, knowledge, sources of information utilized, innovative orientation and progressive values were positively and significantly related with the entrepreneurial characteristics of farmers.

Deivasenapathy (1986) opined that educational level and the family background of entrepreneurs did not influence their career success. Family support and previous job experience influenced their entrepreneurial success.

Scott (1986) stated that work experience of any kind in the public domain was an asset for an individual in launching into self employment. Women with some experience out side their private world had been successful in their own enterprise. This indicated that some previous experience (exposure) gave women more confidence. Exposure to outside work also made women more aware of the opportunities in their environment, and of the facilities and incentives offered by the government.

According to Singh (1986) entrepreneurship is a function of several factors. These factors are:

1. The individual
2. Environment
3. Socio-cultural factors.
4. Support system.

Individual

Environment $\longrightarrow$ Entrepreneurship $\longrightarrow$ Socio-cultural factors

Support system

Thus, while the individual, the environment and the support system directly influence entrepreneurship, the socio-cultural milieu contributes through the individual and the support system. Broadly speaking, support system and socio-cultural factors also constitute the environment. Thus, there are only two factors, on which entrepreneurship development depends; these are the personal characteristics of the individual and the environment.

1. The individual factor:

   Three main factors, which influence the individual behaviour are his

   A. Motivational Factors

      (a) Entrepreneurial motivation Achievement motivation

         Power motivation

         Extension motivation

      (b) Personal efficiency

      (c) Coping capability

   B. Factors concerning various skills.

      (a) Project development skills.

    (b) Enterprise management skills. It includes accounting and financial control, marketing, production inventory control, skill for managing people.

    (c) Knowledge -Includes knowledge about economic political environment, relevant information, raw material and labour, various assistance sources, various industries, technical aspects, cost and benefits of various technologies.

2. Environmental factors.

3. Socio-cultural factors.

    (a) Family background

    (b) Norms and values

    (c) Normative behaviour (Norms of behaviour): The following aspects of normative behaviour are relevant for entrepreneurship is family expectations and pressure, Risk taking, independence, work, socialisation, and training.

4. Support System -Includes financial institutions, extension service of the department of industries, non-government organisations of small industries or entrepreneurs, development administration in the district, corporations specially set up to develop entrepreneurship and small industries in a region.

Singh (1986) studied that following factors responsible for the initiation and subsequent success of an entrepreneur. Factors were educational background, age, friends, community, prior occupation status, family size, parents, marital status, wives, joint family status, migration, peer group, association membership, financial status of the family, economic environment, availability of the material and technology.

Arundhati (1987) concluded that 96 per cent of the DWACRA beneficiaries were illiterate having low to medium urban contact, medium extension contact and low to medium mass media exposure. The suggestions expressed by respondents for the improvement of the DWACRA were permission to get raw material, provision of child care facilities, creation of marketing facilities, Quick release of funds and training in the trade.

Mancuso and Moscolo (1987) found out that formal education was positively correlated with entrepreneurship.

Swamy (1988) stated that the farmers who were more innovative and enterprising migrated early by identifying the existence of income generating opportunities outside the village and provide information for their success.

Scherer *et al.* (1989) stated that education was an important factor in determining the entrepreneurial orientation in individuals. Education might be informal or formal. The informal form of learning stressed the importance of early role models and reinforcement patterns on the acquisition and maintenance of entrepreneurial behaviour.

Thagamuthu and Manimekhalai (1989) suggested that steps f must be initiated for regular supply of raw material and provision of common work shed. They further suggested that if more financial~ infrastructural and marketing facilities were even at the right time and place, these self employed units might become highly enterprising units.

Jones and Cohen (1990) stated that "innovation" was the key factor in entrepreneurship. Innovation, the creation of new products, markets, services, sources of supply, or foffi1S of individual organisations, was viewed as the dynamic force that moved the capitalist system.

Singh (1990) suggested that the groups should be allowed to choose the activity freely, proper training in the trade, raising of loan amount, creation of marketing linkages were essential measures for the successful functioning of DWACRA programme.

Bhagat and Singh (1995) concluded that training intervention had certainly played an overwhelming effect to make the trainees to learn and acquire necessary knowledge/skill required for successful business.

Cooper *et al.* (1995) revealed that on an average, more infoffi1ation was sought by those with no entrepreneurial experience and in experienced entrepreneurs entering known fields, with regards to sources, the latter engaged in more intensive search in known fields through largely personal sources.

Manjula (1995) stated that majority of the participant women had medium value orientation (61.67 per cent), whereas non participant women had medium value orientation (56.67 per cent). Value orientation was positively and significantly related to entrepreneurial behaviour of participant and non participant women's. Socio-economic status was having positive significant correlation with entrepreneurial behaviour of participant women but socio-economic status was found to be positive but non significant correlation with entrepreneurial behaviour of non participant women. Majority of the participant women belonged to medium socio-economic status followed by high and low. In case of non participant women three- fourth of them had low socio-economic status followed by medium. However, high entrepreneurial behaviour had high socio-economic status. Mass media exposure was positive and significant correlation with entrepreneurial behaviour participant women but it was positive, non significant correlation with entrepreneurial behaviour of non participant women. Training received had positive and significant relationship with entrepreneurial behaviour of participant women's which might be due to participant women's newly acquired knowledge which was gifted by' training might have helped her to utilize different sources of employment. An individual who was trained could manage her enterprise better to get more profits and there by expanded her unit which in turn contributed for additional income. Marketing facilities had found to be positive but not significant correlation with entrepreneurial behaviour of participant and non participant women.

Mishra (1996) concluded that training in entrepreneurship provided an impetus to the potential and budding entrepreneurs to acquire a new identity about himself.

This was perceived as an approach towards transforming people who served the purpose of making people aware about their own identity, helping them accept new identify and finally establishing such identify for entrepreneurial pursuit. In order to take of such task of transformation it could find entrepreneurship training serving the purpose of stimulation, orientation, participation and implementation in a sequential order.

Sinha (1996) stated that belief in fundamental values and positive work value played a great role in the healthy growth of the enterprise, reflecting with entrepreneurial effectiveness. Human factors, back ground and demographic characteristics of entrepreneurs, their beliefs, value's orientation and manipulative skill and style of leadership factors together played a critical role in birth and development of an enterprise. 72 per cent of the successful entrepreneurs were those who had some technical qualifications. Whereas most of the unsuccessful entrepreneurs (67 per cent) did not have a technical background. The technical back ground of the entrepreneurs was found to have a positive association with their success as entrepreneurs. Successful entrepreneurs had beliefs which had a functional value in the present social system. He also suggested that successful entrepreneurs had a strong work value, whereas unsuccessful entrepreneurs had a weak work value, which might contribute to the successful or unsuccessful running of their enterprises. Successful entrepreneurs were relatively younger in age, technically educated and had some business background in the family. There was significant positive association between age of the entrepreneurs and that of the criterion variable of the health of the enterprise.

Sinha (1996) found that majority of the entrepreneurs (81 per cent) were first generation entrepreneurs. They did not come from families with business backgrounds. Agriculture and service of some kind figured quite prominently as their father occupation, while business hardly accounted for (19 per cent) of the family background. However, the business background of the entrepreneurs played a positive role in successful running of their enterprises as the association between the two variables was found significant and positive. Younger entrepreneurs were likely to be more effective than older ones. He further stated that majority of the entrepreneurs of both healthy (74 per cent) and unhealthy (71 per cent) enterprises hailed from high income groups (annual income of above 2 lakhs). The economic background did not seem to have a significant relationship with the successful or unsuccessful running of the small enterprises. Entrepreneurs with business and technical education were in a better position compare to others.

Stoner *et al.* (1996) stated that certain psychological and sociological factors were influencing the entrepreneurship. The psychological factors were need achievement, locus control, tolerance of risk, tolerance of ambiguity and type behaviour

Banerjee and Talukdar (1997) stated that two variables namely age and education were not correlated with extent of entrepreneurship which implied that respondents did not differ in their entrepreneurship behaviour due to personality variables like age and education.

Mohanty *et al.* (1997) concluded that training of vegetable growers must have top priority for the production technology components relating to plant protection and storage practices.

Patel (1997) revealed that socio-economic status and education were the major factors showing the higher direct effect on entrepreneurial behaviour of sugarcane growers.

Patel and Sanoria (1997) found out that entrepreneurship was influenced by various socio personal and economic factors. The important factors were caste, age, education, land holding, annual income, social participation, socio-economic status, farm mechanization, cropping intensity, and irrigation potentiality.

Raju (1997) stated that vocational training could play an important role in promoting the enterprises. Agricultural entrepreneurship was to be encouraged among the educated youth by providing the required training on the latest methods of farming with the latest technology, loans at reasonable interest rates *etc.*

Richart and Maurer (1997) stated that the entrepreneurial orientation of an entrepreneur -led firm was directly related to its marketing orientation.

Lekha *et al.* (1998) concluded that statistically significant association was found between economic statuses and time spent on unit-related work as well as training. Women with low income spent more time on business related work compared to women with high and middle income. High and middle income groups were better trained compared to low -income groups. On the whole, the socio- economic status did not show any significant association with the overall improvement of women entrepreneurs. Majority of the women entrepreneurs (73.3 per cent) did not have any business experience prior to starting their own business. Only 25 per cent of the respondents received appropriate training to start their business ventures. 24 per cent of the women entrepreneurs expressed the need for training *i.e.* mainly in marketing, new business opportunities and finding sources of investment and loan facilities. More than half of the respondents (51 per cent) did not express any need for training.

Ghosh *et al.* (1998) found out that majority of women had one to four years of work experience before they started their own enterprise; about 18 per cent had more than 5 years experience. Prior education had definitely contributed towards the development of entrepreneurship among the majority of women. Yet the level of education was, for the majority, matriculation which gave enough knowledge in reading and writing, and created a general awareness about entrepreneurship.

Entrepreneurial Behaviour of Farmers. An Axiomatic Theory Kumar (1998) suggested certain directions that need to be followed while training entrepreneurs. The training requirements in marketing were different for an enterprise before it was set up and for an ongoing enterprise. The focus in the training of potential entrepreneur required conceptualization that helped in the understanding of buyer behaviour, process of benefit segmentation, target -market selection and product strategy for efficient and deployment of resources. A training programme that helped creative thinking and pattern identification along with development of analytical skill was required. The impact of local cultural traditions and social values were critical factors in the development of entrepreneurial talent in the milieu. The presence of conflicting values could easily produce "role strain" in the entrepreneurs self image. The suitable pedagogy would be the case and discussion methods that expose the potential entrepreneur to a process of systematic problem solving. Such methods also developed the ability to apply concepts to real life situations.

Kumar (1998) conducted a survey of 109srnall enterprises in Kerala in 1995. The specific observations made by the study were as follows.

1. Branding is not systematically done by small enterprise.
2. Only a few enterprises performed market research.
3. Small enterprises preferred to sell their product by direct sales.
4. Promotion efforts were taken up, but this did not take into account seasonality in sales.
5. Pricing method used by small enterprise was the cost-plus method.
6. There was no variation in marketing programmes for different customer group in the markets served by these small enterprises.

Kumar (1998) stated that many small enterprises experienced difficulty in marketing their products and consequently their financial performance suffers. This created the need for developing marketing skills among entrepreneurs.

Manimekalai and Abdullah (1998) stated that less experienced entrepreneurs produce Charcoal more economically than the more experienced ones.

Nair *et al.* (1998) found out that education was an important factor in determining the entrepreneurial orientation in individuals., Education might be informal or formal. The informal form of learning stresses the importance of early role models and reinforcement patterns on the acquisition and maintenance of entrepreneurial behaviour. He further stated that successful I entrepreneurship crucially depended on both entrepreneur related, and environmental contingency factors. The relative role of the of two sets of factors in enterprise creation in Kerala, suggested that the environmental contingency factors played a dominant role relative to entrepreneur related factors. Training had definitive role; in enhancing entrepreneurship in the context of a developing country by enlarging the pool of entrepreneurs in society.

Paul (1998) found out that, majority of small farmers (45 per cent) belonged to low category whereas (67.5 per cent) of medium and (52.2 per cent) of big farmers belonged to medium marketing orientation category. Paul (1998) found out that lack of regulated market was a problem expressed by 15 per cent of small, 72.5 per cent of medium and 80 per cent of big farmers. Due to high degree of perishability of vegetables, the farmer had to unload the produce for whatever price offered while a glut in tile market because of high production and lo\v demand also likely to affect the farmers adversely. In adequate transportation and storage facilities were also expressed as a problem in vegetable cultivation by 72.5 per cent small, 55 per cent medium and 60 per cent of big farmers. Ranking the entrepreneurial characteristics by small, medium and big farmers showed that ability to co-ordinate farming activities was given first rank. Majority of the respondents in all the categories, small, medium big farmers had medium ability to coordinate farming activities (55.8 per cent) followed by low (24.2 per cent) and high categories (20 per cent). This indicated that irrespective of farm size all the respondents had medium ability to coordinate farming activities.

According to Ramana and Papaiah (1998) the factors that contribute to the development of entrepreneurship would essentially be a suitable environment and the intuition of an individual in grasping the essential facts.

De and Rao (2001) reported that higher the value orientation, marketing facilities, education and socio-economic status lead to higher entrepreneurial behaviour of farmers.

From the foregoing discussion it can be concluded that entrepreneur is a person of dynamic agent of change, entrepreneurship is a process of creative and innovative response to the environment. These responses can take place in a variety of fields of social endeavour, business, industry, agriculture, education, social work and the like. By a simple definition doing new things or doing things that are already done in a new way is a part of entrepreneurial behaviour. The key factors contributing to the entrepreneurial behaviour of farmers are value orientation, marketing facilities, education and socioeconomic status which are to be taken into account by the change agents for entrepreneurial development programmes in agriculture.

Framework of Entrepreneurial Venture

Entrepreneurship process starts when the entrepreneur recognizes an opportunity in the environment. There are certain words used in this statement which requires elaboration. The words are Entrepreneur, Entrepreneurship and Entrepreneurial Behaviour. For a long time there was no equivalent term for 'Entrepreneur' in English language. Three words were commonly used to connote the sense the French term carried; adventurer, undertaker and projector; these were used interchangeably and lacked the precision and characteristics of a scientific person (Gopakumar, 1995). Hence the term entrepreneur did not find any prominence in the history of economic thought. Ramana (1999) defined entrepreneurs as those people who work for themselves. The word entrepreneur derived from French word "Entreprendre" meaning to "Undertake". The entrepreneur is thus a person who organizes and manages an activity/organization, undertaking the risks for fulfilling some of his needs. His job involves the quality of boldness, courage, dynamism and risk taking in sufficient measure. The entrepreneur in this context is defined as one who could start new activity or a new enterprise which is a deviation from his traditional family occupation or profession. De (1986) stated that a farmer does not become an entrepreneur only by adopting a new agricultural technology but he becomes an entrepreneur only when he comes to be an operator of a farm business. A business involves rational decision on investment after assessing risk, other alternatives and possibilities or profit and loss. An entrepreneur is a dynamic agent of change or the catalyst who increasingly transforms physical, natural and human resources into corresponding production possibilities.

Entrepreneurship is that factor which urges an individual to take advantage of favourable situations by understanding innovative practices with a concern for excellence and assessment of self and the environment. Entrepreneurship is a way of life, a thought process, to bring any sustainable change; effort has to be more broad based. Cole (1949) stated that entrepreneurship comprises any purposeful activity that initiates, maintains or develops a profit oriented business in interaction with the internal situation of the business or with the economic, political and social circumstances surrounding the business. Hegan (1998) stated that entrepreneurship is not only conceiving the idea behind a venture but also designing and maintaining the organization for carrying it out. According to Schumpter (1970) entrepreneurship is a function of group level pattern, a function of managerial skill and leadership, an organizational building function, a function of high achievement, input complementing and gap filling, a function of status withdrawal and a function of social, political and economic structure.

Singh (1986) stated that Entrepreneurial Behaviour was a function of an individuals personality characteristics and environmental factors, it was represented as

$$EB = f (PE)$$

where,

EB = Entrepreneurial Behaviour

P = Personality characteristics

E = Environmental factors.

Phases of Entrepreneurial Venture

Framework for entrepreneurial venture has been developed in line with Ramachandran and Ray (1998). Entrepreneurial activities originate from individuals; the framework of entrepreneurial venture is described as under:

☆ **PHASE – I Identification of Enterprise:** An entrepreneur may come from unsatisfied need of people as well as unsatisfied personal need. This is reflected by the alertness of entrepreneur him/herself, his ability to identify opportunity, family and educational background, professional experience, formal and informal networks *etc*.

☆ **PHASE – II Initiation/Creation of Enterprise:** The entrepreneur needs to create an organization for transforming the concept into a marketable product by accumulating and combining physical and other resources. The ideas of entrepreneurs and environment interact to turn the ideas into reality. The new venture takes away large portion of time and attention of entrepreneur from the management of environment to the management of organization. Interaction with several components like bankers, regulatory agencies, experts and advisors, suppliers and so on entrepreneur alone has to constantly interact with them to have the control over the enterprise that he/she created.

☆ **PHASE – III Nurturing of Enterprise:** Organiosation translates the business concept into marketable product and offers it to the customer. The entrepreneur gets feedback on market response in terms of profitability, sales *etc*. The feedback is measured in terms of efficiency, competitiveness, effectiveness, innovativeness, flexibility *etc*.

☆ **PHASE – IV Concluding/Transition Phase:** Entrepreneurial process does not end with achieving stability and reaches success. At this stage organizations require different managerial style as venture competition gradually builds up more and more.

Typology of Entrepreneur

Ramachandran and Ray (1998) described the typology of entrepreneurs based on the outcome of entrepreneurial ventures. Empirical evidence suggests the existence of following four types of entrepreneurs.

(1) **Mobile entrepreneurs**: They are the persons who leave the ventures as soon as venture is created. They are the true Schumpeterian entrepreneurs as according to Schumpeter (1934) the entrepreneurial activity ends as soon as the venture is created. Ted Nieren berg, founder of Dansk designs in the US, Mohan of Good knight and Vikram Sarabhai both from India comes closer to this type.

(2) **Managerial entrepreneurs**: They are the persons who prefer to continue in the same venture by transforming themselves to fit the changing demands. As for example persons like Henry Ford of Ford Motors (Lacey, 1986; Pollarq, 1996), Pierre S. Du Pont (Chandler and Salsburry, 1971), and George Eastman of Kodak (Brooke ball, 1996).

(3) **Innovative entrepreneurs**: They are the persons who create an organization and remain engaged in their pursuits of innovation and creation of novel product's and technology. Walt Disney, Ibuka and Akio Morita of Sony Corporation (Morita, 1987), Sochiro llinda of llinda motor Company (Pewins, 1967)"and Bill Gates of Microsoft (Marshall, 1996) are some of the notable entrepreneurs belonging to this type.

(4) **Empire Builders**: They are the persons engaged in creating chain of new veptures having an ownership. They have the qualities of vast vision, flair of innovation al1d managerial capability to build an empire for themselves. John D. Rockfeller of the US (Abels, 1965), J.N. Tata (Harris, 1958) and Ghanshyam Birla (Jaju, 1985) both from India, Konosuke Matsushita of Japan (and wins, 1967) -and Chung Ju Yung of Hyundai and Kim Woo Chong of Daewoo both from South Korea are some of examples of this type.

Models of Entrepreneurship

It is supported by research that, entrepreneurship plays an important role for positive development in society by creating jobs and well being of consumers. There is no one best way for a country to promote entrepreneurship. There are several models country can adopt relevant to its cultural context. Dana (1998) has explained different models of entrepreneurship, they are as under.

1. Laissez-faire model
2. Limited environmental model
3. Strategic interventionist model
4. The subsidized interest rate model
5. The egalitarian model
6. The Trade facilitation model
7. The yogopluralist model
8. The top-down reform model
9. The open door and reform model
10. The doi-moi model
11. Culturally sensitive models
    a. Entrepreneurship among the amiss
    b. Entrepreneurship among indigenous people in alaska
    c. Entrepreneurship in the canadian subarctic
    d. Entrepreneurship in laos (Asia)
    e. Entrepreneurship in the kingdom of lesotho (Africa)

1. Laissez-Faire Model

The lack of intervention from the government permits entrepreneurs to become competitive and to become well. In other worlds govt interference and regulation obstruct economic development.

2. The Limited Environmental Model

Govt. can contribute in creating environment by having adequate infrastructure, low levels of taxation, minimum paper work, stable currency *etc.* it will encourage the business sector and this will be limited to providing a positive environment. Deregulation and privatization of enterprise is the example of this type of model.

3. Strategic Interventionist Model

This model indicates that the govt. promotes business by intervening the market place through a strategy and policy formulation. Entrepreneurship is encouraged by providing, training, research, finance, marketing know-how and support to entrepreneurs.

4. The Subsidized Interest Rate Model

High interest rate may cause a heavy debt burden, decreasing the rate of new venture formation and rate of increasing bankruptcy, subsidized interest rates will create many jobs this will help entrepreneurs to develop enterprises into chaebols *i.e.* mega-conglomerates. As for example the South Korean economy is dominated by a small number of highly diversified mega-conglomerates (Dana 1999). Subsidized interest rate model resulted in a chaebol dominated economy.

5. The Egalitarian Model

In egalitarian model all borrowings by entrepreneurs have to be at relatively high interest rate. High interest rate is an incentive to be thrifty and save rather than borrow. The implementation of egalitarian model in Taiwan resulted in a nation of small business (Dana, 1994).

6. The Trade Facilitation Model

In this model govt. facilitates paper work requirements and reduces bureaucratic constraints rather than tangible end. In Kenya there is no grant available for new venture. Loans are few in quantity and minimal in value. Govt. provides free consulting services to entrepreneurs and help them step by step from the time of their inception. Its policy is one with emphasis on trade facilitation (Dana, 1993).

7. The Yogopluralist Model

This model is unique to Yugoslavia and a function and a function of its pluralism. According to Dana (1994) the yogopluralist model was a result of weak central govt. and the cultural heterogeneity of its constituent republic. Culture was an important determinant of differences in economic policy which in turn contributed to regional disparity. Reform was initiated in Yugoslavia at local level and subsequently caused changes at federal level.

8. The Top-down Reform Model

This is by decree from the top level of govt. to down to the people. Several govts. declared the change from a centrally planned economy to one driven by market forces. The German democratic republic followed the top-down reform model. It does not take into account the social conditioning (Dana, 1994).

9. The Open Door and Reform Model

The peoples republic of China has followed an open door policy coupled with major reforms of the completely planned economy. The utilization of open door reform model is consistent with Chinese culture. Land is not privately owned, but long term lease arrangement give the feeling of ownership to the tenants. This arrangement leads to strong incentives to produce. Chinese farmer are kept out side the planned sector of the economy for promoting entrepreneurship in them (Dana, 1998).

10. The Doi-Moi Model

The doi moi is a Vietnamese word it means 'renovation' or 'new thinking'. This model is comfortable with restoring prestige of the ruling communist party. According to Dana (1994), this model is a unique blend of socialist and free enterprise policy in a socialized state. Here is complementarities and harmony between government firms operating under a system of centralization and the business sector operating independently.

11. Culturally Sensitive Models

a. Entrepreneurship among the amiss

Amiss is a religious group evolved in Europe spread to America. These people value asceticism, frugality, thrift, work and humility. They do not prefer to work companies in main stream society. They prefer to be self employed, they believe that a community of believers is the context of life, self employment is perceived as a social activity and as an economic activity. They continue to exhibit a propensity for entrepreneurial behaviour. Their society is self sustaining with no unemployment.

b. Entrepreneurship among indigenous people in Alaska

According to Dana (1995), Eskimos in Nome, Alaska have a lower tendency to become entrepreneurs than do non natives. The majority of the entrepreneurs were not actively seeking opportunity. Eskimo entrepreneurs tend to seek and respond to opportunity less than their non native counter parts. The study suggests that entrepreneurship is not the function of opportunity but as a function of cultural perceptions of opportunity.

c. Entrepreneurship in the Canadian sub arctic

According to Dana (1996), aboriginal tended to express entrepreneurship in the form of informal self employment, reflecting traditional activities such as hunting and fishing. In contrast entrepreneurship among non-aboriginals was found to be the result of opportunity identification. Entrepreneurship is not merely a function of opportunity but rather a function of one's cultural perception of opportunity.

d. Entrepreneurship in Laos (Asia)

The establishment of enterprise in Laos by the entrepreneurs is due to the impact of religion and culture. The national religion of Lao kingdom is Theravada Buddhism, wherein monks have been given higher status than entrepreneurs. A respectable man should not work towards the satisfaction of materialistic desires. He should strive to eliminate desire itself. In this culture, Entrepreneurship may be perceived as a means to satisfy desire, therefore this behaviour is ignored. Lao men are thus excluded from the entrepreneurship. This culture dampens the Entrepreneurial spirit among Lao men. The economy in Laos is growing. There are several hundred foreign entrepreneurs in this country, they are ethnic Chinese, Muslims, Australians and Thais own enterprises in Laos. Numerous Lao women have stands at local market but for Lao men cultural values make Entrepreneurial work a taboo (Dana, 1995).

e. Entrepreneurship in the kingdom of Lesotho (Africa)

In Lesotho culture dictates that certain property can not be bought or sold for cash. This is because of difference of culture in western society and in Lesotho. As for example a cow is not treated as commodity by the people of Lesotho. There are two types of properties in Lesotho one, property for personal consumption, and two, property with social value. Property for social value can not be exchanged for cash, although it dictates to trade cattle in exchange for bride. So cattle are property with social value. Property with social value should not be consumed because such assets ensure wealth in terms of long term security and prestige. If an individual is to sale an ox for personal gain it is interpreted as hostility against society. Real state is also considered as social value. Fields can not be bought or sold, home can not be bought or sold. Right to fields and homes are acquired through allocation, inheritance or bribery. A man can build a fancy home on the land on which he has rights, if he becomes poor and might not have money for food, culture prevents him to sale his house, which comes under property for social value.

Different countries have implemented a variety of models to help their entrepreneurs. An entrepreneurship model should not be copied into a new environment without verifying appropriateness. Identification of opportunity and response to it by entrepreneurs is culture bound.

Theories of Entrepreneurship

Schumpeter Theory of Entrepreneurship

Schumpeter (1949) proposed a dynamic theory of entrepreneurship. It breaks up the stationery circular flow of economy resulting in initiation and sustenance of development process. Entrepreneur activates the economy through new combinations, factors of production known as innovation. This takes place in five functions:

1. Introduction of a new good.
2. Introduction of new method of production.
3. Opening of a new market.
4. Conquest of a new source of supply of raw materials and
5. Carrying out of new organizations of any industry. Innovations create new demand and entrepreneur brings these innovations to the market. Entrepreneurs are major mechanisms of wealth creation and distribution in capitalism.

Classical Entrepreneurship Theory

This theory is explained with the help of classical economic theory. In classical economic theory there are two basic issues: 1. wealth creation in society, and 2. wealth distribution in society. It is based on private ownership of the society the creation and distribution of the wealth are done through the exchange of goods and services under open uncontrolled markets. It is open to all buyers and sellers. Adam Smith developed classical capitalism as an economic theory.

Neoclassical Entrepreneurship Theory

Neoclassical entrepreneurship theory is explained with the help of neoclassical economic theory. The key concept of this theory is that there are many buyers and sellers in the market. They interact and insure that supply equals demand. Neoclassical theory proposes that capitalized market create wealth and distribute wealth among buyers and sellers. It is based on the concept of "economics of scale" the cost of production per unit decreases with the increase of the size of the firm. This theory suggests that large firms are more profitable than small firms.

Mark Christopher Casson's Economic Theory

According to Casson, Individuals differ in taste and access to information. The entrepreneur proceeds on the basis of unique information available to him. There are inherent difficulties (transaction costs) in organising markets. The entrepreneur often must create market institutions. According to this theory entrepreneurs take judgemental decision regarding the coordination of scarce resources. Cassons theory emphasises coordination as a problem and a process, private or social. Contract and conjuncture are two mechanisms of allocation of scarce resources. The state of equilibrium reaches through bargaining. The occurrence of coordination is crucial through the assessment of a situation. According to Casson "The theory of entrepreneurs is related to the theory of market making firm; the entrepreneur operates in a market economy through the firm, of which the entrepreneur is the founder or owner manager"

Leibenstein's X-efficiency Theory (http://www-rohan.sdsu.edu/~frantz/docs/Hosseini.pdf)

X-efficiency refers to the degree of inefficiency in resource use within the farm. It measures the failure of harvesting the productive potential of resources by the farm. X-efficiency is due to wrong use of farm resources or due to not used at all. Leibenstein's points out two main roles for the entrepreneur one is input completion and other is gap filling.

Input Completion: Entrepreneurs plays an important role to improve the flow of information in the market. Inputs are made available to improve the efficiency of production methods or help in introducing the new one.

Gap filling: Arbitrage function has been emphasized in gap filling. The economy is made up of nodes and path ways in a network.

Harvard School Theory: Entrepreneurs are engaged in purposeful activity of developing a profit oriented business in interaction with internal and external forces. Internal forces are the characteristics' of entrepreneurs *viz.*, risk taking, innovativeness, achievement orientation personal resourcefulness, opportunity seeking, strategic reasoning and external forces include surrounding the business, economic, social political, cultural and legal framework. This theory emphasise two types of activities. 1. The organisation or coordination activity and 2. Environmental characteristics' affecting decision making.

Axiomatic Theory of Entrepreneurship

De and Rao (2001) proposed an axiomatic theory of entrepreneurship in agriculture. Axiomatic theory consists of a propositional statements ranging from postulates to hypothesis with a built-in logical system for explaining and predicting the entrepreneurial behaviour of farmers. According to them Axiomatic theory of entrepreneurship is higher the value orientation, marketing facilities, education and socioeconomic status lead to the higher entrepreneurial behavious of farmers.

Empirical Research on Entrepreneurial Behaviour of Farmers

A study was conducted on entrepreneurial behaviour of vegetable growers in Andhra Pradesh by Rao and De (2000) they found out the following package of personality characteristics (predictor) of entrepreneurial farmers after reviewing the past researchers in the field. They also found out the different dimensions of entrepreneurial behaviour which have been presented in the tabular form below.

Dimensions Related to Entrepreneurial Behaviour

Sl.No.	Name of the Dimension	Number of Studies	Number of Studies Significant	Number of Studies Non Significant
1.	Innovativeness	9	8	1
2.	Not to discourage by failure	1	1	–
3.	Future oriented	3	2	1
4.	Hard working	1	1	–
5.	Persistent	1	1	–
6.	Takes personal responsibility	2	2	–
7.	Risk taking ability	8	7	1
8.	Set goals and realistic sub goals	1	1	–
9.	Drive for power but not excessive	3	2	1
10	Achievement motivation	14	12	2
11.	Desires feed back and learns from experience	1	1	–
12.	Desires, feed back and learns from experience recognising error	1	1	–
13.	goal oriented	2	2	–
14.	Ability to exploit situation	4	2	2
15.	Willing to learn and does not suffer from complex	1	–	1
16.	Has self confidence	7	5	2
17.	Constantly under stress	1	–	1
18.	Knowledge of farming	9	9	–
19.	Person of integrity	1	1	–
20.	Like to excel in work	1	1	–
21.	Pleasant personality	3	2	1
22.	Success oriented	4	3	1
23.	Egoist	1	–	1
24.	Competitive	2	2	–
25.	Dreamer	1	–	1
26.	Time is important	3	2	1
27.	Imagination	1	1	–
28.	Courageous	1	1	–

Contd...

Contd...

Sl.No.	Name of the Dimension	Number of Studies	Number of Studies Significant	Number of Studies Non Significant
29.	Self starter	2	1	1
30.	Failure as step toward success	2	1	1
31.	Individualist	1	–	1
32.	Multifaceted interests	3	2	1
33.	Likes challenges	3	3	–
34.	Dynamic	5	4	1
35.	Leadership ability	7	7	–
36.	Takes initiative	4	3	1
37.	Higher energy level	1	1	–
38.	Impatient	1	1	–
39.	Sensitive and perceptive to people and environment	2	2	–
40.	There is always better way	1	1	–
41.	High tolerance and ambiguities	2	2	–
42.	Good communication with people	6	5	1
43.	Mentally hyper active	4	3	1
44.	Intuitive	1	–	1
45.	Ability to grasp quickly	2	2	–
46.	Belief in god	3	3	–
47.	Enjoys living, work and play	1	1	–
48.	Dissatisfied with general situation	4	2	2
49.	At least average intelligence	5	3	2
50.	Want to make lot of money	4	3	1
51.	Personal resourcefulness	1	1	–
52.	Originality	1	1	–
53.	Task result oriented	3	3	–
54.	Utilization of available assistance	6	5	1
55.	Management orientation	4	3	1
56.	Ability of organization and administration	3	2	1
57.	Searching environment and time bound planning	4	4	–
58.	Hopeful about future	7	6	1
59.	Is good sales man of idea	2	2	–
60.	Strategic vision	1	1	–
61.	Likes to excel in his work	1	1	–
62.	Managerial competence	2	2	–

Contd...

Contd...

Sl.No.	Name of the Dimension	Number of Studies	Number of Studies Significant	Number of Studies Non Significant
63.	Tactful	1	1	–
64.	Farm decision making	14	12	2
65.	Progressivism	2	2	–
66.	Creative	6	5	1
67.	Locus control	2	1	1
68.	Tolerance of ambiguity	2	1	1
	Total = 68			

Sl.No.	Name of the Predictor Variable	Number of Studies	Number of Studies Significant	Number of Studies Non Significant
1.	Age	5	3	2
2.	Education	9	8	1
3.	Family size	2	2	–
4.	Parents	1	–	1
5.	Wife	1	1	–
6.	Friends	1	1	–
7.	Joint family	1	1	–
8.	Migration	3	2	1
9.	Peer group	1	–	1
10.	Caste	6	4	2
11.	Community	1	1	–
12.	Socio-economic status	14	11	3
13.	Income	2	1	1
14.	Sources of information utilization	6	5	1
15.	Training received	8	7	1
16.	Marketing facilities	7	6	1
17.	Attitude towards farming	3	2	1
18.	Farm size	3	1	2
19.	Experience	4	2	2
20.	Avalailability of material	2	2	–
21.	Family support	2	1	1
22.	Community support	1	–	1
23.	Availability of technology	1	1	–
24.	Social participation	7	5	2
25.	Ability to co-ordinate farming activities	6	5	1

Contd...

Contd...

Sl.No.	Name of the Predictor Variable	Number of Studies	Number of Studies Significant	Number of Studies Non Significant
26.	Economic resources	1	1	–
27.	Inter-personal competence	1	1	–
28.	Knowledge of results	2	1	1
29.	Adaptability	1	1	–
30.	Birth order	3	2	1
31.	Independence	1	1	–
32.	Cosmopoliteness	4	3	1
33.	Adoption	3	2	1
34.	Level of aspiration	1	1	–
35.	Occupation	3	2	1
36.	Financial status of family	5	4	1
37.	Value orientation	6	4	2
38.	Lack of employment opportunities	1	–	1
39.	Little opportunity for advancement	1	1	–
40.	Cropping intensity	1	1	–
41.	Irrigation potentiality	1	1	–
	Total = 41			

Number of predictor variables related to entrepreneurial behaviour of farmers is 41. Number of postulates and hypotheses were formulated and tested and concluded that "higher the value orientation, marketing facilities, education and socioeconomic status lead to higher entrepreneurial behaviour of farmers in the social system".

Conclusion

The conceptual framework presented here identifies several key components of entrepreneurs. Entrepreneurship process starts when the entrepreneur recognizes an opportunity in the environment. Entrepreneurship Development process goes through i. identification of enterprise ii. Creation of enterprise iii. Nurturing of enterprise and iv. Concluding/transition phase of enterprise. Empirical research in entrepreneurial behaviour of farmers has found out 41 predictor variables related to entrepreneurial behaviour and 68 dimensions related to entrepreneurial behaviour. This will help to identify relevant and significant variables and dimensions and their complex interrelationships. Postulates and hypotheses can be developed from these variables and dimension for subsequent empirical verifications and generalizations. This will also identifies to address the question: Why do they act as entrepreneurs? *i.e.,* the cause of entrepreneurial actions of individuals?

References

Agarwal, R. (1997). Innovations in entrepreneurship: major issues in new era of gloalisation. *Abhigyan.* Vol (15), pp. 65-71.

Aitken, H.G.J. (1963). The future of entrepreneurial research, explorations in entrepreneurial history. *Second series,* Vol (1), pp. 3-9.

Akhouri, M.M.P. (1975). Planning and monitoring entrepreneurship. *Developing entrepreneurship. A hand book* (Ed. Rao, T. V and Pareek, U). Learning system, New Delhi. pp. 336-347.

Akhouri, M.M. (1996). Conceptual background and rationale for entrepreneurship development training programmes. Paper presented in orientation course on entrepreneurship development in agriculture. December 17-31.IARI, New Delhi. pp. 10.

Ali, H. (1995). Gateways to market entry. *The Journal of entrepreneurship.* Vol (4), No. (I), pp. 49-69.

Appley, L.A. (1956). Management in action. *American management association,* New York. pp.19.

Atkinson, J.W. and Litwin, G.H. (1960). Achievement motive and test anxiety conceived motive to approach success and pp. 105- motive to avoid failure. *Journal of abnormal and social psychology,* Vol (60), No. (1), pp. 52-63.

Banerjee, M. and Talukdar, R.K. (1997). Variables influencing entrepreneurship of women entrepreneurs. *Indian Journal of Extension Education.* Vol. (33), No. (1 and 2). pp. 25-36.

Bankston, C.L. and Zhou, M. Q 996). Go fish: The Louisiana vietnamese and ethnic entrepreneurship in extra active industry. *National Journal of Sociology.* Vol (10), No. (1), pp. 37-55.

Basavanna, M. (1971). A study of self-confidence as an attitude of self-concept. *Ph.D. Thesis (Unpublished).* Srivenkateswara University, Tirupathi, Andhra Pradesh.

Basu, S.K. and Moulik, T.K. (1979). Preference oriented selection criteria for rural entrepreneurs, identification and selection of small scale entrepreneurs. (Ed. Rao, T.V. and Moulik, T.K.). Indian Institute of Management, Ahmadabad, Gujarat.

Battacharjee, S.K. and Akhouri, M.M.P. (1975). Profile of small industry entrepreneurs. *SEDME* Vol. (2), No. (1), pp. 73-85.

Battacharya, S.N. (1983). Entrepreneurship development in India and the South East Asian countries in agricultural and industrial sectors. Metropolian book company, New Delhi.

Bema, J. (1960). Industrial entrepreneurship in Madras State. Asia Publishing House, London.

Bethal, L.L. Atwar,F.S. Smith, G.B.E. and Stockman, H.A. (1930). Industrial Organization and Management. Mc Graw Hill Book Company, New York, pp. 126.

Bhagat, G.R. and Singh, Y.P. (1995). Effectiveness of entrepreneurial development training, on the knowledge level of farm entrepreneurs. *Indian Journal of Extension Education*. Vol. (31), No. (1-4), pp. 102-104.

Bhatt, R.S. (1974). Growth of entrepreneurship in small and medium sectors. The *Indian Journal of Public Administration*. Vol (20), No. (3), pp. 453-465.

Bhattacharya, D. Sarkar, S. and Gangopadhyay, P. (1997). Decision making role of women in rural development. A quantitative study. Social science perspectives in agriculture. Edited by Das Gupta, D. and Mukhopadhyay, S.D. Santineketan, W.B. pp: 143-156.

Bisht, N.S. and Sharma, P.K. (1991). *Entrepreneurship expectations and experience*. Himalaya publishing house, New Delhi. pp. 3-4.

Bradford, L.A. and Glenn, L. J. (1953). *Farm management analysis*. John wiley and sons, Inc. New York.

Bull, I. and Willard, G. E. (1993). Towards a theory of entrepreneurship. *Journal of Business Venturing*. Vol. (8), pp: 183-195.

Busenitz, L.L. (1996). Research on entrepreneurial alertness: Sampling measurement and theoretical issues. *Journal of Small business management*. Vol. (34), No. (4), pp. 35 44.

Bygrave, W.D. and Hofer, C.F. (1991). Theorizing about entrepreneurship. *Entrepreneurship Theory and Practice*. Vol. (6), No. (2), pp. 13-22.

Cantillon, R. (1955). *International encyclopedia of social sciences*. Quoted by Cochran, T.C. (1968) under entrepreneurship. MacMillan, New York. Vol. (5), pp. 88-90.

Carland, J. W. (1984). Differentiating entrepreneurs from small business owners. *Academy of Management Review*. Vol. (9), No. (2), pp. 354-359.

Carrier, C. (1996). Entrepreneurship in small business: An exploratory study. *Entrepreneurship Theory and Practice*. Vol. (2), No. (1), pp. 5-20.

Christopher, K.J. (1969). Socio psychological factors influencing the adoption of innovations of starting a small industry unit. SIET Institute, Hyderabad, pp. 68-71. Christopher, K.J. (1971). Identification and detection of latent entrepreneurship. Productivity. Vol. (12), No. (1), pp. 78- 83.

Cohen, N. (1980)"The Five Ages of the Entrepreneur." *Venture*: 40– 42.

Cole, A.H. (1949). Entrepreneurship and entrepreneurial history. In change and the entrepreneur. Prepared by the research centre in entrepreneurial history. Harward University Press, Cambridge.

Cole, A.H. (1959). *Business enterprise in it social setting*. Harward University Press, Cambridge.

Cole, A.H. (1968). Meso-economics: A contribution from entrepreneurial history. *Explorations in Entrepreneurial History*. Vol. (6), No. (1), pp. 78-86.

Cooper, A.C. Timothy, B and Woo, C. (1995). Entrepreneurial information search. *Journal of Business Venturing*. Vol. (10), No. (2), pp. 107-120.

Covin, J.G. and Slevin, D.P. (1991). A conceptual model of entrepreneurship as firm behaviour. *Entrepreneurship Theory and Practice*. Vol. (16), No. (1), pp. 7-25.

Dasgupta, A. (1969). Indian business management. Deptt. Of business management and industrial relations, University of Delhi. pp. 60.

De, D. (1985). Status symbol and innovative entrepreneurship are predictors of farmers progressivism. *Journal of Extension System*. Vol. (1), No. (1), pp. 81-84.

De,D. (1986). Factors affecting entrepreneur characteristics of farmers. *Indian Journal of Social Work*. Vol. (XLVI), No. (4), pp. 541-546.

De D and Rao M. S. (2001) Entrepreneurial behaviour of farmers: An Axiomatic theory. Ganga Kaveri Publishing House, D.35/77 Jangamwadi Math Varanasi.

Deivasena Pathy, P. (1986). "Entrepreneurial success: Influence of certain personal variables". *The Indian Journal of Social Work*. Vol. (XLVI), No. (4), pp. 547-553.

Derossai, F. (1971). The maxican entrepreneur. Development centre of the organisation for economic co-operation and development. Paris.

Desai, G.R. (1981). A critical analysis of the contribution of Extension guidance to economic performance of cotton farmers of Karnataka State. Thesis (Unpublished), U.A.S., Banglore.

Downing, J. (1991). Gender and growth of enterprises: Small Enterprise Development. Vol. (2), No. (1), pp. 4-12.

Drucker, P. (1954). *The practice of management*. Harper and Row, New York. pp. 17.

Drucker, F.P. (1985). *Innovation and Entrepreneurship: Practice and Principles*. Heinemann, London. pp. 31.

East West Centre, (1977). Entrepreneurial discovery and development progress of action research. East –West technology and development institute, Honolulu, Hawai. pp. 82 83.

Encyclopedia Britanica: Micropedia Vol. III: Ready reference and index.

Encyclopedia Britanica, Inc. Benton, W. Publishers, 1943-1973. Helen Hemingway, Chicago.

Encyclopedia of psychology, Search press, Ed. Eysenck, H.J. and Arnold, W. and Meili, R. London. Vol (2).

Encyclopedia of social sciences, (1933). Mc Millan Company, New York. Vol. (20), pp. 76 77.

English,H.B. and English, A. (1958). A comprehensive dictionary of psychological and psychoanalytical terms. Longmans. Green and Company, New York.

Fishbein, M. (1965). Attitude and the prediction of behaviour. In Fishbein, M (Ed.), *Readings in attitude theory' and measurement*. John Wiley and Sons, New York. pp. 477-492.

Fraser, T.M. (1961). Achievement motivation as a factor in rural development, a report on research in Western Orissa. Haverford College, Haverford, Paris.

Gaikwad, J.G. Inamke, N.M. and Kadam, K. R. (1997). Involvement of farm women in decision making. *Maharashtra.J. Extn. Edn.* Vol. (16), pp. 359-361.

Gaikwad, V.R. (1978). Entrepreneurship: The concept and social context in developing entrepreneurship. A hand book of Rao, T. V. and Pareek, U. in collaboration with Deshpande, M. V Mehta, P. and Nadkarni. M.S. Learning systems, New Delhi.

Gartner, W.B., Bird, B.J. and Starr, J.A. (1992). Acting as if Differentiating entrepreneurial from organizational behaviour. *Entrepreneurship theory and Practice.* Vol. 16 (3). pp. 13-31.

Ghosh, R. Gupta, M. and Dhar, S.N. (1998). Women and entrepreneurship in India. *Entrepreneurship innovation* edited by Kanungo, R.N. Sage Publications, New Delhi. pp.156-175.

Gogoi, M. and Talukdar, R.K. (1997). Determining achievement motivation of agricultural scientists. *Indian Journal of Extension Education.* Vol. (33), No. (1 and 2), pp. 64-67.

Gopakumar K (1995) Entrepreneurship in economic thought: A thematic review. *Journal of Entrepreneurship,* 4 (1) Pp 1-17.

Gordon, L. (1961). Private enterprise and international development. *The American Review.* Vol. (6), No. (1), pp. 5-13.

Gupta, C.B. and Srinivasan, N. P. (1992). Entrepreneurial development: Second edition. Sultan Chand and Sons Educational Publishers, New Delhi.

Harbison, F.N. and Hyers, C.A. (1964). Education man power and economic growth. In Nodder, B.W. (Eds.). *Economic development in the topics,* Meuthen, London.

Haredero, J.M. (1979). *Agricultural entrepreneurship. Identification and selection of small scale entrepreneurs.* (Ed. Rao. T.V. and Moulik, T.K.). Indian Institute of Management, Ahmadabad. pp. 151-159.

Harper, M. and Vyakaranam, S. (1988). Rural enterprise: Case studies for developing countries. Intermediate technology publications, London.

Hawley, F.B. (1882). Entrepreneurship in interaction. In *Encyclopedia of the social sciences* (1968). Mac Millan, New York. Vol. (5), pp. 87-90.

Hegan (1968) *The economics of development.* Irwin RD Inc. Illinois, New York

Hepple, L.M. (1959). Group organization and Leadership in rural life. Lucas bros, Columbia, Missouri.

Himachalam, A (1990). Developing entrepreneurial industrialist - An alternate approach. *Productivity.* Vol. 34 (2). pp.281-287.

Homans, G.C. (1970). The relevance of psychology to the explanation of social phenomena. Borger, R. Cioffi, F. (Ed.) *Explanation in the behavioural sciences.* Cambridge University Press, Cambridge. pp. 313-326.

Hornaday, J.A. and Aboud, J. (1971). Characteristics of successful entrepreneurs. *Personal psychology.* Vol. (14), No. (6), pp. 66-78.

Hoselitz, B.F. (1957). Non-Economic factors in economic development in Tandoil, B.C. (Eds.). *Environmental and entrepreneur.* Chugh publications, Allahabad, India.

Hundal, P.S. and Singh, S. (1975). Structure of personality characteristics and motive pattern of farmers. *Indian Journal of Psychology:* Vol. (50), No. (1), pp. 33-43.

Jaiswal, N.K. (1965). A study of factors associated with low level of adoption of improved agricultural practices. *Ph.D. thesis,* Division of agricultural extension, IARI, New Delhi.

Jenkins; J. C. (1983). A resource mobilization theory and the study of social movements. *Annual review of sociology.* Vol. (9), pp. 527-533.

Jhamtani, A. (1996). Dynamics of entrepreneurship. Orientation course on entrepreneurship development in agriculture. December 17-31, JARI, New Delhi, pp 5-6.

Jones, S. and Cohen, M.B. (1990). The emerging business. Ronald Press, New York.

Joshi, S.S. and Kapur, T.R. (1973). *Fundamentals of farm business management.* Kalyani publishers, New Delhi.

Jyothi, V. (1990). Entrepreneurship development among women - study of entrepreneurs in Guntur district. M. Phil dissertation submitted to Nagarjuna University. Guntur, Andhra Pradesh.

Jyothi, V. and Prasad, G. (1993). A profile of potential rural women entrepreneurs. *Small enterprises development management and extension Journal (SEDME).* Vol. (20), No. (1), pp. 23-38.

Kanungo, R.N. and Bhatnagar, J. (1978). Achievement orientation and occupational values: A comparative study of Young French and English Canadians. Canadian Journal of Behavioural Science. Vol. (10), No. (3), pp. 202-213.

Kanungo, R.N. and Conger, J.A. (1995). Model orientation in leadership research and their implications for developing countries. New approaches to employee management. Vol (3), pp. 155-170.

Kanungo, R.N. and Mendonca, M. (1994). *Fundamentals of organizational behaviour.* Kendall Hunt publishing company, Iowa.

Keshari, P.K. and Kota, R. (1995). Determinants of the propensity to export: The case of small firms in the diesel engine and chemical industries in Gujarat. *The Journal of entrepreneurship.* Vol. (4), No. (1), pp. 35-48.

Khan, N.A. (1998). Agricultural marketing: Present position and future strategy, *Kurukshetra.* Vol. (47), No. (3), pp. 11-13.

Khan, W, M. (1998). Technological innovation in SMEs in Pakistan. *Entrepreneurship and innovation* edited by Kanungo, R.N. Sage publications, New Delhi. pp. 287-309.

Khanka, S.S. (1998). Making the entrepreneurial society. *Yojana.* Vol. (42), No. (2), pp. 9 11.

Khemmani, M. (1983). Turning thoughts towards training, Manita, phillippines, Asia and Pacific programme for development training and communication planning (DTCP), UNDP. Training tips No.1 Editor: Andrew P. Bartlett.

Kilby, P. (1971). *Entrepreneurship and economic development.* Free press, New York.

Kirzner, I.M. (1979). *Perception, opportunity and profit: Studies.in the theory of entrepreneurship.* The University Chicago press, Chicago.

Kittur, M.M. (1976). A study of adoption behaviour of marginal farmers in relation to their characteristics and value orientation in Bizapur district of Karnataka state. *M.Sc. (Ag.) Thesis.* Division of Agricultural Extension, U.A.S. Dharwar.

Knudson, W., Wysocki A., Champagne J. and H. Christopher Peterson, (2004). Entrepreneurship and innovation in the Agri-food System. *Amer. J. Agr. Econ.* 86:1330-1336.

Kogan, N. and Wallach, M.A. (1964). *Risk taking. A study in cognition and personality.* Holt, Rinehart and winston, New York.

Kotler, P and Levy, S.J. (1969). Broadening the concept of marketing. *Journal of marketing.* Vol (16), No. (8), pp.63-64.

Koontz, H. (1994). *Essentials of management.* (I edition). Mc. Graw. Hill publishing company, New York. pp. 39.

Kourilsky, M. (1980). Predictors of entrepreneurship in stimulated economy. *Journal of creative behavioul:* Vol. (14), No.3. pp. 76-86.

Krishnan, S.B. Kumar, K. G. and Kanungo, R.N. (1998). *Entrepreneurship development concept and context. Entrepreneurship and innovation.* Sage publications, New Delhi. pp. 21.

Kumar, E.S. (1.998). Theorizing in entrepreneurship: A review, critique and future directions. *Entrepreneurship innovation.* Edited by Kanungo, R.N. Sage publications, New Delhi, pp.90-115.

Kumar, G.K. (1995). The entrepreneur in economic thought: A thematic overview. *The Journal of entrepreneurship.* Vol. (4), No. (1), pp. 1-15.

Kumar, K.G. (1998). Management skills for rural enterprises. *Entrepreneurship innovation.* Edited by Kanungo, R.N. Sage publications, New Delhi, pp. 338-357.

Kumar, V.S. (1998). Small business marketing: Need for specilized skills. *Entrepreneurship innovation.* Edited by Kanungo, R.N. Sage publications, New Delhi, pp. 325-337.

Lamb, R.K. (19652). The entrepreneur and the community in men in business. In Baumback, C.M. and Mancuso, JoR. (Eds.). *'Entrepreneurship and venture management.* Prentice Hall Inc, Engle wood cliffs, New Jersey.

Learned, K.E. (1992).What happened before the organization? A model of organization formation. *Entrepreneurship Theory and Practice.* Vol. 17(1). pp. 39-48.

Leeds, C.N. and Stainton, R.S. (1978). *Management and business studies (Second edition).* Macdonald and Evans.

Leibenstein, H. (1976). *Beyond economic man: A new foundation for micro economics.* Harward University, Cambridge, Massachusetts.

Lekha, K.C. Kalyan, W and Lavoie, D. (1998). Micro enterprise management by women in India. *Entrepreneurship innovation.* Edited by Kanungo, R.N. Sage publications, New Delhi, pp. 176-199.

Lim, C. (1977). Small industry in Malaysia: A socio economic profile of entrepreneurs. *Social action.* Vol. (27), No. (2), pp. 22-36.

Lumpkin, G.T. and Dess, G.G. (1996). Clarifying the entrepreneurial orientation construct and linking it to performance. *Academy of management review.* Vol. (21), No. (1), pp. 135-172.

Ma, Hao and Tan, Justin, (2006). Key components and implications of entrepreneurship: A 4 P framework, *Journal of Business Venturing.* 21:704-725.

Mall, C.P. (1993). Impact of financial incentives on entrepreneurial development in backward regions: A case study of Deoria district of V.P. *The Journal of entrepreneurship.* Vol. (2), l No. (2), pp. 199-207.

Mancuso, J. and Mascolo, M. (1987). Recognising achievement motivation. *Motivation and emotion.* Vol. (11), No. (3). pp. 323-330.

Manimala, M.J. (1998). Net working for innovation: Anecdotal evidences from a large sample study of innovative enterprises. *The Journal of entrepreneurship.* Vol. (7), No. (2), pp. 153-169.

Manimekalai, N. (1998). Gem cutting and rural entrepreneurship development. *Kurukshetra.* Vol. XLVI, No. (5), pp. 22-26.

Manimekalai, N. and Abdullah, M. (1998). Charcoal production and rural entrepreneurship development. *The Indian Economic Journal.* Vol. (45), No. (2), pp. 159-169.

Manjula, S. (1995). A study of entrepreneurial behaviour of rural women in Ranga Reddy district of Andhra Pradesh. *M. Sc. (Home Science). Thesis.* ANGRAV, Rajendra Nagar, Hyderabad.

Massie, J.L. (1982). *Essentials of management (Third edition).* Prentice Hall of India private limited, New Delhi.

Mcclelland, D.C. (1953). *Achievement motive.* Appleton, New York.

Mcclelland, D.C. (1965). Need achievement and entrepreneurship: A longitudinal study. *Journal of personality and social psychology.* Vol. (1), pp. 389-392.

Mcclelland, D.C. (1969). *Motivating economic achievement.* Free Press, New York.

Mishra, S.P. (1979). A study of farm entrepreneurship in a backward district of Bihar, *Ph.D. Thesis* (Unpublished), IARI, New Delhi.

Mishra, S.P. (1996). Training in entrepreneurship -A conceptual background. Paper presented in orientation course on entrepreneurship development in agriculture. December 17-31. IARI, New Delhi.

Mokry, B.W. (1988). Entrepreneurship and public policy. Quorum books, New York.

Nair, K.B. Sivaraman, M. and Kanungo, R.N. (1998). Rural entrepreneurship in a developing economy: A contingency approach. *Entrepreneurship innovation*. Edited by Kanungo, R.N. Sage publications, New Delhi. pp. 78-89.

Nandapurkar, G.G. (1980). A study of the entrepreneurial behaviour of small farmers. Unpublished *Ph.D. thesis*. U.A.S., Bangalore.

Nandi, A. (1973). Motives, modernity and entrepreneurial competence. *Journal of social psychology*. Vol. (91), No. (3), pp. 127-136.

Neill, R.E. and Rogers, E.M. (1963). Measuring achievement motivation among farmers. Department series A.E. 346. Ohio Agricul. Expt. Stn. Columbus.

Niles, M.C. (1956). *The essence of management*. Orient longmans, Bombay. pp. 20.

Palmer, B.l. Ahiamadu, R.K. (1997). Leadership style and performance of public enterprises. A Nigerian study. *Productivity*. Vol. (38), No. (1), pp. 54-64.

Pandey, VK. and Sharma, K.C. (1996). Crop diversification and self sufficiency in food grains. *Indian Journal of agricultural economics*. Vol. (5), No. (4), pp. 644-651.

Pareek, U. and Nadkarni. M. (1978). Development of entrepreneurship. A conceptual model. In Rao, T.V and Pareek, U. (Eds.), *Developing entrepreneurship: A hand book, learning systems*, New Delhi.

Pareek, U. and Rao, T.V. (1995). Counseling and helping entrepreneurs. *The Journal of entrepreneurship*. Vol. (4), No. (1), pp. 19-34.

Patel, D.R. (1995). Entrepreneurship and small business development for women. *Kurukshetra*. Vol. (XLIII), No. (11), pp. 65-68.

Patel, G. V. (1970). A successful ventures in developing entrepreneurs. Lessons for Gujarat. A symposium paper on development of entrepreneurial talent. New Delhi.

Patel, M.M. Sanoria, Y.C. (1997). Correlates of entrepreneurial behaviour of sugarcane growers. *Maharashtra J Extn. Edn*. Vol. (16), pp. 344-346.

Paul, K.C. (1998). A study on entrepreneurial behaviour of vegetable growers in Krishna district of Andhra Pradesh. *M.Sc. (Ag.) thesis*, ANGRAU, Rajendra Nagar, Hyderabad.

Pearce, I.A. Kramer, T.R. and Robbins, K. (1997). 'Effect of managers' entrepreneurial behaviour on subordinates. *Journal of business venturing*. Vol. (12), No. (2), pp. 147 160.

Pollo, L.V. (1996). On economic theory of entrepreneurship: The strategic role of the entrepreneurial function. International review of sociology. Vol. (6), No. (2), pp. 191-213.

PRADAN (2000). System setting document. Self Help Groups Pradan Programme, New Delhi

Prakasam, D.S. (1998). Entrepreneur: An important change agent in every society. *Kurukshetra*. Vol, (46), No. (10), pp. 51- 52.

Prasad, R.M. (1983). Comparative analysis of achievement motivation of rice growers in three states in India. *Ph.D. thesis (unpublished)*, U.A.S., Banglore.

Prasad, S.K. (1996). Promoting an entrepreneurship development programme. *Laghu Udyog*. Vol. (21), No. (2), pp. 3-5.

Rajak, M.P. (1998). Craft technology and entrepreneur. *Kurukshetra*. Vol. (XLVI), No. (5), pp. 17-19.

Ramachandran K and Ray Sougata (1998) A framework and entrepreneurial typology for developing a comprehensive theory of entrepreneurship. In *Entrepreneurship and innovation: Models for development*. Ed Kanungo R. N. Sage Publications New Delhi. Pp.40-63.

Ramana, A. V. and Papaiah, R. (1998). Nuances of entrepreneurship development. *Indian management*. Vol. (37), No. (2), pp. 48-53.

Ramakrishnan, K. (1979). Interview techniques as a method of selection in identification and selection of small entrepreneurs (Eds. Rao, T.V. and Moulik, T.K.) Indian Institute of Management, Ahmedabad.

Ramana, A. V. (1999). Entrepreneurship and economic development. *Kurukshetra*. Vol. (48), No. (2), pp. (8 to 10 and 45).

Rani, C. (1986). Potential women entrepreneurs. A study. *SEDME*. Vol. (13), No. (3), pp. 13 32.

Rao, G.K.V. (1985). A prediction analysis of farming performance of farmers through their entrepreneurial behaviour factors. *Ph.D. thesis*, A.P.A.U. Rajendra Nagar, Hyderabad.

Rao, M.S. (2000). Axiomatic approach to theory building in entrepreneurial behaviour of farmers. *Ph.D. thesis*, Institute of Agricultural Sciences, B.H.U., Varanasi.

Rao, M.S. and Mishra, S.K. (1997). Farmers perceptions of environmental risk in pesticide use. Maharashtra Journal of Extension Education. Vol. (16), pp. 214-219. Rao, T. V (1975). Development of an entrepreneur: A behaviouristic model. *SEDME*. Vol. (1), No. (4), pp. 325-345.

Rao, T.V. (1975). Characteristics of entrepreneurs as perceived by entrepreneurial development agents. Paper presented at the work shop on identification and selection of entrepreneurs. Indian Institute of Management, Ahmedabad.

Rao, T.V. and Mehta, P. (1978). Psychological factors in entrepreneurship. In Rao, T.V. and Pareek, U. (Eds.). *Developing entrepreneurship*. A hand book. Learning systems, New Delhi.

Rao, T.V.S. (1989). A study on entrepreneurial characteristics of vegetable growers in Guntur district of Andhra Pradesh. Unpublished *M.Sc. (Ag.) thesis*, A.P.A.U., Rajendra Nagar, Hyderabad.

Ray, S. and Ramachandran, K. (1996). Towards a frame work for a comprehensive theory of entrepreneurship. *The Journal of entrepreneurship*. Vol. (5), No. (1), pp. 1-22.

Razin, E. and Langlois, A. (1996). Metropolitan characteristics and entrepreneurship among immigrants and Ethnic groups in Canada. *International Migration Review*. Vol. (30), No. (3), pp. 703-727.

Reddy, Y.V.R. (1992). Economics of Agro forestry system with particular reference to agro horticultural systems on dry land region. *Int. sym. on natural resource management for a sustainable agriculture*, New Delhi. Feb. 6-10.

Richart, C.B. Maurer, J.G. (1997). The moderating effect of environmental variables on the entrepreneurial and marketing orientation on entrepreneur-led firms. *Entrepreneurship Theory and Practice*. Vol. (22), No. (1), pp. 47-58.

Ricky, W.G. (1992). *Essentials of management*. Prentice Hall of India Private Lmited, New Delhi. pp. 22-24.

Ripsas, S. (1998). Towards on interdisciplinary theory of entrepreneurship. *Small business economics*. Vol. (10), No. (2),pp.l03-115.

Rogers, E.M. and Svenning, L. (1969). *Modernization among peasants: The impact of communication*. Hort, Rinehart and Winston, New York.

Rogers, E.M. and Shoe maker, F.F. (1971). *Communications of innovations, A cross cultural approach*. The Free Press, A division of Macmillan publication company, Inc. New York.

Rogers, F.M. and Neill, R.E. (1966). *Achievement motivation among columbian peasants*. Michigan state University, East I Lansing, Michigan.

Rousmasset, J.A. (1977). *Risk and uncertainty in agricultural development*. Seminar report No. 15. Agricultural Development Council, New York.

Samanta, R.K. (1977). A study of some agro economic, socio psychological and communication variables associated with repayment behaviour of agricultural credit users of nationalised banks. *Ph.D. Thesis*, Bidanchandra Krishi Vidyalaya, Nadia, West Bengal.

Samanta, R.K. (1989). *Management in agriculture and rural development*. UDH publishing house, Naisarak, New Delhi.

Samuelson, P.A. (1963). Discussion -problems of methodology. *American economic review*. Vol. (53), pp. 236.

Sara, C. (1998). Portfolio entrepreneurship in the farm sector: Indigenous growth in rural areas? *Entrepreneurship and regional development*. Vol. (10), No. (1), pp. 17-32.

Say, J.B. (1815). Quoted under entrepreneurship in international encyclopedia of the social sciences (1968). Macmillan, New York. Vol. (5), pp. 87-90.

Schumpter, J.A. (1961). *The theory of economic development: An enquiry into profits, capital, interest and the business cycle.* Cambridge Harward University Press. Cambridge.

Schumpter, J.A. (1970). The entrepreneur as innovator. Reading in management (Second edition). Mc. Graw Hill, New York. Schwartz, E.A. (1979). Entrepreneurship: New female frontier. *Journal of Contemporary business winter issue.* Vol. (15), pp.150-163.

Scott, C.E. (1986). Why more women are becoming entrepreneurs. *Journal of small business management.* Vol. (24), No. (4), pp. 37-44.

Scott, M. (1997). Reengineering the rapidly growing entrepreneurial company. *Industrial management.* Vol. (39), No. (3), pp. 21-23.

Sengupta N (1998). Empowerment: A socio-psychological approach to Self Help Group formation. *Prajnan.* XXVI (4): 525-531

Shailendra, V. (1990). When the harvest is in developing rural entrepreneurship. Oxford and IBH publishing Co. Pvt. Ltd., New Delhi

Sharma, A Mehta: P. and Sha S.K.(1996). Evaluation of task m diversified crop farming of Himanchal Pradesh,. A case study of vegetable farmers; *Indian Journal of Agricultural Economics;* Vol. (51), No. (4); pp. 688.

Sharma, K.L. (1973). Growth of small scale entrepreneurship in India: A theoretical frame of analysis. *Interdisciplinary.* Vol. (10), No. (3), pp. 97-108.

Sharma, S. V.S. (1980). Characteristics of entrepreneurs. In developing entrepreneurship: Issue and problems. SIET (Small industry extension training) Institute, Hyderabad.

Singh, D.N. (1986). *Developing entrepreneurship for economic development.* Vikas publishing house private limited, New Delhi. pp. 83-92.

Singh, K. (1992). *Women entrepreneurs.* Ashish publishing house, New Delhi.

Singh, K.N. and Pal, K. (1974). A study on achievement motivation of farmers. *Interdiscipline.* Vol. (II), No. (I), pp. 31-40.

Singh, N.P. (1979). Achievement motivation among agricultural and business entrepreneurs of Delhi. *Journal of Social Psychology.* Vol. (81), pp. 146-149.

Singh, N.P. (1979). Tribal entrepreneurship, efforts and achievements. In Rao, T.V. and Moulik, T.K. (Eds.), *identification and selection of small scale entrepreneurs.* Indian Institute of Management, Ahmedabad.

Singh, N.P. and Gupta,R.S. (1985). *Promotion women entrepreneurs.* NIESBUD, New Delhi.

Singh, N.P. and Singh, K. (1971). Motivational components of agricultural and business entrepreneurs in India. *Indian Journal of Industrial relations.* Vol. (7), No. (1), pp.31 52.

Singh, P:N. (1986). *Factors influencing entrepreneurship. Developing entrepreneurship for economic development.* Vikas publishing house private limited, New Delhi. pp. 46-92.

Singh, 'S. (1978). Achievement motivation, decision making orientation and work values of fast and slow progressing fam1ers in India. *The Journal of social psychology*. Vol.:(.106), pp. 153-160.

Singh, S. (1979). Motives, Personality characteristics and work values of fast and slow progressing farmers. *Behavioural science and rural development*. Vol. (2), No. (2), pp.71-77.

Singh, T.P. and Kumar, K. (1975). Achievement motivation, socio- economic status and green revolution. *Indian Journal of Extension Education*. Vol. (1.1), No. (10), pp. 1-7.

Sinha,B.P. and Mehta, P. (1972). Size and land holding and farmers achievement motivation and change proneness. *Behaviour science and community development*. Vol. (6), No. (2), pp.244-25.1.

Sinha, T.N. (1996). Human factors in entrepreneurship effectiveness. *The Journal of entrepreneurship*. Vol. (5), No. (1), pp. 23-39.

Sisodia, G.S. (1976). A study of the achievement motivation of farmers for mechanization. *Indian Journal of Extension Education*. Vol. (13), No. (3 and 4), pp. 72-75.

Stevenson, H.H. and Jari110, I.C. (1990). A paradigm of entrepreneurship: Entrepreneurial management. *Strategic management Journal*, 10th summer special issue. Vol (11),pp. 17-27.

Strayton, R. (1986). The organization of course and conferences., in *ILO* (1986b) pp. 1-27.

Sullivan, A.O. Munir, K. and Dougherty, D. (1998). Building organizational capacities for product innovation. *Entrepreneurship innovation* edited by Kanungo, R.N. Sage publications, New Delhi.

Sundari, S.S.K. (1995). Entrepreneurship development. for rural women. *SDEM*. Vol. (1)., pp. 15-30.

Suresh, B. Kumar, G.K. and Kanungo, R.N. (1998). Entrepreneurship development: Concept and context. *Entrepreneurship innovation*. Edited by Kanungo, R.N. Sage publications, New Delhi.

Sutton, R.I. and Staw, B.M. (1965). What theory is not. *Administrative science*, Vol. (40), No. (3), pp. 371-384.

Swamy, B.A. (1988). Agricultural entrepreneurship in India. *Ph.D. thesis*, Andhra University, Waltair, Andhra Pradesh.

Terry, G.R. (1968). *Principles of management*. Irwin, R.D. Inc., Homewood, Illinois.

Thagamuthu, and Manimekhalai, N. (1989). Generation of employment for women through DWACRA. *Journal of rural development*. Vol. (8), No. (40), pp. 431-438.

Thomas, T. T. and Padmakar, K. (1993). Developing entrepreneurial industrialists. An alternative approach. *Productivity*. Vol. (34), No. (2), pp. 281-287.

Timmons,l.A. (1996). Characteristics and role demands of entrepreneurship. Paper presented in orientation course on entrepreneurship development in agriculture. December 17 31, IARI, New Delhi.

Tripathi, D. (1985). An integrated view of entrepreneurship. Economic and political weekly review of management. Vol. (20), No. (48), pp. 62-69. *Administrative science*. Vol. (31), No. (3), pp. 439-465.

Tyson, S. and York, A. (1989). Personal management. Made simple books, an imprint of Heinemann professional publishing limited. Oxford, England.

Vplankar, A. T. (1977). A study of self employed industrial entrepreneurs. *The Indian Journal of Social Work*. Vol. (38), No. (2), 137-147.

Vijaya, V. Kamalanabhan, T.J. (1998). A scale to assess f entrepreneurial motivation. *The Journal of; entrepreneurship*. Vol. (7), No. (2), pp. 183-198.

Vinze, M.D. (1987). *Women entrepreneurs in India*. Mittal publications, New Delhi.

Vyas, V.S. (1996). Diversification in agriculture: Concept, rationale and approaches. *Indian Journal of Agricultural Economics*. Vol. (51), No. (4), 635-643.

Weber, M. (1930). *Entrepreneurship and economic development*. The Free Press, New York.

Webster, S. (1968). *New world dictionary of the American language*. Cleveland.

Westermark, N. (1951). The human factor and success in farming. *Acta agriculture scandanavica*. Vol. (1), pp. 146.

Westermark, N. (1961). The human factors and economic progress. *Indian Journal of Agricultural Economics*. Vol. (16), pp. 12-17.

Wick, J.H. (1998). Understanding entrepreneurial intentions: The attitudinal determinants. *Entrepreneurship and innovation*. Edited by Kanungo, R.N. Sage Publications, New Delhi, pp. 119-138.

Wilken, P.H. (1979). *Entrepreneurship. A comparative and historical study*. Ablex publishing, Norvood. pp. 60.

Yasunobu, K. (1997). Entrepreneur fanners in peninsular Malaysia: profile of Peladang Jaya. *JIRCAS News letter. (December)*. No. (13). pp. 4-5.

Zahir, M.A. (1994). Entrepreneurship in Punjab. A case study. *Yojana*. Vol. (38), No. (8), pp. 24-28.

Chapter 2
Social Entrepreneurship
Dipak De, Basavaprabhu Jirli and Kirti
Department of Extension Education, Institute of Agricultural Sciences,
Banaras Hindu University, Varanasi – 221 005, U.P.

Introduction

It is now well established fact that Entrepreneurship is one of the important factors for social inclusion and economic growth leading to development and well being of the society. It helps to overcome the problems of unemployment by increasing the adoption of innovations and accelerating changes in economic structure. It creates healthy economic environment in the society. In the present context it has become increased interest to understand value creation in the field of entrepreneurship. This has helped academicians to develop a field of enquiry known as social entrepreneurship.

Whenever society is in stuck or has an opportunity to seize a new opportunity, it needs an entrepreneur to see the opportunity and then to turn that vision into a realistic idea and then a reality and then, indeed, the new pattern all across society. We need such entrepreneurial leadership at least as much in education and human rights as we do in communications and hotels. This is the work of social entrepreneurs (Bill Drayton Founder of Ashoka: Innovators for the Public).

Concept of Social Entrepreneurship

Social entrepreneurship is the field in which entrepreneurs tailor their activities to be directly tied with the ultimate goal of creating social value. In doing so, they often act with little or no intention to gain personal profit. A social entrepreneur "combines the passion of a social mission with an image of business-like discipline, innovation, and determination commonly associated with, for instance, the high-tech pioneers of Silicon Valley" (Dees, 1998; tinyurl.com/86g2a6).

Social entrepreneurship is the product of individuals, organizations, and networks that challenge conventional structures by addressing failures–and identifying new opportunities–

in the institutional arrangements that currently cause the inadequate provision or unequal distribution of social and environmental goods. Social entrepreneurship can further be defined as any action that displays three key characteristics: sociality, innovation, and market orientation (Nicholls and Cho, 2006). The notion of "sociality" entails a context, process, and/ or set of outputs that might reasonably be considered to be in the public benefit. "Innovation" indicates the creation of new ideas and models that address social or environmental issues. Socially entrepreneurial innovation can be manifested in three ways: in new product and service development (institutional innovation); in the use of existing goods and services in new, more socially productive, ways (incremental innovation); in reframing normative terms of reference to redefine social problems and suggest new solutions (disruptive innovation). Finally, "market orientation" here suggests that social entrepreneurship exhibits a performance-driven, competitive, outlook that drives greater accountability and cooperation across sectors. Social enterprises specifically address conventional competitive markets, but elsewhere social entrepreneurship broadens the conception of a "market" beyond the merely neo-liberal to suggest that markets establish exchange value and that this is inevitably socially embedded. Thus, market orientation here includes ideas of reciprocity and the common good, as well as the rational, utility maximizing individual. Social entrepreneurship can be the product of for-profit or not-for-profit organizations, as well as of many hybrids in between these two poles. It is also found in the public sector (Nicholas, 2009).

Bornstein and Davis (2010) define three historical phases that the idea has passed through in the last three decades.

1. The first phase was to "identify people with innovative ideas and practical models for achieving major social impact," spotlight their work and provide a support system.
2. The second phase was to help social entrepreneurs build sustainable and high impact organizations through better processes of management, finance and outcome evaluation. The focus at present,
3. In the third phase, is to help promote all people as potential change makers in whatever form that someone can manage, much more in the style of a social movement.

According to Pontus Braunerhjelm and Ulrika Stuart Hamilton Social entrepreneurship – a survey of current research working paper 2012:09 Swedish Entrepreneurship Forum (http://entreprenorskapsforum.se/wp-content/uploads/2013/03/WP_09.pdf) three decisive macro-dynamics have been identified in the emergence of social entrepreneurial activities around the world. The first of these is the slowdown of the public offering of products and social services, which has contributed to an increase in needs not fulfilled (Light, 2008). This is especially true for social welfare, with regards to which public sector involvement is rather limited (Sharir and Lerner, 2006).

Second, the existing disequilibrium in the distribution of income level in both developing and developed countries has increased the need for a new paradigm and new business strategies (Bornstein, 2004). Over the past two decades, social entrepreneurs have also increasingly employed business strategies to address problems and generate revenues.

Finally, the increased competition within the non-profit sector to achieve donations and grants has led to the need to professionalize the activities undertaken with the objective of reducing financial dependence and thus ensure their economic stability for the development of their social mission (Perrini, 2006). Social movements that over the last couple of decades have begun promoting social entrepreneurship comprise *e.g.* Ashoka Foundation (Bill Drayton), the Skoll Foundation (Jeff Skoll), and Schwab Foundation (Hilde and Klaus Schwab).

The Grameen Bank and the Bangladesh Rural Advancement Committee (BRAC) in Bangladesh and the USA-based Ashoka organization are the examples of social entrepreneurship. Beginning in the late 1970s in Bangladesh, Grameen and BRAC helped to organize millions of the poor. Ashoka, founded in 1980, helped to identify people on the ground who had already begun promising social change projects, and provided them with interim support as they attempted to scale their work to reach a critical mass. Ashoka's founder Bill Drayton proposed the term **"social entrepreneur"** just as Grameen had popularized the term "micro-lending."

The practice of foundations like Skoll2, Echoing Green3 and the Geneva-based Schwab Foundation for Social Entrepreneurship4, which have funded and helped to define practice over the last decade. The practice of social entrepreneurship is not recent. Its roots go back to the 1970s, when applications of social organizing took root in Bangladesh without the benefit of theory. In what is now part of the "creation myth" of the idea of microlending and microfinance, Muhammad Yunus (1999) recounts how during a severe famine in his country he and some of his economics students at a university went out to distribute small amounts of money to surrounding villages as unsecured loans and found an astoundingly high level of repayments. This grew from a handful of recipients in 1974 to 7,000,000 by 2007, and the establishment of a large set of institutions that did everything from telecommunications to manufacturing of yogurt as well as acting as a bank for the poor (Yunus, 2007). A similar history of growth and success is the parallel history of BRAC, also in Bangladesh. The kind of organizations that BRAC and Grameen represented were different from usual development efforts at that time, and the difference suggested an innovation in thinking about development and social change.

There were some characteristics that Ashoka looked for in their recruitment of these change makers whom they would support, which suggest a vision of social change. These were:

☆ **Creativity/social innovation**: A new and original idea about how to solve a significant social problem that had been field tested.

☆ **Entrepreneurial quality**: Someone attuned to a particular context and not satisfied until their idea works locally and is spread to as many people in the region or nation as possible (sometimes termed "scaling up" in development terminology).

☆ **Social impact of the idea**: Two aspects of the idea are included: the idea has some kind of measureable impact (not necessarily quantitative– more below on this) and it is adopted by others in different contexts for application and spread (replication).

☆ **Ethical fiber:** The change maker inspires trust in others (a concept so important in Drayton's mind, that a unanimous vote was required by the interviewing group) (Bornstein, 2007).

"Someone who targets an unfortunate but stable equilibrium [in society] that causes the neglect, marginalization or suffering of a segment of humanity; who brings to bear on this situation his or her inspiration, direct action, creativity, courage and fortitude; and who aims for and ultimately affects the establishment of a new stable equilibrium that secures permanent benefits for the targeted group or society at large".

Social Entrepreneurship relates to non profit organisation (Lasprogata and Cotton, 2003), It is a profit making company operated by non profit organisation (Wallace 1999); It is an organisation that create a firm at a financial loss (Baron, 2007); it is a philanthropic act (Ostrander, 2007), It relates to individuals or organisation engaged in entrepreneurship activity with a social goal (Cesrto and Miller, 2008, Vanden Ven, Sapienza and Villa Nueva, 2007).

Definitions of social entrepreneurship range from broad to narrow. In the former, social entrepreneurship refers to innovative activity with a social objective in either the for-profit sector, such as in social-purpose commercial ventures (*e.g.*, Dees and Anderson, 2003; Emerson and Twersky, 1996) or in corporate social entrepreneurship (*e.g.*, Austin, Leonard, Reficco, and Wei Skillern, 2004); or in the nonprofit sector, or across sectors, such as hybrid structural forms which mix for-profit and nonprofit approaches (Dees, 1998). Under the narrow definition, social ntrepreneurship typically refers to the phenomenon of applying business expertise and market-based skills in the nonprofit sector such as when nonprofit organizations develop innovative approaches to earn income (Reis, 1999; Thompson, 2002).

James Austin *et al.*, 2006 define social entrepreneurship as innovative, social value creating activity that can occur within or across the nonprofit, business, or government sectors.

As innovative, social value creating activity that can occur within or across the nonprofit, business, or government sectors.

Difference between Social Entrepreneurship and Entrepreneurship

According to Austin (2006) difference between social entrepreneurship and entrepreneurship:

☆ **Market failure:** One theory behind the existence of social-purpose organizations is that they emerge when there is social-market failure, *i.e.*, commercial market forces do not meet a social need, such as in public goods (Weisbrod, 1975, 1977) or in contract failure (Nelson and Krashinsky, 1973). This is often due to the inability of those needing the services to pay for them.1 A problem for the commercial entrepreneur is an opportunity for the social entrepreneur. Our proposition here is: *Market failure will create differing entrepreneurial opportunities for social and commercial entrepreneurship.*

☆ **Mission:** The fundamental purpose of social entrepreneurship is creating social value for the public good, whereas commercial entrepreneurship aims at creating profitable operations resulting in private gain. This contrast is, of course, overstated. Commercial entrepreneurship does benefit society in the form of new and valuable goods, services, and jobs, and can have transformative social impacts. Such transformations can even be a driving motivation for some commercial entrepreneurs. Nonetheless, the differences in purpose and reward are useful for our comparative analysis. Our proposition is: *Differences in mission will be a fundamental distinguishing feature between social and commercial entrepreneurship that will manifest itself in multiple areas of enterprise management and personnel motivation. Commercial and social dimensions within the enterprise may be a source of tension.*

☆ **Resource mobilization:** The nondistributive restriction on surpluses generated by nonprofit organizations and the embedded social purpose of for-profit or hybrid forms of social enterprise limits social entrepreneurs from tapping into the same capital markets as commercial entrepreneurs. Additionally, the economics of a social entrepreneurial venture often make it difficult to compensate staff as competitively as in commercial markets. In fact, many employees in social entrepreneurial organizations place considerable value on nonpecuniary compensation from their work. Our proposition is: *Human and financial resource mobilization will be a prevailing difference and will lead to fundamentally different approaches in managing financial and human resources.*

☆ **Performance measurement:** The social purpose of the social entrepreneur creates greater challenges for measuring performance than the commercial entrepreneur who can rely on relatively tangible and quantifiable measures of performance such as financial indicators, market share, customer satisfaction, and quality. Additionally, the various financial and nonfinancial stakeholders to which a social entrepreneurial organization are readily accountable to are greater in number and more varied, resulting in greater complexity in managing these relationships (Kanter and Summers, 1987). The challenge of measuring social change is great due to nonquantifiability, multicausality, temporal dimensions, and perceptive differences of the social impact created. Our proposition is: *Performance measurement of social impact will remain a fundamental differentiator, complicating accountability and stakeholder relations.*

Leadbeater (1997) Social entrepreneurs are entrepreneurial, innovative, and "transformatory" individuals who are also: leaders, storytellers, people managers, visionary opportunists and alliance builders. They recognize a social problem and organize, create, and manage a venture to make social change.

Bornstein (1998) A social entrepreneur is a path breaker with a powerful new idea who combines visionary and real-world problem-solving creativity, has a strong ethical fiber, and is totally possessed by his or her vision for change.

Dees (1998) Social entrepreneurs play the role of change agents in the social sector by:

☆ Adopting a mission to create and sustain social value;

☆ Recognizing and relentlessly pursuing new opportunities to serve that mission;

☆ Engaging in a process of continuous innovation, adaptation, and learning;

☆ Acting boldly without being limited by resources currently in hand;

☆ Exhibiting a heightened sense of accountability to the constituencies served for the outcomes created.

Flower (2000) Social entrepreneurship is the creation of viable (socio) economic structures, relations, institutions, organisations and practices that yield and sustain social benefits.

Thompson *et al.* (2000) Social entrepreneurs are people who realize where there is an opportunity to satisfy some unmet need that the state welfare system will not or cannot meet, and who gather together the necessary resources (generally people, often volunteers, money, and premises) and use these to "make a difference".

Alvord *et al.* (2004) Social entrepreneurship that creates innovative solutions to immediate social problems and mobilizes the ideas, capacities, resources and social arrangements required for sustainable social transformations.

Austin *et al.* (2006) define social entrepreneurship as innovative, social value creating activity that can occur within or across the non profit, business or government sectors.

Mair and Marti (2006) define Social entrepreneurship broadly as a process involving the innovative use and combination of resources to pursue opportunities to catalayse social change and/or address social needs.

Peredo and McLean (2006) Social entrepreneurship is exercised where some persons or group:

1. Aim(s) at creating social value either exclusively or at least in some prominent way;

2. Show(s) a capacity to recognise and take advantage of opportunities to create that value (envision);

3. Employ(s) innovation, ranging from outright invention to adapting someone else's novelty, in creating and/or distributing social value

4. Is/or willing to accept an above-average degree of risk in creating and disseminating social value, and

5. Is/or unusually resourceful in being relatively undaunted by scarce assets in pursuing their social venture.

Weerawardena and Mort (2006) defines Social entrepreneurship as a behavioural phenomenon expressed in a NFP organisation context aimed at delivering social value through the exploitation of perceived opportunities.

Sharir and Learner (2006) define Social entrepreneurship as to apply business strategies for the purpose of more effective confrontation with complex social problems.

Zahra *et al.* (2008) Social entrepreneurship encompasses the activities and processes undertaken to discover, define, and exploit opportunities in order to enhance social wealth by creating new ventures or managing existing organizations in an innovative manner.

Brinckerhoff (2009) A social entrepreneur is someone who takes reasonable risk on behalf of the people their organization serves.

The social entrepreneur is a mission-driven individual who uses a set of entrepreneurial behaviours to deliver a social value to the less privileged, all through an entrepreneurially oriented entity that is financially independent, self-sufficient, or sustainable. This definition combines four factors that make social entrepreneurship distinct from other forms of entrepreneurship.

Social Entrepreneurs

1. Are *mission-driven*. They are dedicated to serve their mission of delivering a social value to the underserved.

2. Act *entrepreneurially* through a combination of characteristics that set them apart from other types of entrepreneurs.

3. Act within *entrepreneurially oriented organizations* that have a strong culture of innovation and openness.

4. Act within *financially independent organizations* that plan and execute earned-income strategies. The objective is to deliver the intended social value while remaining financially self-sufficient. This is achieved by blending social and profit-oriented activities to achieve self-sufficiency, reduce reliance on donations and government funding, and increase the potential of expanding the delivery of proposed social value (Bacq *et al.*, 2011; tinyurl.com/7nry6jp).

Ashoka (2012) Social entrepreneurs are individuals with innovative solutions to society's most pressing social problems. They are both visionaries and ultimate realists, concerned with the practical implementation of their vision above all else.

Going through all the definitions proposed by different authors the central theme of social entrepreneur is 'social mission'.

Differences between Entrepreneurship, Business, Social Entrepreneurship and Non Governmental Organisations

The basic differences between entrepreneurus and businessman include Entrepreneurs create needs while businessmen satify needs. Businessmen follow others but entrepeneurs create path (to see the world differently). Entrepreneurs capture opportunities while Businessman unnotice opportunities.

Sl.No.	Entrepreneur	Businessman
1.	Create Needs	Satisfies Needs
2.	Create Path	Follow others
3.	Capture opportunities	Un-notice opportunities

Differences between Social Entrepreneurship and Entrepreneurship

Sl.No.	Social Entrepreneurship	Entrepreneurship
1.	Satisfy needs	Creates Needs
2.	Knows exactly what the problem is and struggles with how to create a solution to fix them	Struggles with what the real problems are and build solutions that look for a problem to solve
3.	In social good before they become entrepreneurial first. They understand the social problems better than any other entrepreneur	
4.	Do care about making revenue and turning a profit in a much different way. Revenue is the life blood of any non profit or profit company regardless of social mission	It is very difficult to draw conclusion on social good first and make money secondary or vice-versa
5.	Really entrepreneurs (Truly) with a different mindset	
6.	Improves social conditions	Improves commercial markets
7.	Value lies in social benefit to a community or transformation of a community that lacks resources to fulfil its own needs	Value lies in profit
8.	Creates wealth for effecting social change by changing minds of profit, alleviating suffering and by reversing injustice	Benefits are driven to innovate within a commercial market. The innovation creates wealth.
9.	Older the organisation or initiative the more entrepreneurial skill. Skill are the key to the long term success of the social enterprise	
10.	Less ambitious to grow their employee basis	more ambitious to grow their employee basis
11.	More interested in understanding the social, economic, political and cultural context of the problem they are trying to solve	Not much interested in understanding the social, economic, political and cultural context of the problem they are trying to solve
12.	Delivers social value	Creates economic wealth
13.	Social returns	Economic returns
14.	Comamnd regards in the society	Demand regards in the society
15.	Mission leader	High achiever
16.	Emotionally charged	Risk bearer
17.	Organiser	Change agent
18.	Opinion Leader	Strategic thinker
19.	Visionery	Arbitrageur (Investor/Trader)
20.	Socially alert	Economically alert

According to Abu-Saifan (2012) common to both qualities include:Innovator, dedicated, initiative, leader, opportunity alert, persistent and committed.

The Differences between Non-governmental Organisations and Social Entrepreneurship

Non Governmental Organisations	Social Entrepreneurship
Purely Social Purpose	Driven by social as well as financial goal
Source of funding donations, charitable contributions	Earnings from the business
Selection of employees: people participate, voluntary	People are selected based on skill and performance
Performance measurement: on the basis of social values delivered.	On the basis of social values delivered alongwith financial returns.
Runs on social plan for fund raising, social out reach, selling services or products true of cost	Runs on enterprise setting using market tools of commercial and social advertisement for selling products and services
Risk of non sustainability	Highly sustainable
EX: Helpage India, CRY, AID *etc.*	SKS Microfinance, Rang De, Arvind Eye Care, *etc.*

Examples of Social Entrepreneurship in India

SKS Microfinance

Microfinance is an effective tool that can help reduce poverty and spread economic opportunity by giving poor people access to financial services, such as credit and insurance. SKS distributes small loans that begin at Rs. 2,000 to Rs. 12,000 (about $44-$260) to poor women so they can start and expand simple businesses and increase their incomes. Their micro-enterprises range from raising cows and goats in order to sell their milk, to opening a village tea stall. SKS uses the group lending model where poor women guarantee each other's loans. Borrowers undergo financial literacy training and must pass a test before they are allowed to take out loans. Weekly meetings with borrowers follow a highly disciplined approach. Re-payment rates on our collateral-free loans are more than 99 per cent because of this systematic process. SKS also offers micro-insurance to the poor as well as financing for other goods and services that can help them combat poverty (http://www.sksindia.com/).

Rang De

The idea of Rang De was sown in the year 2006, the same year that Professor Muhammad Yunus was awarded the Nobel Peace Prize for his work with Grameen Bank in Bangladesh. The motivation for starting Rang De was the belief that the peer-to-peer lending model could be leveraged to lower the cost of microcredit. Rangde.org went live on 26 January 2008 and this is the belief system that has shaped Rang De.

The vision of Rang De is to make poverty history in India by reaching out to underserved communities through microcredit. We are striving to do this through a network of committed field partners and social investors, by offering microcredit that has a positive impact on business, education, health and environment of the communities we work with. The values adopted by Rang De include Empathy, Creativity, Transparency, Accountability and Tenacity are the core values that guide Rang De's actions. We believe that we are not just a non-profit peer-to-peer lending platform. We are as much about social change as we are to do with technology and as much about investing time as we are about investing money. Rang De believes that it needs to be sustainable and scale meaningfully. We get a nominal cut of 2 per cent on all the loans repaid by our borrowers. We believe this will help us in achieving our goals and fulfilling our vision in the long run(https://www.rangde.org/).

Conclusion

The concept of social entrepreneurship finds a place in the curriculum of academic field. It is exercised by persons to create social value of some kind through recognising and exploiting opportunities; employing innovation and refusing limitations in available resources for creating social values. Social entrepreneurs deal with complex social needs specially when there is a less public funding for doing social welfare deeds/works. Social entrepreneur exhibits in the social arena. It has been tried to distinguish social entrepreneurs from other entrepreneurs and a contribution to understand the complex phenomenon. It improves large scale social development by reducing poverty.

References

Armonk, NY: M.E. Sharpe. Thompson, J. (2002). The world of the social entrepreneur. *International Journal of Public Sector Management, 15*(5), 412–431.

Ashoka Official Website. www.ashoka.org (December 5, 2005).

Austin, J.E. (2000). *The collaboration challenge: How nonprofits and business succeed through strategicalliances.* San Francisco: Jossey-Bass Publishers.

Austin, J.E. and Harmeling, S.S. (1999, October 13). *Women's World Banking: Catalytic change through networks.* HBS Case No. 9-300-050. Boston: Harvard Business School Publishing.

Austin, J.E. and Porraz, J.M. (2002, September 19). *KaBOOM!* HBS Case No. 9-303-025. Boston: Harvard Business School Publishing.

Austin, J.E., Leonard, H., Reficco, E., and Wei-Skillern, J. (2004). Corporate social entrepreneurship: A new vision of CSR. Harvard Business School Working Paper No. 05-021. Boston: Harvard Business School.

Bornstein, D. (2004). *How to Change the World: Social Entrepreneurs and the Power of New Ideas.* Oxford: Oxford University Press.

Bornstein, D. and S. Davis (2010). *Social Entrepreneurship: What Everyone Needs to Know.* New York: Oxford University Press.

Borntsein, D. (2007). *How to Change the World*. New York: Oxford University Press.

Bradach, J.L. (2003). Going to scale. *Stanford Social Innovation Review, 1*, 18–25.

Bradach, J.L. and Tempest, N. (2000, October 13). *New schools venture fund*. HBS Case No. 9-301-038. Boston: Harvard Business School Publishing.

Burgelman, R.A. (1983). Corporate entrepreneurship and strategic management: Insights from a process study. *Management Science*, 29, 1349–1364.

Burgelman, R.A. (1984). Designs for corporate entrepreneurship in established firms. *California Management Review*, 26, 154–166.

Cardoso, F.H. and E. Faletto (1979). *Dependency and Development in Latin America*. Berkeley CA: University of California Press (Portuguese edition 1969).

Center for Effective Philanthropy Official Website. http://www.effectivephilanthropy.com/about/about_timeline.html (January 26, 2006).

City Year. (2002). About City Year. Available at http://www.cityyear.org, accessed 7 August 2002.

Clark, C., Rosenzweig, W., Long, D., Olsen, S. (2004). Double Bottom Line Project Report: Assessing Social Impact in Double Bottom Line Ventures. *The Rockefeller Foundation*.

Colby, S., Stone, N., and Carttar, P. (2004). Zeroing in on impact. *Stanford Social Innovation Review*, 2, 24–33.

Cole, B. (1982). Compensation. *Modern Healthcare*, 12, 67–90.

College Summit Official Website: Results and Awards. http://www.collegesummit.org/resres.html (December 6, 2005).

Collins, O.F. and Moore, D.G. (1964). *The enterprising man*. East Lansing, MI: Michigan State University.

Cooper, A.C. and Bruno, A.V. (1975). Predicting performance in new high-technology firms. *Academy of Management, Proceedings of the 35th Annual Meeting*, August (pp. 426–428).

Dees, G. (1998/2001). "The meaning of 'Social Entrepreneurship'" www.fuqua.duck.edu/centers/case/docu-ments/Dees_SEdef.pdf

Dees, J. G. (1998). The Meaning of Social Entrepreneurship. [Online] http://www.fntc.info/files/documents/The per cent 20meaning per cent 20of per cent 20Social per cent 20Entreneurship.pdf.

Dees, J.G. (1998). *The meaning of "social entrepreneurship."* Comments and suggestions contributed from the Social Entrepreneurship Founders Working Group. Durham, NC: Center for the Advancement of Social Entrepreneurship, Fuqua School of Business, Duke University. Available at http://faculty.fuqua.duke.edu/centers/case/files/dees-SE.pdf

Dees, J.G. and Anderson, B.B. (2003). For-profit social ventures. *International Journal of Entrepreneurship Education* (special issue on social entrepreneurship), 2, 1–26.

Dees, J.G. and Oberfield, A. (1991, March 5). *Steve Mariotti and NFTE*. HBS Case No. 9-391-169. Boston: Harvard Business School Publishing.

Dees, J.G., Anderson, B.B., and Wei-Skillern, J. (2004). Scaling social impact. *Stanford Social Innovation Review, 1*, 24–32.

Dervin, B. and R. Huesca (1999). "Participatory Communication and Development Narrative: An Examination of the Meta-Theoretical Implications and their Impacts." In T. Jacobson and J. Servaes (eds.) *Theoretical Approaches to Participatory Communication*. Cresskill,NJ: Hampton.

Dollinger, M.J. (1984). Environmental boundary spanning and information processing effects on organizational performance. *Academy of Management Journal, 27*, 351–368.

Dorfmann, A. and A. Mattelart (1975). *How to read Donald Duck: Imperial Ideology in the Disney Comic*. New York: International General, (Spanish edition 1971).

Drucker, P. F. (1990). Managing the Non-Profit Organization. New York: Harper Collins.

Drucker, P.F. (1989). What business can learn from nonprofits. *Harvard Business Review, 67*, 88–93.

Ducks Unlimited. (2002). Fact sheet. Available at http://www.ducks.org, accessed 7 August 2002.

Easterly, W. (2006). *The White Man's Burden: Why the West's Efforts to Aid the Rest has Done so Much Ill and so Little Good*. New York: Penguin Press.

Emerson, J. and Twersky, F. (Eds). (1996, September). *New social entrepreneurs: The success, challenge and lessons of non-profit enterprise creation*. San Francisco: Roberts Foundation, Homeless Economic Development Fund.

Emile G. McAnany (2012). Social Entrepreneurship and Communication for Development and Social Change *Rethinking Innovation Nordicom Review* 33 (2012) Special Issue, pp. 205-218

Foster, W. and Bradach, J. (2005). Should nonprofits seek profits? *Harvard Business Review, 83*, 92–100.

Frank, A.G. (1969). *Underdevelopment or Revolution: Essays on Development of Underdevelopment and the Immediate Enemy*. New York: Monthly Review Press.

Fraser, C. and S. Restrepo-Estrada (1998). *Communicating for Development: Human Change for Survival*. London/New York: I.B. Tauris Publishers.

Freeman, R.B. (1979). The job market for college faculty. In D.R. Lewis and W.E. Becker (Eds), *Academic rewards in higher education* (pp. 63–103). Cambridge, MA: Ballinger.

Freire, P. (1970). *Pedagogy of the Oppressed*. New York: H and R Paper Books.

Freire, P. (1975). *Extension or Communication: The Conscientization of the Rural Countryside*. Mexico City: Siglo XXI, 1975 (Spanish edition).

Gerace, F. (1973). *Horizontal Communication*. Lima: Libreria Studium, (Spanish).

Golding, P. (1974). "Media Role in National Development." *Journal of Communication*, 24, 39-53.

Grantmakers for Effective Organizations Official Website. http://www.geofunders.org/index.cfm?fuseaction=Page.viewPage and pageId=2 (January 26, 2006).

Grønbjerg, K.A., Martell, L., and Paarlberg, L. (2000). Philanthropic funding of human services: Solving ambiguity through the two-stage competitive process. *Nonprofit and Voluntary Sector Quarterly*, 29(Supplement 1), 9–40.

Grossman, A. and McCaffrey, A. (2001, May 7). *Jumpstart*. HBS Case No. 9-301-037. Boston: Harvard Business School Publishing.

Grossman, A., Wei-Skillern, J., and Lieb, K.J. (2003, January 9). *Guide Dogs for the Blind Association*. HBS Case No. 9-303-006. Boston: Harvard Business School Publishing.

Hart, M.M., Stevenson, H.H., and Dial, J. (1996). Entrepreneurship: Adefinition revisited. In W. Bygrave (Ed.), *Frontiers of entrepreneurship 1995: Proceedings of the 15th Annual Entrepreneurship Research Conference* (pp. 75–89). Babson Park, MA: Babson College.

Hornik, R. (1988). *Development Communication: Information, Agriculture and Nutrition in the Third World*. New York: Longman.217

Huesca, R. (2003). "Participatory Approaches to Communication for Development" B. Mody (ed.) *International and Development Communication; 21ˢᵗ Century Perspective*. Thousand Oaks: Sage.

Jacobson, T. (1985). "Epistemological Shift in Development Communication Theory" in B. Dervin and M. Voigt, eds. *Progress in Communication Sciences*. Norwood NJ: Ablex.

James Austin, Howard Stevenson and Jane Wei-Skillern Social and Commercial Entrepreneurship: Same, Different, or Both? Entrepreneurship Theory and Practice Volume 30, Issue 1, pages 1–22, January 2006.

Jeremy C. Short, Todd W. Moss, and G. T. Lumpkin Research In Social Entrepreneurship: Past Contributions And Future Opportunities *Strategic Entrepreneurship Journal Strat. Entrepreneurship J.*, 3: 161–194 (2009) Published online in Wiley InterScience (www.interscience.wiley.com). DOI: 10.1002/sej.69

Kanter, R.M. and Summers, D. (1987). *Doing well while doing good: Dilemmas of performance measurement in nonprofit organizations and the need for a multiple-constituency approach*. In W.W. Powell (Ed.), *The nonprofit sector: A research handbook* (pp. 154–166). New Haven: Yale University Press.

Kiflemariam Hamde Akrem Abdu and Erik Johansson Social Entrepreneurship - A Case Study of SIFE Umeå University

Kramer, M. (2005). Measuring Innovation: Evaluation in the Field of Social Entrepreneurship. *A Publication for Skoll Foundation by Foundation Strategy Group*.

Lerner, D. (1958). *The Passing of Traditional Society: Modernizing the Middle East*. New York: The Free Press.

Letts, C., Grossman, A., and Ryan, W. (1999). *High performance nonprofit organizations: Managing upstream for greater impact*. New York: Wiley.

Letts, C., Ryan, W., Grossman, A. "Virtuous Capital: What Foundations Can Learn from Venture Capitalists." *Harvard Business Review*. March, 1997.

Lowell, S., Silverman, L., Taliento, L. Not-for-Profit Management: The gift that keeps on giving. *McKinsey Quarterly*. Number 1, 2001.

Lowell, S., Trelstad, B. and Meehan, B. The Ratings Game. *Stanford Social Innovation Review*, Vol. 3, No. 2, Summer 2005.

Martin, L. and S. Osberg (2007). "Social Entrepreneurship: The Case for Definition." *Stanford Social Innova-tion Review*, Spring, 29-39.

McAnany, E. (2009). "Incentives and Participation in Development Communication: Evidence from Sixty-three Recent Projects." Paper presented at The World Congress on Communication for Development, Rome, 2006. (Also in edited online form in *Glocal Times*, 13 Nov. 2009).

McAnany, E. (2010). "Communication for Development and Social Change: New Millennium." *Communica-tion Research Trends*, 29, 3, pp. 3-17.

McAnany, E. (2012). *Saving the World; A Brief History of Communication for Development and Social Change*. Champaign IL: University of Illinois Press.

McClelland, D.C. (1961). *The achieving story*. Princeton, NJ: D. Van Nostrand.

McFarlan, F.W. and Elias, J. (1996, October 9). *Mt. Auburn Hospital*. HBS Case No. 9-397-083. Boston:Harvard Business School Publishing.

Morely, E., Vinson, E., Hatry, H.P. (2001). Outcome Measurement in Nonprofit Organizations: Current Practices and Recommendations. A project collaboration between Independent Sector and The Urban Institute. Washington, D.C.: Independent Sector.

Moyo, D. (2009). *Dead Aid: Why Aid is Not Working and How There is a Better Way for Africa*. New York: Farrar. Strauss, Giroux.

National Organization for Rare Diseases. (2002). About NORD. Available at http://www.rarediseases.org, accessed 7 August 2002.

Nelson, R. and Krashinsky, M. (1973). *Two major issues of public policy: Public policy and organization of supply*. In R. Nelson and D. Young (Eds), *Public subsidy for day care of young children* (pp. 47–69). Lexington, MA: DC Heath and Co. *The new nonprofit almanac and desk reference*. (2002). San Francisco: Jossey-Bass.

Newschools Venture Fund. (2002). Our Strategy. Available at http://www.newschools.org, accessed 7 August 2002.

Nichols, A. (ed.). (2006). *Social Entrepreneurship: New Models of Sustainable Social Change*. New York: Oxford University Press.

Noga Leviner Leslie R. Crutchfield and Diana Wells Understanding The Impact of Social Entrepreneurs: Ashoka's Answer To The Challenge Of Measuring Effectiveness (https://www.ashoka.org/resource/4784).

Oster, S.M. (1994). *Modern competitive analysis*. New York: Oxford University Press.

Oster, S.M. (1995). *Strategic management for nonprofit organizations: Theory and cases*. New York: Oxford University Press.

Patrizi, P., McMullan, B. (1998). Evaluation in Foundations: The Unrealized Potential. Battle Creek, MI: W.K. Kellogg Foundation Evaluation Unit.

Pontus Braunerhjelm and Ulrika Stuart Hamilton Social entrepreneurship – a survey of current research Working Papers Series from Swedish Entrepreneurship Forum

Porter, M.E. (1980). *Competitive strategy: techniques for analyzing industries and competitors*. New York:Free Press.

Porter, M.E. (1985). *Competitive advantage: creating and sustaining superior performance*. New York: Free Press.

Preston, A. (1989). The nonprofit worker in a for-profit world. *Journal of Labor Economics, 7*(4), 438–463.

Quarry, W. and R. Ramirez (2009). *Communication of Another Development: Listening before Telling*. London/New York: Zed Books.

Quinn, R.E. and Cameron, K. (1983). Organizational life cycles and shifting criteria of effectiveness. *Management Science, 29*, 33–51.

REDF Official Website. www.ref.org (December 5, 2005).

Reis, T. (1999). *Unleashing the new resources and entrepreneurship for the common good: A scan, synthesis and scenario for action*. Battle Creek, MI: W.K. Kellogg Foundation.

Rogers, E. (1962). *Diffusion of Innovations*. New York: The Free Press.

Rogers, E. (1976). *Communication and Development: Critical Perspectives*. Beverly Hills CA: Sage.

Sahlman, W.A. (1996). Some thoughts on business plans. In W.A. Sahlman, H. Stevenson, M.J. Roberts, and A.V. Bhide (Eds), *The entrepreneurial venture* (pp. 138–176). Boston: Harvard Business School Press.

Salamon, L. (2002). *The State of Nonprofit America*. Washington, DC: Brookings Institution Press.

Samer Abu-Saifan Social Entrepreneurship: Definition and Boundaries *Technology Innovation Management Review* February 2012 www.timreview.ca Pp:**22-27**

Sawhill, J. (1999). Mission Impossible? Measuring Success in Nonprofit Organizations. Washington, D.C.: The Nature Conservancy.

Schramm, W. (1964). *Mass Media and National Development: The Role of Information in Developing Coun-tries*. Stanford/Paris: Stanford University Press and UNESCO Press.

Schumpeter, J.A. (1934). *The theory of economic development*. Cambridge, MA: Harvard University Press.

Servaes, J. (1989). *One World Multiple Cultures: Towards Another Paradigm on Communication for Develop-ment*. Leuven: Acco.

Sesame Workshop (2002). Annual report. Available at http://www.sesameworkshop. org/aboutus/pdf/Sesame-

Singhal, A. and E. Rogers (1999). *Entertainment-Education: A Communication Strategy for Social Change*. Mahwah NJ: LEA.

Singhal, A., M. Cody, E. Rogers, M. Sabido (eds.) (2004). *Entertainment-Education and Social Change: His-tory, Research and Practice*. New York: Routledge,.

Sparks, C. (2007). *Globalization, Development and the Mass Media*. New York: Sage.

Special Edition for the Skoll World Forum (2009). Social Entrepreneurship: Shifting Power Dynamics http://www-tc.pbs.org/now/shows/537/Shifting-Power-Dynamics.pdf

Stevenson, H.H. (1983). *A perspective on entrepreneurship*. Harvard Business School Working Paper No. 9-384-131. Boston: Harvard Business School.

Stevenson, H.H. (1985). The heart of entrepreneurship. *Harvard Business Review*, *63*, 85–94.

Stevenson, H.H. and Jarillo, J.C. (1991). A new entrepreneurial paradigm. In A. Etzioni and P. Lawrence (Eds), *Socio-economics: Toward a new synthesis* (pp. 185–208).

Timmons, J.A. and Bygrave, W.D. (1986). Venture capital's role in financing innovation for economic growth. *Journal of Business Venturing, 1*, 161–173.

Venture Philanthropy 2000: Landscape and Expectations (2000). Produced for the Morino Institute Youth Social Ventures by Community Wealth Ventures, Inc. http://www.vppartners.org/learning/reports/report2000/report2000.html (December 5, 2005).

Weisbrod, B. (1975). Toward a theory of the voluntary sector in a three-sector economy. In E.S. Phelps (Ed.), *Altruism, morality and economic theory* (pp. 171–195). New York: Russell Sage Foundation.

Weisbrod, B. (1977). *The voluntary nonprofit sector*. Lexington, MA: DC Heath and Co.

Weisbrod, B. (1983). Nonprofit and proprietary sector behavior: Wage differentials among lawyers. *Journalof Labor Economics, 1*(3), 246–263.

Wei-Skillern, J. (2002, March 26). *The September 11ᵗʰ fund: The creation*. HBS Case No. 9-302-083. Boston: Harvard Business School Publishing.

Workshop2002.pdf, accessed 7 August 2002.

Yunus, M. with A. Jolis (1999). *Banker to the Poor: Micro-Lending and the Battle Against World Poverty*. New York: Public Affairs.

Yunus, M. with K. Weber (2007). *Creating a World Without Poverty: Social Business and the Future of Capi-talism*. New York: Public Affairs.

Zadek, S. and Thake, S. (1997, June 20). Send in the social entrepreneurs. *New Statesman, 26*, 31.

Chapter 3

Entrepreneurship

P.K. Mishra

Professor, Department of Chemical Engineering
Indian Institute of Technology, Banaras Hindu University,
Varanasi, U.P.
E-mail: drpkm18@gmail.com, pkmishra.che@itbhu.ac.in

"Entrepreneur and Entrepreneurship" have become popular and respectable words today. We use the term entrepreneur to describe men and women who set up and manage their own business. The process involved in this is called entrepreneurship.

The real meaning of an entrepreneur is becoming more and clearer with the passage of time. Certain special characteristics, roles and economic functions have begun to be attributed to him/her. In this lecture, personality characteristics (traits) and functions will be discussed in detail.

You may have realized that first of all an entrepreneur has to be a highly motivated person. He or she will have to take initiative at various stages in organizing and reorganizing resources like money, materials and manpower to produce goods or services. In this process he/she will have to be continually innovative. The entrepreneur also has to take risks in terms of money, time, technology, career commitment and social relationships. The monetary reward for these activities comes in the form of profit. But there is a certain amount of uncertainty about this reward.

You must try and remember the key expressions given below:

☆ Being motivated
☆ Taking initiatives
☆ Organizing resources
☆ Taking risks
☆ Innovating

I will now begin to explain some of the difficult but nevertheless important expressions here:

Motivation

Motivation comes from the word 'motive' (or goal). It means the urge in an individual to achieve a particular goal. In other words, it is the need to achieve that motivates a person. You may find many people with sufficient financial resources and family support, who are interested in independent ventures. Yet, very few actually start their businesses. Why is it so? Because they are comfortable in doing routine jobs and no higher goals in life. They lack 'motivation'.

Entrepreneurs generally are highly active. They struggle constantly to achieve something better than what they already have. They like to be different from others and are ready to work hard to reach their goal.

Persons experiencing constant need to achieve always try to understand their strengths and weaknesses. This enables them to seek external help whenever needed.

> *Let us take a look at the 'Vadilal' group, which is a households name today in Gujrat. 'Vadilal ice cream' is a premier brand in the consumer market. You may be surprised to know that Ramachandra Gandhi and Laxman Gandhi, the two brothers, who founded the Vadilal empire, could not even complete their school education. They started in a small way by selling homemade ice cream in the city of Ahmedabad. Now, Vadilal is the largest ice cream company in the country.*

How did they do it? They did not stop thinking big. They had the courage to do what they wanted to. Entrepreneurial persons seek rewards or returns earned through their own efforts and do not depend upon 'luck'. They do not like to be idle.

The following case is a very good example:

> *Balachandran is a commerce graduate and belongs to a lower middle class family in Tamil Nadu. Even as a student, his ambition was to become his own boss one day. With this desire, he attended an Entrepreneurship Development programme (EDP) at Karaikudi, in Tamil Nadu. This strengthened his self confidence. His idea was to produce high molecular high density shopping bags. But he could not manage to procure the finances and hence dropped the idea. However, he did not lose hope. With just Rs. 500/- in his pocket, he left for Madras. He started selling used packing cartoons and earned some money. In the meantime he happened to meet a commercial artist, who used to work in the field of advertising. Balachandran became his assistant and started earning Rs. 300/- a month. The salary was too small and therefore he also began working independently. He started getting closer to his dream of having an enterprise of his own.*

> By December 1988, he had setup his own outdoor advertising unit and had quit his job. The first order came for Kwality Ice Cream. He has thereafter been manufacturing acrylic glow sign boards and also doing engraving work and screen printing. His clients include big names like Cadbury India and Pepsi. By 1993, he had set up three enterprises of his own.
>
> For Balachandran, there is no looking back now. And he continues his journey towards greater heights.

Risk-taking

Risk-taking implies taking decisions under conditions, where the reward on a certain action is known, but the occurrence of the event is uncertain. While doing so, an entrepreneur becomes responsible for the result of the decision. This responsibility however cannot be insured against failure.

Imagine that you are a qualified Pharmacist and that you have got a large sum of money from your parents. Which of the following options would you choose?

a) Invest in a bank deposit with 8 percent annual interest;

b) Invest in a company with a possible return of 15 percent;

c) Start a medical shop in your locality (because people there have to travel a long distance to get medicines) with a fairly good chance of making an immediate return of around 10 percent. (You are also aware that the business is sustainable and can bring in more return (20,30 or 50 per cent) in future if you put in your time and effort);

d) Try your luck in the share market.

Clearly, option 'c' calls for an entrepreneurial quality. Remember that successful entrepreneurs usually choose the moderate or middle path. They are not 'gamblers'. At the same time, they are not afraid of taking risk, if there is a reasonable chance of success.

You must be aware that businessmen and women spend considerable amount of time planning their enterprises. They study the market, technology, examine and re-examine the demand, the prices, machinery and processes involved, make detailed enquiries about sources of finance and think about other business lines. Why do you think they go through all these processes? In order to minimize the risks involved and avoid difficulties that may arise in the future, as far as possible.

Innovation

The process of commercializing an invention is innovation. For example, steam as an alternative source of energy was invented as early as AD 100. Later, in 1712, when it was used to run engines then the process was called innovative. In simple words, it means introduction of 'newness' or 'novelty'. When it is used in the context of a business activity, novelty may take any one or a combination of the following:

a) New products;
b) New methods of production;
c) New markets;
d) New sources of raw material; or
e) New forms of organization.

Innovation is a critical aspect of entrepreneurship. Entrepreneurs always try to create new and different values and get satisfaction in doing so. They try to convert a material into a resource or, combine the existing resources in a new and more productive manner. The act of innovation thus provides resources with a new capacity to create wealth.

The story of Penicillin, the popular antibiotic that is used against a variety of diseases and infections explains, what innovation can do in our society. Penicillin is made of a particular fungus or 'mould'. Till the beginning of 1920's scientists considered this fungus a nuisance, as it used to destroy their bacteria cultures. They were desperately trying to get rid of it. A doctor from London, Alexander Fleming, realized that this 'nuisance' was nothing but the 'bacteria killer', the scientists were searching for! All of a sudden Penicillin became a valuable 'resource'.

Innovation and change are in fact the most distinguished features of entrepreneurship. It needs not only your ability to 'create', but also the ability to understand your surroundings.

He/she

1. Searches for and discovers economic opportunities,
2. Evaluates the economic opportunities,
3. Arranges for financial resources necessary for the enterprise,
4. Makes time bound arrangements,
5. Takes ultimate responsibility for management of the enterprise,
6. Is the ultimate uncertainty/risk bearer,

Mr. Karsanbhai Patel, the proud founder of 'Nirma' washing powder belongs to a family of farmers in a small village in Gujrat. After having completed B.Sc. in chemistry, he began working as a laboratory technician. But soon, with limited know-how acquired from his work, karsanbhai began to manufacture a detergent in the backyard of his house. Finally, in 1969 he had set up 'Nirma Chemical Works' with an investment of Rs. 1000/- obtained from friends and relatives. He gave up his job in 1972 and ventured into full time manufacturing. Ever since then, the 'Nirma' empire has been growing and is now one of the world's largest detergent producers.

7. Is responsible for providing motivation within the enterprise,
8. Searches for and discovers new economic informations,
9. Translates these informations into new markets, techniques and goods,
10. Provides leadership to the workers.

Charms of Being Entrepreneur

1. You become your own boss. You are independent. You no longer work for somebody. Hence you face no frustration.
2. You no longer waste your energy and time looking through job advertisements. You create job for others.
3. You can use your talent, skills and knowledge for your own benefit and, also for your nation's prosperity.
4. You can earn a healthy income. Your successful enterprise can bring you unlimited returns.
5. Above all, you prove to the world that you are an 'achiever', because you seek nothing less than excellence.

Chapter 4

Entrepreneurship Development: Concept and New Approaches

S.P. Mishra

*Former Vice-Chancellor, Dev Sanskriti Vishwavidyalaya,
Gayatrikunj-Shantikunj, Haridwar – 249 411, U.K.*

The Underlying Concept

The growing awareness of the need for and urgency of, building "entrepreneurs" for the attainment of accelerated and self sustained economic growth could be ascribed to two factors, first, the belief is gaining ground that economic growth in the advanced countries appears to be attributable to entrepreneurial awareness in the community rather than to capital. In fact, several of the empirical studies recently conducted show that entrepreneurs or the 'human capital' has grown in the western societies at much faster rate than conventional capital and has made a much larger contribution to economic growth that non human capital. Second, investment in human resources has directly contributed to economic development and growth, by promoting the knowledge and application of science and technology to production process, developing innovations and research, training the workers in different technical skills needed for modern production and building up of the right type of attitude, values and interests conducive to right output. The fundamental problem in developing countries is, therefore, not so much the creation of wealth but the capacities to create wealth towards strengthening widening and improving the absorptive capacity of the country.

In other words, entrepreneurship development is an approach of developing human resources. It is concerned with the growth and development of people towards high level of competency, creativity and fulfillment. This approach helps people to grow in self control, responsibility and other abilities and then try to create a climate in which all clients may contribute to the limits of their improved abilities. It is assumed that expanded capabilities and opportunities for people will lead directly to improvement in operating effectiveness. Thus, Entrepreneurship Development as an approach does not confine itself in setting up enterprises but surpasses this limit in creating conducive climate for optimum utilization of limited and scattered resources and making people functional in all walks of life.

Entrepreneurship is the creative response to an environment that combines innovativeness, readiness to take risk, sensing opportunities, heightened initiative, standard of excellence, persistence in achieving the goal, positive orientation to problem solving and constant striving for growth and excellence. When all those attributes are developed in once person, the person can be found in any field of activity such as industry, business, education, public or professional bodies *etc.* But wherever they go they create a landmark, turn the direction of the side and attain heights.

The Experiences

The concept of entrepreneurship has undergone remarkable changes during the recent decades. Professionals of different disciplines contributed largely to the growth and maturity of the concept and application for economic growth of the society. During the last three decades, academicians and practioneres contributed significantly theorizing and applying Entrepreneurship concept, which resulted into few important generalizations:

☆ There is positive linkage between Entrepreneurship and economic prosperity.

☆ Entrepreneurship is instrumental in achieving progress all walks of life.

☆ Entrepreneurship can be developed through planned training intervention/efforts.

☆ Entrepreneurship is a process consisting of stimulatory, support and sustaining activities.

☆ It requires to be viewed in terms of integration of activities and coordination among different departments.

☆ Entrepreneurial culture need to be in built mechanism of development process.

These findings helped acceptance of different approaches to develop Entrepreneurship. One of them is the development of human factor – the entrepreneur himself, another major factor is the development of environment through which entrepreneurial activities can flourish and grow. The human factor refers to the, values, attitude, desire and motivation of individual, his capability to perceive the environmental changes and opportunities as well as his ability to solve the problem which entrepreneur is likely to face. All these qualities differ on the average from country to country or culture to culture and these differences are related to economic growth. These qualities known to be entrepreneurial qualities are not inherited by a person rather acquired in the process of socialization in a particular culture. Emergence, growth and development of entrepreneurial qualities takes root in childhood and continue to get sharpened with conscious attempt.

Many attempts in promoting Entrepreneurship in different countries have yielded variety of experiences. Some were highly successful other disappointing and perhaps considerable success. Analyzing these past experiences mainly from the perspective of the client system the whole process of Entrepreneurship development can be seem as a process having three distinct phases:

1. Stimulatory
2. Support
3. Sustaining

The stimulatory phase would include all activities that create willingness amongst the specific target groups such as generating entrepreneurial awareness in the community through planned campaign, identifying and selecting potential entrepreneurs, helping them raise their motivation level through training, improving their skills in modern management methods, developing technical competence, helping entrepreneurs select new products, evolving new products and processes *etc*. These stimulatory activities help in the emergence of entrepreneurs in the society. They prepare the background from where Entrepreneurship can sprout. People start looking for entrepreneurial pursuits. It generates the initial motivation, helps people perceive opportunities and incentives that await them besides acquiring relevant information and skills. All these taken together stimulate entrepreneurship in a society.

Support phase includes all such activities that help entrepreneurs in establishing and running their enterprises like registration of the unit, arrangement of finance, helping in purchase of plant and machinery, providing land, shed, power and water *etc* for establishing the unit, guidance for selecting and obtaining plant and machinery lay out, Issue of licenses for scarce raw materials, providing common facilities *etc*.

This group of activities provide nurturance and help the already stimulated entrepreneurs to move ahead in achieving his immediate goal of setting up and running his enterprise. These activities remove many hurdles which are likely to cause sickness to the unit or which discourage the new entrepreneurs.

Activities in sustaining phase are those that help the entrepreneur in continued, efficient and profitable running of his enterprise. After the stimulation of entrepreneurship and subsequent adequate support the entrepreneur succeeds to starting his unit. Quite often, however, an entirely new set of problems crop up before him once he commences production. He may not be fully prepared. For some of them it may be non-viable due to many unforeseen and uncontrollable factors. The small industries are susceptible to a variety of vagaries. Even well managed units may suffer closure for want of raw materials, lack of availability of substitute, change of skilled hands, *etc*. Such causalities which are comparatively higher in the small units require a lot of sustaining activities by the promotional agencies.

The sustaining activities may include- helping in modernization diversification/ expansion/product substitution, Additional financing for full capacity utilization, Deferring repayment/interest depending on the situation *etc*.

Each phase of activities mentioned above is highly interactive, supplementary and crucial to the other phases. Often it is found that the promotional agencies place excessive emphasis on support activities such as arranging finance, providing plant and machinery on long term basis, and establishing industrial estates to provide the infrastructure like land, shed and power *etc*. However, in models where stimulatory phase is neglected and it begins with support activities, the outcome can be of two types:

If outsiders are brought in, the lack of acquaintance with the locality may render the industry sick which may further be aggravated if the program does not have sustaining activities. Stimulatory activities are sometimes over emphasized

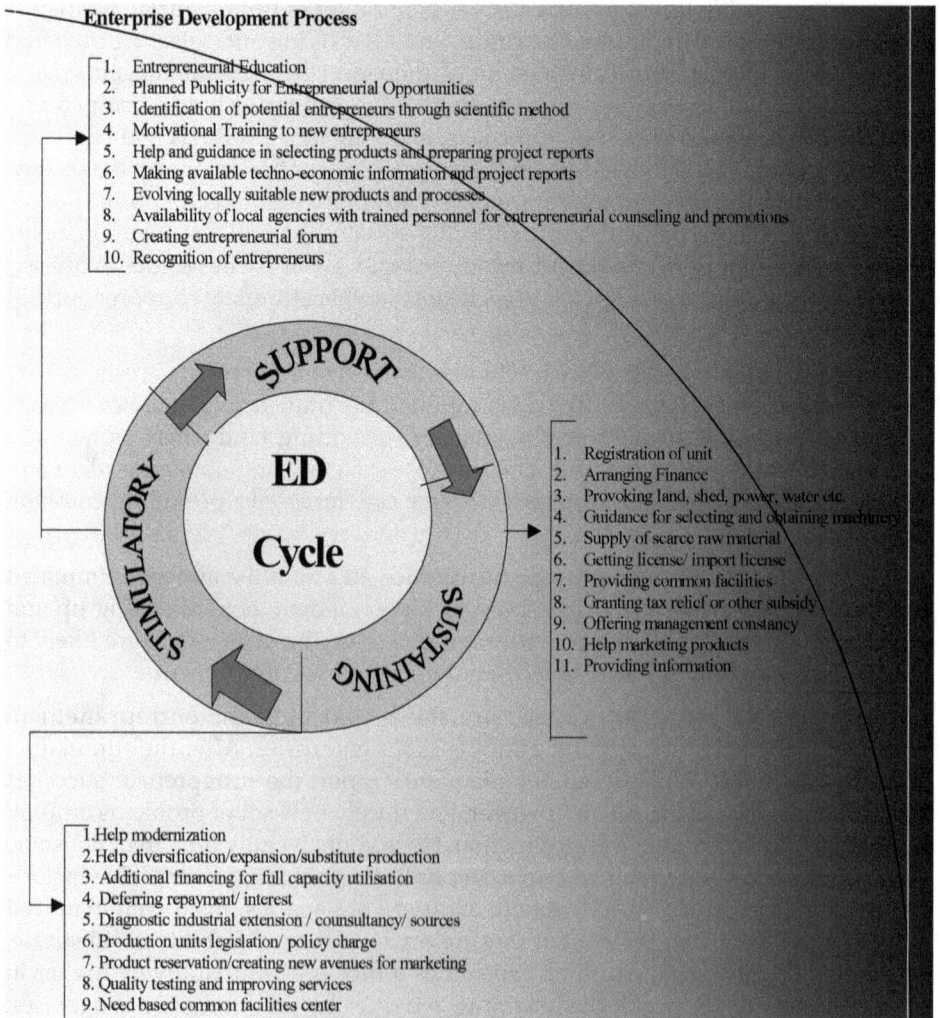

Enterprise Development Process

1. Entrepreneurial Education
2. Planned Publicity for Entrepreneurial Opportunities
3. Identification of potential entrepreneurs through scientific method
4. Motivational Training to new entrepreneurs
5. Help and guidance in selecting products and preparing project reports
6. Making available techno-economic information and project reports
7. Evolving locally suitable new products and processes
8. Availability of local agencies with trained personnel for entrepreneurial counseling and promotions
9. Creating entrepreneurial forum
10. Recognition of entrepreneurs

ED Cycle — SUPPORT / STIMULATORY / SUSTAINING

1. Registration of unit
2. Arranging Finance
3. Provoking land, shed, power, water etc.
4. Guidance for selecting and obtaining machinery
5. Supply of scarce raw material
6. Getting license/ import license
7. Providing common facilities
8. Granting tax relief or other subsidy
9. Offering management constancy
10. Help marketing products
11. Providing information

1. Help modernization
2. Help diversification/expansion/substitute production
3. Additional financing for full capacity utilisation
4. Deferring repayment/ interest
5. Diagnostic industrial extension / counsultancy/ sources
6. Production units legislation/ policy charge
7. Product reservation/creating new avenues for marketing
8. Quality testing and improving services
9. Need based common facilities center

at the cost of other activities which may result in an over abundance of budding entrepreneurs but there may be a lack of timely support from concerned agencies. This will not only demotivate entrepreneurs, this may also have a negative multiplier effect. The ultimate casualty in this process is the credibility of the agencies concerned with entrepreneurship development. Similarly if provision is made for sustaining activities without arranging a matching degree of stimulatory or support activities, there may not be sufficient number of units to be sustained. Thus balanced emphasis is required on all the three phases to set desired result on continued basis.

Recent Approaches for Entrepreneurship Training Programme

With the increased emphasis of developing entrepreneurs, witnessed a phenomenon change has been witnessed in its approach to create entrepreneurial awareness and generate entrepreneurship in the society. What appears to be most significant is the realization that no approach can meet all requirements in dealing with diverse population scattered over widely dispersed area of any country. Consequently, different types of entrepreneurship development training programme took a shape in making it as an effective preposition for industrial development. Recent approaches to entrepreneurship training may be seen as:

☆ **Target oriented**

☆ **Location specific**

☆ **Product and Process Oriented**

Target Oriented

Such EDPs are directed and planned for a specific group which has distinguishing features as compared to other groups. It is because of the fact that target audience vary in terms of their background, experiences, training and exposure to business world. Moreover, the size of the proposed enterprise may also vary from one target group to others. These variations demand for the matching training inputs with varying degree of intensity to bring these entrepreneurs to the threshold from where they can launch their enterprise and be able to manage them successfully. For example, a group of MBA entrepreneurs may require less intensity on management skill development input as compared to other aspects. Similarly, for a group of entrepreneurs who want to set-up considerably larger units may require high intensity on management skills rather than those who are going to set up comparatively smaller units. For this purpose, various target groups may be identified like (i) General Entrepreneurs (ii) Science and Technology Entrepreneurs, (iii) Women Entrepreneurs (iv) Educated unemployed entrepreneurs (v) Agril. Entrepreneurs *etc.*

Location Specific

In order to reduce regional imbalances and the imbalances that may exist between progressive and non-progressive areas, it is of considerable importance to plan EDPs depending on the characteristics of the area itself. Location in terms of urbanization, concentration of specific target group in a particular area *etc.* may be considered under such specific EDPs.

Product and Process Oriented EDPs

The late eighties is experiencing a shift from Target Oriented EDP. Such EDPs are organized for a group of prospective entrepreneurs who opt for enterprises having set product line or process such as plastic or electronics or construction materials of food technology *etc*. It seems to have a great future, but some of the efforts made recently were found suffering from the lack of appreciation of the role played by inputs, other than product and process orientation. In the absence of such comprehensive program, the acquired knowledge and skills about product and process will not automatically result in development of entrepreneurial quality, competence for enterprise launching and ability to manage which are crucial to start and ensure success.

All these approaches need to be viewed interims of certain activities which also in a way, indicates as how such programmes could be conducted. These activities are as follows:

Documentation of Entrepreneurial Opportunities in the Area

Before launching the entrepreneurship development training it is considered necessary to document the industrial opportunities in the area. It is expected that in selected districts some techno economic survey indicating availability of raw material, human, infrastructural facilities, potential demand of goods *etc*. have already been conducted. Similarly, Bank report may also be considered to substantiate the potentiality of developing a particular area. The training institutions, therefore, need to collaborate with local bank and other promotional organizations to prepare project profile based on such documents. These documents may be made available to the participants during the program with an objective to help them identify a suitable project for themselves.

Information about the Possible Supply of Entrepreneurs

Availability of resources and demand of products in a particular area is one aspect of development but equally important aspect of the availability of potential people capable of transforming the resource into production possibilities. Therefore, it is important to ascertain the availability of potential entrepreneurs in the area. For this, a detailed information regarding availability of potential people may be obtained through technical institutions, universities, financial institutions *etc*. This may even help us deciding the target group for the programme.

Motivational Campaign for Attracting and Identifying Potential Entrepreneurs

Based of several experiences it is considered necessary to organize a well-planned and effective campaign in the area in order to create awareness about the entrepreneurial opportunities and the role and support of various agencies. The motivational company will include preparation of some pamphlets to be distributed through various channels, educational institution and other promotional organizations. Announcement in the local newspaper, circular to various offices *etc*. may be taken up seriously for giving wide coverage to the programme and attract potential people for the programme. Response to such organized campaign itself is an indication of interest a person might be having towards entrepreneurship. Such identified people, in fact, form the base of selection of a limited number of participants for the training programmes.

Selection of Potential Entrepreneurs

The task of the training institution becomes vary crucial for this aspect, it has been established through researches and experience that the proper selection of participants for the training programme yields better result. But the scientific selection is possible and meaningful only when good response is obtained through motivational campaign. Whenever emphasis is only on selection without motivational campaign, the result is seen it terms of limited candidates taking on selection test and therefore, minimizing the scope and choice of selection.

Fortunately a battery of selection tests is available whose reliability and validity is beyond any doubt. The training institution, therefore, needs to be aware about such test and use it with sincerity of purpose to finally select a group of candidates suitable for such programme.

Identifying Training Needs to the Target Group

The candidates could be selected either as preconceived target group like science and technology group, women, educated unemployed, *etc* or it could be mixed group entirely left on selection criterion. In both the cases, it is desirable to assess the training needs of the candidates based on their backgrounds, experiences, aptitude and entry behavior in terms of their awareness, understanding, competencies and skill in diverse areas. This provides a base to formulate a training strategy for the identified group. The assumption is that all the target groups are not the same and the training treatment requires different approach to help achieve the basic objective of 'Transformation'.

Training

The training institution is expected to take the full responsibilities of planning, designing and conducting the training programme. The detailed activities, and process of designing in a broader perspective. However, some of the activities related to this phase may be enlisted here for illustration only. Such activities can be divided into two parts: (a) Academic and (b) Non-academic. The academic activities comprise of: (a) designing the training programme, specifying content coverage is sequence, mode of content coverage, faculty requirement and availability (b) arrangement of training aids (c) preparation of information material for the participants (d) arrangements for market survey (e) in-plant attachment (f) interface with promotional agencies (g) interface with existing entrepreneurs *etc*. In case of non-academic activities the institute is required to pay attention towards, making arrangements for (a) training venue and class room facilities (b) Board and lodge (c) Inaugural and valedictory normally for long duration ranging from 4 to 12 weeks on full time basis. Therefore, the involvement and commitment of training institution plays a significant role for the success of the programme. At any point of time, even little slackens and unconcern shown by the training institution makes the programme rote learning instead of facilitating learning process. Any kind of commercial approach by the institution-conducting programme can only fulfill the target of training but remains devoid of developing entrepreneurship through training intervention. It is, therefore, important for the training organization to understand the delicacies of undertaking entrepreneurship training programme.

Entrepreneurship Development through Planned Training Intervention

Training in entrepreneurship provides an impetus to the potential and budding entrepreneurs to acquire a new identity about himself. This is perceived as an approach towards transforming people which serves the purpose of making people aware about their own identity, help them accept a new identity and finally establish such identity for entrepreneurial pursuit. In order to take up such task of transformation we can find entrepreneurship training serving the purpose of stimulation, orientation, preparation and implementation in a sequential order. In order to apply the model, one requires an understanding about approaches to entrepreneurship training programme.

Entrepreneurial Approach in Training

The entrepreneurship training differs with other type of training in terms of its nature, scope, result, target group, post training activities *etc*. Promoter of training to the small business requires undertaking a business like approach. This business like approach to training must also, however, be entrepreneurial in delivery and marketing. Following are the salient points of entrepreneurial approach to training:

Entrepreneurial Approach

a) Major focus on process of delivery
b) Ownership of learning by participant
c) Trainer as fellows learner/facilitator
d) Emphasis upon know-how and know-who
e) Participants generating knowledge
f) Sessions flexible and responsive to needs
g) Learning objectives negotiated
h) Mistakes to be learned from
i) Emphasis upon practice
j) Problem/multi-disciplinary

The entrepreneurial approach in training can be immense use in visualizing and immolating the training process in entrepreneurship.

Post-Training Support/Follow-up

Post training support by the institutions organization entrepreneurship training programme is as important as providing a well-knit package of inputs to help the individual in his transition from an individual to an entrepreneur. Often the potential entrepreneur, after undergoing training, is still riffled with myriads problems, both personal and institutional. While he has chosen a product line and prepares a project report, he still needs some one to whom he can go for clarifications and help in many steps, before he can successfully ground the project. This much needed support and counseling, after training, has to be given by the faculty concerned for him to sustain his enthusiasm and efforts to become an entrepreneur. Monthly review with the trained entrepreneurs and keeping contact with them by the coordinating faculty has proved beneficial to them. It has been felt that the counseling and support makes their movement much faster in reaching their goal.

Emerging Trends

The scope of wage employment is not only limited but decreasing very fast. Even the nature of wage-employment is undergoing drastic transformation from a security oriented job to performance based contractual assignment. Such a situation has virtually compelled people in general to think more for entrepreneurship/ self employment. In this process, an analysis of the situation indicates clearly that today's environment is pointing concern towards the aspects like:

- ☆ Highly Competitive Market
- ☆ Competition between public and private sector
- ☆ Quality consciousness and standardization of products
- ☆ Security/guarantee
- ☆ Fast changing Technology
- ☆ Technological Collaboration
- ☆ Liberalizaton in policy and action
- ☆ Health/Hygiene
- ☆ NRIs readiness to invest in India
- ☆ Privatization of public sector
- ☆ Increased export opportunities and mental readiness of other countries to buy products from India
- ☆ Emphasis on self sustenance of supporting/promotional organization leading to better services from them.

Therefore, the environment in totality appears to be quite challenging. In such a direction the question arises as how to help people sense business opportunities which fulfills the global expectations in general and local needs in particular? For this an attempt is made here list out possible models in operation in one hand and emerging demands on the other hand.

Models

In order to operate effectively the above approaches a few models in operation are as follows:

Market Driven Model

Demand estimation and demand forecasting are the best ways through which demands of various products are reflected in the market. While the demand is estimated on the basis of the gap between the demand and supply, the forecasting of the demand is made on the basis of future projection of possible needs of the concerned products. Therefore economic activity is generated related to this model to fulfill the requirement of the market.

Need based Specific Model

Generally, we find two types of needs *i.e.* felt and unfelt needs. It is possible understand the felt need on the basic requirements and the gap but it seems difficult to visualize the unfelt needs *i.e.* need below the surface. In other words, unfelt need to indicate the need in reality but in latest form. This is one area on the basis of which various products find place in the market. Related to this is another concept which is known as need creation in the society. There are number of examples where in different products have been introduced through this process in a particular social system like emergence of fast food in traditionally strong society in African and South Asian countries.

Uniqueness Model

A variety of products have been introduced in the market by projecting unique features of particular product for specified segment of society. This may be seen quite clearly in automobile industry *i.e.* cars, two wheelers *etc.*

Based on these models and some more conceptual models already in operation, the following emerging demands have come in to focus.

Conclusions and Suggestions

In this context, it seems important to take advantages of the earlier experiences and research findings which may be summarized as follows:

☆ Promotion and development of entrepreneurship is more developmental in nature and is not administrative or target-oriented

☆ Entrepreneurship development is not a programme but a process whereby people in general are stimulated, supported and their enterprising activities are sustained.

☆ Accumulated experience has brought maturity in approaches and strategy of developing entrepreneurship.

☆ Experience suggests that entrepreneurs could be developed through a well planned integrated approach

☆ A set of competencies is already identified to help people become entrepreneurs. These competencies could be developed through training intervention.

☆ Competencies could be broadly grouped as:

  ❑ Behavioral

❑ Enterprise launching and resourcing

❑ Enterprise management.

☆ New economic opportunities are identified and available which may be integrated in entrepreneurship development programmes.

☆ Special attention may be made to promote "marketing entrepreneur" and R&D entrepreneur.

☆ Emergence of women entrepreneur through women empowerment seems to be an established fact.

☆ Micro enterprise creation and development through micro-finance has shown a very significant result.

These findings could be used to promote entrepreneurship professionally in all the sectors including agriculture. This deserves a mention here that farm entrepreneurship need to be promoted keeping in view in shaping the farmers as farm entrepreneurs capable of looking farming not as hereditary vocation but as business venture.

In view the restricted environment in the area of wages employment and potentiality of developing agriculture as an enterprise, it is perhaps important to look at the following aspects for consideration:

☆ Entrepreneurship curriculum may be devised especially in the context of Agri. education and be introduced as part of the total curriculum.

☆ The institute may initiate establishing Entrepreneurship Development Cell for providing modular training to the students and farmers.

☆ Technology Incubator especially focused on agriculture technology and R and D for value added agriculture products may be established for promoting agriculture based enterprises.

☆ Attempt may be made to organize training on market entrepreneurs and promote such entrepreneurs.

It is expected that a modest beginning will be made through deliberating and discussing about the conceptual aspects of entrepreneurship development and its possible integration for promoting agriculture based enterprises in the country.

Chapter 5

Conceptual Developments in Research on Entrepreneurship: An Agri-business Perspective

P.S. Badal

Professor, Department of Agricultural Economics,
Banaras Hindu University, Varanasi, U.P.

The reasonable man adapts himself to the world; the unreasonable persists in trying to adapt the world to himself. Therefore, all progress depends on the unreasonable man.

– *George Bernard Shaw*

The term "entrepreneurship" originates from the French term "entreprendre" and the German word "unternehmen", both mean "to undertake". The earliest aatempt to define entrepreneurship was by the French writer Bernard F. de Belidor, who defines entrepreneurship "as buying labour and materials at uncertain prices and selling the resultant output at contracted prices". However, the concept gathered prominence in economics literature mainly through the writings of Richard Cantillon (1680-1734), who gave the concept some analytical treatment and assigned the entrepreneur an economic role by emphasising on 'risk' as a prominent entrepreneurial function (Gopakumar, 1995). Despite a number of developments in the later years (Table 5.1), a dynamic theory of entrepreneurship was first developed by Schumpeter (1934). He saw innovation and the entrepreneur as the central figure in economic growth and development. Innovations introduced by the entrepreneur are a source of creative disequilibrium. As other firms adopt the innovation, less efficient firms are forced out of business until a new equilibrium is reached. The post-Schumpeterian developments have proceeded along two different schemes – the Harvard tradition and the neo-Austrian School, the former is portrayed as an extension of Schumpeterian view (Liebesnstien, 1968), while the latter is represented as an alternative approach.

Definitions of Entrepreneurship

Table 5.1: Definitions of Entrepreneurship

Author	Definition
Richard Cantillon (1730)	Entrepreneurship is defined as self-employment of any sort. Entrepreneurs buy at certain prices in the present and sell at uncertain prices in the future. The entrepreneur is a bearer of uncertainty.
Jean Baptiste Say (1816)	The entrepreneur is the agent "who unites all means of production and who finds in the value of the products.the reestablishment of the entire capital he employs, and the value of the wages, the interest, and rent which he pays, as well as profits belonging to himself."
Frank Knight (1921)	Entrepreneurs attempt to predict and act upon change within markets. Knight emphasizes the entrepreneur's role in bearing the uncertainty of market dynamics. Entrepreneurs are required to perform such fundamental managerial functions as direction and control.
Joseph Schumpeter (1934)	The entrepreneur is the innovator who implements change within markets through the carrying out of new combinations. The carrying out of new combinations can take several forms; 1) the introduction of a new good or quality thereof, 2) the introduction of a new method of production, 3) the opening of a new market, 4) the conquest of a new source of supply of new materials or parts, 5) the carrying out of the new organization of any industry. Schumpeter equated entrepreneurship with the concept of innovation applied to a business context. As such, the entrepreneur moves the market away from equilibrium. Schumpeter's definition also emphasized the combination of resources. Yet, the managers of already established business are not entrepreneurs to Schumpeter.
Penrose (1963)	Entrepreneurial activity involves identifying opportunities within the economic system. Managerial capacities are different from entrepreneurial capacities
H. Leibenstein (1968)	An entrepreneur fills market deficiencies through input-completing activities. Entrepreneurship involves "activities necessary to create or carry on an enterprise where not all markets are well established or clearly defined and/or in which relevant parts of the production function are not completely known.
Israel Kirzner (1979)	The entrepreneur recognizes and acts upon market opportunities. The entrepreneur is essentially an arbitrageur. In contrast to Schumpeter's viewpoint, the entrepreneur moves the market toward equilibrium.
Bygrave and Hofer (1991)	They defined the entrepreneurial process as 'involving all the functions, activities, and actions associated with perceiving of opportunities and creation of organizations to pursue them'.
Kao (1993)	Entrepreneurship is the process of doing something new and something different for the purpose of creating wealth for the individual and adding value to the society

Proponents of the Austrain School emphasise the enetrepreneur's ability to take advantage of imperfections in information to make innovations. The entrepreneur uses superior information to introduce an innovation and earn profits. Unlike Schumpeter, the Austrians see a market before innovation as being in disequilibrium. The introduction of innovation increases the amount of knowledge in the market moving it toward new equilibrium (Kirzner, 1979).

In summary, entrepreneurship is often viewed as a function which involves the exploitation of opportunities that exist within a market. Such exploitation is most commonly associated with the direction and/or combination of productive inputs. Entrepreneurs usually are considered to bear risk while pursuing opportunities, and often are associated with creative and innovative actions. In addition, entrepreneurs undertake a managerial role in their activities, but routine management of ongoing operations is not considered to be entrepreneurship. An individual may perform an entrepreneurial function in creating an organization, but later he is relegated to the role of managing it without performing an entrepreneurial role.

In this sense, many small-business owners would not be considered to be entrepreneurs. Finally, individuals within organizations (*i.e.* non-founders) can be classified as entrepreneurs since they pursue the exploitation of opportunities. This form of entrepreneurship is called intrapreneurship.

In case of agriculture a farmer does not become an entrepreneur only by adopting a new agricultural technology but he becomes an entrepreneur only when he comes to be an operator of a farm business. A business involves rational decisions on investment after assessing risk, other alternatives and possibilities of profit and loss (De, 1986).

Elements of Entrepreneurial Disposition

There are certain key elements that define the entrepreneurial disposition namely: personal resourcefulness, achievement orientation, strategic vision, opportunity seeking and innovativeness (Balakrishnan *et al.*, 1998; Knudson at al. 2004).

Personal Resourcefulness

Personal resourcefulness is the belief in one's own capability for initiating actions directed towards creation and growth of enterprises. Such initiating process requires cognitively mediated self-regulations of internal feelings and emotions, thoughts and actions (Kanungo and Misra, 1992).

Strategic Vision

Milton (1989) suggests that entrepreneurs have a knack for looking at the usual and finding the unusual. This would lead to what is commonly known as entrepreneurial vision. This entrepreneurial vision or the abstract image of the kind of business they wish to create is one that guides their own intentions.

Achivement Orientation

Entrepreneurs have a need for achievement, or a strong ego-drive. Entrepreneurs strive to make a difference in their own lives and in the lives of others. They are determined, persistent and committed to a job until it is finished.

Opportunity Seeking

Opportunity seeking involves one's ability to see situations in terms of unmet needs, identifying markets or gaps for which product concepts are to be evolved, and the search for creating and maintaining a competitive advantage to derive benefits on a sustained basis.

Innovativeness

While a capitalist is one who seeks market return for the capital he invests an entrepreneur is one who invests skills, assumes responsibilities and controls the activities of the enterprise to seek residual surplus for the 'human industry' and 'coordination' he puts in for the creation and survival of the enterprise. Schumpeter (1949) went on to conceptualise entrepreneurs as persons who are not necessarily capitalists or those having command over resources, but as ones who create new combinations of the factors of production and the market to derive profit.

It is evident that each of the above given disposition can be found in each and every individual with varying intensities. Therefore, it is assumed that a person with a very high intensity of the above predispositions could emerge as an entrepreneur. This gives rise to a process view of entrepreneurship *i.e.* entrepreneurship is more a state of becoming than a steady state event. Table 5.2 gives a relationship between Entrepreneurial predispositions and enterprise life cycle.

Table 5.2: Entrepreneurial Dispositions in Enterprise Life Cycle

Entrepreneurial Dispositions	Enterprise Life Cycle		
	Startup	Stabilisation	Diversification
Resourcefulness	High	High	Medium
Achievement-orientation	Medium	High	High
Visioning	Medium	Low	High
Opportunity-seeking	High	Medium	High
Innovativeness	High	Low	Medium

Source: Balakrishnan *et al.* (1998).

Entrepreneurs are Born or Made?

Are certain individuals born with certain characteristics that predispose them to entrepreneurial endeavors? Is there a set of traits that can be attributed to an entrepreneurial personality? Or does environmental context, such as early exposure to entrepreneurialisam makes the entrepreneur? These questions are as old as the study of entrepreneurship itself.

Professor of Psychology Alan Jacobowitz, believes that entrepreneurs are born, not made (Cohen, 1980). Through interviews with over 500 entrepreneurs over a three-year period, Jacobowitz observed that entrepreneurs commonly share certain personality characteristics. These include: restlessness, independence, a tendency to be a loner, and extreme self-confidence.

Other researchers have not reached the same conclusion as Jacobowitz. There are a growing number of researchers who eschew to a more dynamic approach to entrepreneurship in which personality traits and subsequent behavior are shaped by a variety of factors including the interaction between personal characteristics, perceptions, values, beliefs, background, and environment. They conclude that the intention to initiate and continue entrepreneurial behavior is influenced by the interaction of various factors. These include individual characteristics, individual environment, business environment, an individual's personal goal set, and the existence of a viable business idea. Through these interacting factors, individuals make several comparisons between their perceptions of a probable outcome, their intended goals, intended behavior, and actual outcomes (Knudson. *et al.*, 2004).

The authors believe that many individuals are born with "entrepreneurial DNA." *i.e.*, they exhibit a predisposition to entrepreneurship. In fact, early studies found a strong correlation between entrepreneurial predisposition or propensity and firm start-up decisions (Learned, 1992).

There is a distinct difference between possessing entrepreneurial tendencies and acting on them. Many latent entrepreneurs reach a level of comfort in their jobs and careers that may never result in an outward expression of entrepreneurship, while others grow bored of routine and seek out new challenges. Many entrepreneurs are born out of an "event." This event could take many forms including: losing one's job, threat of bankruptcy, loss of a significant other, frustration on the job, and discovering a marketplace gap (Knudson *et al.*, 2004).

While a life-changing event releases the entrepreneurial spirit in some people, for others, unleashing entrepreneurial spirit is a matter of training and education. There is a growing demand for training and educational programs in entrepreneurship, as evidenced by the growing number of academic programs offered in entrepreneurship. Graduates holding agriculture degrees may not be getting enough of the tools needed by entrepreneurs to reach their full potential. College graduates often possess a relatively narrow, but deep amount of knowledge in a given subject area, but they may lack more general information when it comes to securing loans, managing people, understanding the legal process of running a business, or how to segment and target a given market.

The conclusion from the literature and from growing practice favors the concept of "making" entrepreneurs provided they possess the "entrepreneurial DNA." Whatever the entrepreneurial disposition, most individuals can succeed at being an entrepreneur if they are equipped with the proper tools to execute entrepreneurial activity.

Entrepreneurship under Constrained Environment

Absence of resources with poor households of developing countries is considered an insurmountable obstacle to developing an entrepreneurial culture. However, availability of resources under control has not been found to necessarily be an impediment to growth and success of an entrepreneur (Sexton and Bowman-Upton, 1991). Bryant (1989) argues that entrepreneurs are characteristically people who go beyond the limits of resources over which they have direct control. Bygrave

(1994) similarly contends that entrepreneurs find ways to control critical resources without owning them. Indeed, an important quality of entrepreneurs is their ability to be creative with limited resources (Saylor, 1987). According to Saylor (1987), ownership of resources is not a mandatory requirement for entrepreneurs to make use of them. This seems to comply with Kirzner's (1973) argument that ownership of capital is not necessary to provoke its movement or change of application. Thus, it could be concluded entrepreneurship is a process by which individuals pursue opportunities without regard to resources they control (Kodithuwakku and Rosa, 2002).

Interaction between Entrepreneurship and Innovation

The connection between entrepreneurship and innovation has been depicted by Kudson *et al.* (2004) as presented in Table 5.3. It is hypothesized that two drives propel the behaviour of economic actors (firms and individuals): (a) the drive to commercialise an idea, that is, bring it to market realization; and (b) the drive to innovate. These two drives interact to form four possible entrepreneurial/ innovation types: (a) master entrepreneurs who are skilled managers, and risk bearers, but not innovators. (b) innovative entrepreneurs, skilled entrepreneurs, who are also innovators, (c) entrepreneurial innovators, skilled innovators who are also entrepreneurs, and (d) master innovators, who are skilled innovators, but not entrepreneurs.

Table 5.3: The Interaction between Entrepreneurship and Innovation

Type of Entrepreneur/ Innovator	Drive to Bring to Market	Drive to Innovate	Examples
Master entrepreneur	Dominant	Minimal	Individual producer or agri-food processor
Innovative entrepreneur	Primary	Secondary	Innovative farmers
Entrepreneurial innovator	Secondary	Primary	First Generation organic food firms
Master innovator	Minimal	Dominant	Bench engineers and scientists

Master Entrepreneurs

Master entrepreneurs express a dominant entrepreneur trait with little or no desire to innovate. Although decidedly entrepreneurial in their approach (*e.g.* willingness to take risk, high energy, desire to control their environment, and so on) they are not motivated by desire to innovate. Master entrepreneurs see market gaps and fill them with existing business models, products and services. Privately held grain merchandising firms and food manufacturers often are master entrepreneurs. Local grocery store managers fit this description as they often need to be entrepreneurial in their approach to developing local business, but are not innovating as they follow corporate guidelines and make the most out of given resources.

Innovative Entrepreneurs

Innovative entrepreneurs tend to be entrepreneurial first and innovative second. Innovative entrepreneurs constantly seek out new challenges, enjoy taking calculated risks, and are driven by a vision of what future could be. Unlike master entrepreneurs, they also look out for new ways of doing things. Innovative entrepreneurs often grow restless with the pace of change and lack of control over their future. Eventually they start their own ventures. Unlike master entrepreneurs, they are content to use true and tried business models, products and processes. They are driven to improve these products and processes in an effort to carve out a niche in the marketplace. Innovative entrepreneurs see market gaps and are willing to take risk based on their perception of market opportunities. They use a mix of tried and true and new business models, products and processes.

Entrepreneurial Innovators

Entrepreneurial innovators exhibit primary innovator and secondary entrepreneur traits. Such individuals are never satisfied with the status quo. They are constantly developing improvements to processes. Unlike master innovators who are content with innovation in and of itself, entrepreneurial innovators seek out change, grow tired of bureaucracy, and are driven to take risks to see that their innovations reach the marketplace. Entrepreneurial innovators see market gaps and fill them with new business models, products and processes that they are willing to take personally to market place. The first generation firms in the organic food industry started with a conviction about what products to produce. This conviction motivated production innovation that, in turn, triggered the need to market the products so created.

Master Innovators

Master innovators express dominant innovator traits with little or no interest to take these innovations to the marketplace. They are content to innovate in a given area, particularly in the one where they specialize or have expertise. Scientists and research and development engineers spend entire career improving products, processes and formulae with little interest in bringing these innovations tot the marketplace.

All of the above four combinations have to play a major role to play in development of entrepreneurship in the agibusiness sector.Moreover, it is not essential that a particular individual will be stuck in a one of the above given stylized types. Personal dispositions, life events, changes in technology, events affecting the market, and education and training may allow individuals and firms that are flexible enough to switch from one type to another. In order for market realization of an entrepreneur's innovation, barriers to entrepreneurship and innovation must be overcome.

A 4-P Framework of Entrepreneurship

In an effort to present a more integrated and coherent and comprehensive framework of entrepreneurship, Ma and Tan (2006) propose a 4 P framework of entrepreneurship. The four major components of entrepreneurship are – *Pioneer*, denoting entrepreneur as an innovator or champion for innovation; *Perspective*, denoting the entrepreneurial mindset; *Practice*, denoting the entrepreneurial activities; and *Performance*, denoting the outcome or result of entrepreneurial actions and activities. They define entrepreneurship as the process in which pioneers, innovators or champions of innovation, immersed in and guided by the creativity-oriented perspective, engage in the practice of creation and innovation driven activities, which lead to a certain level of performance as indicated by the realized creation and innovation. The specific form of entrepreneurship depends on the patterns of interaction among the pioneer, perspective, and practice, whose effects jointly determine entrepreneurial performance.

Table 5.4: The 4 Ps of Entrepreneurship and Stylized Illustrations

Perspective	Unique mind set for creativity and innovation	There got be a better way!
Purpose	Clear sense of mission and vision	Everyone is on this earth for a reason!
Policy	A wining formula	The strategy matters!
Pioneer	**Relentless Champion for Innovation**	**We can make a difference!**
Passion	Desire to achieve, to create, to make it happen.	Chase your dream!
Perseverance	Mental toughness	Never give up!
Practice	**Action matters**	**Just do it!**
Persuasion	Ability to convince others about your vision	Salesmanship is a natural ingredient to entrepreneurship!
Pursuit	Effort to attract and demand societal resources	God helps those who help themselves!
Performance	**Result driven**	**I did it my way!**
People	Innovation to improve and enrich people's life	Business is about serving people!
Profit responsible!	Innovation pays	Creating economic value is socially

Relationship among 4 Ps

Ma and Tan (2006) identify factors that are necessary but not independently sufficient for entrepreneurship. Thus, it is important to recognize the relationships among these factors that compose entrepreneurship (Shane and Venkataraman, 2000). As discussed earlier, wealth creation characterises the fundamental mission of entrepreneurship and stands as a critical criterion for judging entrepreneurship performance (Ireland *et al.*, 2001). Following the convention in the literature, authors treat performance as the dependent variable in building their models (Figure 5.1). They examine the individual as well as the joint effects of pioneer, perspective, and practice on performance, respectively, in the direct effect model, mediation model, interaction model, and the full model.

Direct Effects

This model depicts the main effects of pioneer, perspective, and practice on performance. In the literature, there exist both theoretical arguments and empirical evidence linking each of the other three Ps to the performance of entrepreneurship.

```
                    ┌─────────────────────┐
                    │      PIONEER        │
                    │    Champion of      │
                    │    Innovation       │
                    │      Passion        │
                    │    Perseverance     │
                    └─────────────────────┘

              ╱─────────────────────────────╲
             ╱        PERFORMANCE             ╲
            │         Pay-off of              │
            │         Innovation              │
            │          People                 │
             ╲          Profit               ╱
  ┌──────────────────┐  ╲─────────────╱  ┌──────────────────┐
  │   PERPECTIVE     │                   │    PRACTICE      │
  │ Unique and Creative                  │ Innovation-driven│
  │     mindset      │                   │      Action      │
  │     Pupose       │                   │      Pursuit     │
  │     Policy       │                   │    Persuation    │
  └──────────────────┘                   └──────────────────┘
```

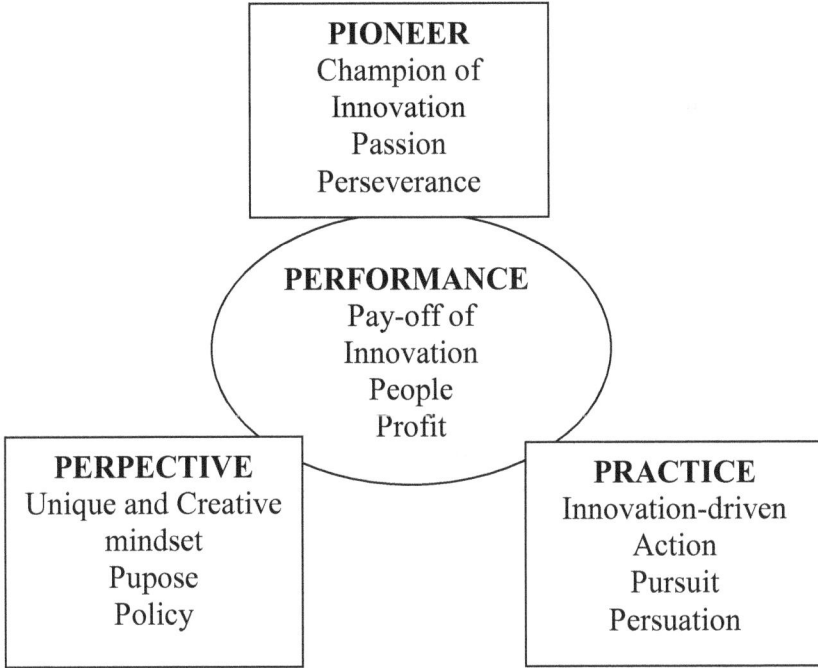

Figure 5.1: A 4-P Definition of Entrepreneurship (Adapted from Ma and Tan 2006).

First, there is a sizable literature on the impact of entrepreneurs' personal characteristics (Herron and Robinson, 1993; Shane, 1994; Baum *et al.*, 2001) on their success in new venture creation and venture growth, and in particular psychological characteristics (Begley and Boyd, 1987; Miner, 1997). Whether in the case of independent entrepreneurs or champions of innovation in corporate settings (Shane, 1994), the tenacity and determination to achieve and the passion to create (Baum *et al.*, 2001) will help pioneers sustain their enthusiasm and commitment and hence reach their goals (Ghemawat, 1991), whether in new venture creation (Vesper, 1990, Stearns and Hills, 1996), product innovation (Jennings and Young, 1990), or corporate venturing (Zahra, 1995; Zahra and Covin, 1995).

Second, the entrepreneurial perspective also has direct effects on entrepreneurial performance (McGrath and MacMillan, 2000). A clear sense of entrepreneurial vision and the strategic intent of being the global leaders of what they do (Hamel and Prahalad, 1989) will likely motivate entrepreneurs to embody their visions into a consistent entrepreneurial orientation (Lumpkin and Dess, 1996) and institutionalized policy or entrepreneurial strategy (Ireland *et al.*, 2001), and the choice and focus of a firm's entrepreneurial strategy will have a definite impact on the firm's wealth creation and performance over time (Tan and Tan, 2005).

Finally, it is entrepreneurial practice, the actual actions and activities pursued, which create the desired results. We expect that entrepreneurial practice will have a direct impact on entrepreneurial performance. McGrath and MacMillan (2000) argue that the adaptive execution in exploiting opportunities plays an important role in entrepreneurial success. How aggressively and intensely pioneers engage in entrepreneurial actions and pursue their entrepreneurial agendas will also affect the success of their endeavor (Morris and Lewis, 1995; Morris and Sexton, 1996). The direct effect model has been summarized in Figure 5.2(a).

Mediation Effects

For practical reasons in research design and data collection, some variables are often difficult and tedious to capture and measure empirically, including the specific amount, scope, and intensity of entrepreneurial actions (Morris and Sexton, 1996). As such, researchers typically use relevant antecedents or consequence variables, such as entrepreneurs' personal traits, either theoretically grounded or empirically derived, as necessary surrogates for theory testing. It is not surprising that there are rarely studies examining the direct relationship between actual entrepreneurial activities and entrepreneurial performance (*e.g.*, Morris and Lewis, 1995; Morris and Sexton, 1996). While pioneers and perspective each may have a direct effect on entrepreneurial performance, it could also be argued that their effects on performance are mediated by the entrepreneurial practice. That is, pioneer and perspective primarily determine entrepreneurial practice, which in turn determines entrepreneurial performance.

Baum *et al.* (2001) theoretically argue for and empirically demonstrate the indirect effect of personal characteristics of entrepreneurs, such as tenacity and passion, on firm performance through other intermediate variables of entrepreneurial action. Practically, it is also reasonable to conjecture that entrepreneurs with strong passion and great perseverance will be more committed to their entrepreneurial pursuit (Ghemawat, 1991; Powell and Ma, 1996).

It is argued that a strong sense of purpose and intention will help the entrepreneurs enlist commitment and support of stakeholders to implement their entrepreneurial agendas (Bird, 1988; Hamel and Prahalad, 1989). Similarly, they also expect that over time, a firm's entrepreneurial orientation (Lumpkin and Dess, 1996, Tan and Tan, 2005) affects the intensity and scope of its entrepreneurial activities. Additionally, in the mediation model, a direct and immediate effect is found between entrepreneurial practice and entrepreneurial performance. The model is presented in panel (b) of Figure 5.2.

It is also theoretically meaningful and practically interesting to examine the interaction effects among pioneer, perspective, and practice on performance. With the vision to create value by targeting over-priced or under-served regions, Rollin King and Herb Kelleher founded Southwest Airlines with a simple perspective: If they could get passengers to their destinations when they wanted to get there, on time, at the lowest possible fares, and making sure they had a good time doing it, people would fly their airline. Unlike many people who enjoy discussing perspectives, these pioneers have turned their simple perspective into practice and served increasing numbers of customers under a simple and consistent business model, still driven by that simple perspective. While doing all this, Southwest

a. Direct Effect Model

b. Mediation Model

c. Practice

d. Performance

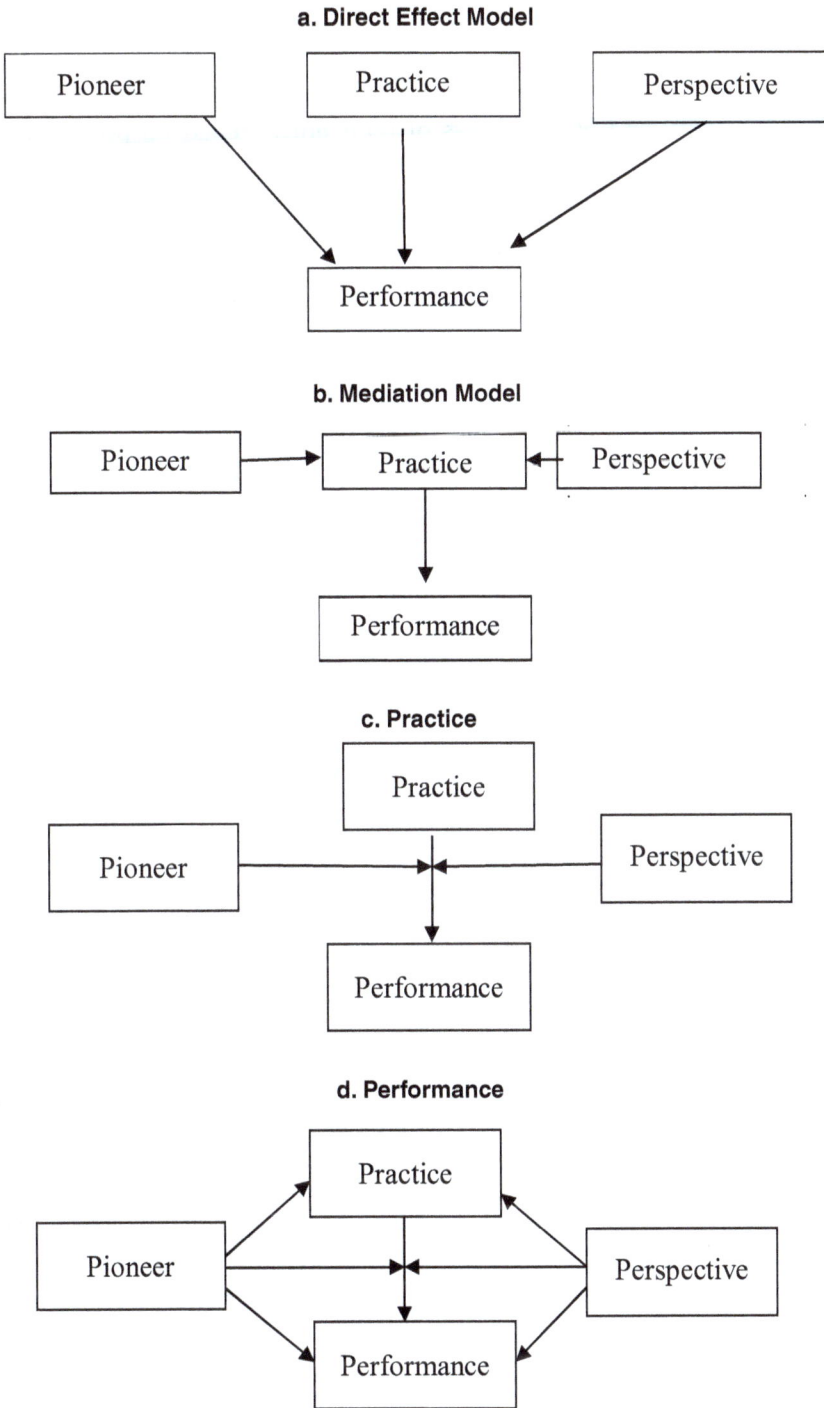

Figure 5.2a-d: A Process-Vased View of Entrepreneurship
(Adapted from Ma and Tan, 2006).

Airlines has also rewarded investors handsomely and made Southwest one of the best places for employees to work.

As advocated by McGrath and MacMillan (2000), the discipline of entrepreneurs affects the quality and effectiveness of their entrepreneurial pursuit in that habitual entrepreneurs pursue only the best opportunities. As such, we expect that the relationship between practice and performance will be qualified by the characteristics of the pioneers.

Similarly, we expect that entrepreneurs who are goal-driven and possessing a clear vision (Collins and Porras, 1996) are better at focusing on the essential entrepreneurial activities that are likely to create superior performance. Lumpkin and Dess (1996), in hypothesizing the interaction effects between entrepreneurial orientation and firm action on firm performance, suggest that firm proactiveness, combined with quick response in the market place, results in high performance. This relationship is depicted in panel (c) of Figure 5.2.

The Full Model

Integrating the above three models into a comprehensive framework, a full model can be established to better represent the various possible relationships among the 4 Ps. Both the main individual effects and joint effects, mediation and interaction, are included in panel (d) in Figure 5.2.

To put the framework in perspective, this approach differs from the micro view of entrepreneurship, which mainly examines personal characteristics that are specific to entrepreneurship. It is contended that the search for a single psychological profile of the entrepreneur is bound to fail, because for each of the traditional definitions of the entrepreneurial type, there are numerous counter examples that disprove the theory. This framework also differs from the macro view, which focuses primarily on environmental factors external to the entrepreneurial business that are capable of creating or destroying entrepreneurship by virtue of the climate they establish (Aldrich and Wiedenmayer, 1993).

Entrepreneurs create and are created by opportunities. Environment and opportunities do not exist in time-space independent of the entrepreneurs. Rather, they have only a virtual existence as they are constituted and changed by entrepreneurial actions. Following the line of reasoning advocated by Shane and Venkataraman (2000), our framework emphasizes entrepreneurship as a process in which pioneers for innovation and change, guided not by social norms but by a unique set of creative perspectives, engage in the practice of wealth creation and convert opportunities into performance. Such a theoretical perspective would view entrepreneurship as a recursive process that evolves as the entrepreneurs interface with resources and opportunities (Shane and Venkataraman, 2000; Venkataraman, 1997) and proactively "enact their own environments" (Levinthal and March, 1993).

Conclusion

The chapter compiles the fragmented literature on conceptual development in entrepreneurship development and research presents some of the most important conceptual models of entrepreneurship. The entrepreneurial disposition, entrepreneurship-innovation continuum, and a 4Ps view of entrepreneurship are the major conceptual developments in this area which is gradually growing in importance.

Given the producer demands and policy imperatives that exist to support value-added ventures, particularly in agri-business development, the concepts presented herein give guidance to researchers and policy makers to foster entrepreneurship and innovation. First, establishment of agri-business incubation centres can play a key role in strengthening the linkages between entrepreneurship and innovation. They can also assist in the creation of business plans, marketing, feasibility studies, and other aspects of product and business development. These centers can also help reduce the potential for market failure. Market failure occurs when would-be entrepreneurs or innovators let good ideas and intentions die or they do not know how to foster their entrepreneurial or innovation abilities. Another market failure occurs when a poorly conceived idea goes forward and fails, leading to lost jobs, the misallocation of capital, and in the worst cases bankruptcy. University programs in entrepreneurship can empower the would-be entrepreneur to answer difficult questions and consider factors that the would-be entrepreneur may be overlooking before committing financial resources. To be effective, these centers and programs need to be geared toward new firms, new products, innovation processes, and entrepreneurial education. Traditional services such as situation and outlook analysis, or farm financial management will not be sufficient in themselves to meet the new needs.

Second, apart from innovation centers, agricultural economists need to develop innovations in curricula and extension activities to more broadly enhance the probability of successful entrepreneurship and innovation. Undergraduate and graduate students need exposure and experience with these concepts. Producers and agri-food managers could likely benefit from education about their entrepreneurial/innovation type and how to transition from one type to another. Also, we may need to create marriages between those who are more strongly entrepreneurs and those who are more strongly innovators. University outreach programs and centers have a key role to play by establishing structures that foster mentorship and or act as a clearing house where entrepreneurs and innovators interact and benefit from one another. Such entrepreneurial/innovation teams may be much more effective than individuals working alone. Such teams may help solve the commercialization gap that exists between our laboratory-based science innovations and getting these innovations (and their potential patent revenues) to market.

References

Aldrich, H.E., Wiedenmayer, G., 1993. From traits to rates: an ecological perspective on organizational foundings. In: Katz, J.A., Brockhaus Sr., R.H. (Eds.), *Advances in Entrepreneurship, Film Emergence, and Growth.* JAI Press, pp. 145– 195.

Balakrishnan, S., Gopakumar, K. and Kanungo, R. N. 1998. Entrepreneurship development: Concept and context,pp. 1-38. In: *Entrepreneurship and Innovation,* (Ed. Kanungo, R. N.), Sage Publications, New Delhi.

Baum, J.R., Locke, E.A., Smith, K.G., 2001. A multi-dimensional model of venture growth. *Academy of Management Journal* 44 (2), 292–303.

Begley, T.M., Boyd, D.P., 1987. Psychological characteristics associated with performance in entrepreneurial firms and smaller businesses. *Journal of Business Venturing* 2, 7993.

Bird, B., 1988. Implementing entrepreneurial ideas: the case for intention. *Academy of Management Review* 13, 442–453.

Bryant, C. R., 1989. Entrepreneurs in the rural environment. *J. Rural Stud.* 5(4), 337-347.

Bygrave, W. D., 1994. *Portable MBA in Entrepreneurship*. John Wiley and Sons, New York.

Bygrave, W.D. and Hofer, C.W., 1991. Theorizing About Entrepreneurship. *Entrepreneurship: Theory and Practice* **16** 2, pp. 13–22.

Cohen, N. "The Five Ages of the Entrepreneur." *Venture* (1980): 40– 42.

Collins, J.C., Porras, J.I., 1996. Building your company's vision. *Harvard Business Review*, 65– 77 (September- October).

De, Dipak, 1986. Factors Affecting Entrepreneur Characteristics of Farmers, *The Indian Journal of Social Work*, Vol. XLVI No. 4 pp. 541-546.

Ghemawat, P., 1991. *Commitment: the Dynamics of Strategy*. Free Press, New York.

Gopakumar, K. 1995. "Entrepreneurship in economic thought: a thematic review", *Journal of Entrepreneurship*, 4(1), 1-17.

Hamel, G., Prahalad, C.K., 1989. Strategic intent. *Harvard Business Review* 67 (3), 63– 76.

Hamel, G., Prahalad, C.K., 1994. *Competing for the Future*. Harvard Business School Press, Boston, MA.

Herron, L.A., Robinson Jr., R.B., 1993. A structural model of the effects of entrepreneurial characteristics on venture performance. *Journal of Business Venturing* 8, 281–294.

Ireland, R.D., Hitt, M.A., Camp, L., Sexton, D.L, 2001. Integrating entrepreneurship and strategic management actions to create firm wealth. *Academy of Management Executive* 15 (1), 49– 63.

Jennings, D.F., Young, D.M., 1990. An empirical comparison between objective and subjective measures of the product innovation domain of corporate entrepreneurship. *Entrepreneurship Theory and Practice* 15 (1), 53–66.

Kanungo, R. N. and Misra, S., 1992. Managerial resourcefulness: A Re-conceptualisation of managerial skills, *Human Relations*, 45(12), 1311-32.

Kao, R.W. Y., 1993. Defining Entrepreneurship: Past, Present and ? *Creativity and Innovation Management*, 2(1):69-70.

Kirzner, I. M. 1973. *Competiton and Entrepreneurship*, University of Chicago Press, Chicago.

Kirzner, I. *Perception, Opportunity and Profit.* Chicago: Chicago University Press, 1979.

Knight, Frank H. 1921. *Risk, Uncertainty, and Profit.* Boston, MA: Hart, Schaffner and Marx; Houghton Mifflin Company, 1921.

Knudson, W., Wysocki A., Champagne J. and H. Christopher Peterson, 2004. Entrepreneurship and innovation in the Agri-food System. *Amer. J. Agr. Econ.* 86:1330-1336.

Kodithuwakku, S. S. and Rosa, P. 2002. The entrepreneurial process and economic success in a constrained environment, *Journal of Business Venturing.* 17(2002):431-465.

Learned, K.E. "What Happened Before the Organization? A Modal of Organization Formation." *Entrepreneurship Theory and Practice* **17**(1992): 39– 48.

Leibesntein, H. 1968. Entrepreneurship and development. *American Economic Review.* 58(2):75.

Levinthal, D.A., March, J.G., 1993. The myopia of learning. *Strategic Management Journal* 14, 95– 112 (Special Issue).

Lumpkin, G.T., Dess, G., 1996. Clarifying the entrepreneurial orientation construct and linking it to performance. *Academy of Management Review* 21 (1), 135– 173.

Ma, Hao and Tan, Justin, 2006. Key components and implications of entrepreneurship: A 4-P framework, *Journal of Business Venturing.* 21:704-725.

McGrath, RG., MacMillan, I.C., 2000. The Entrepreneurial Mindset: Strategies for Continuously Creating Opportunity in the Age of Uncertainty. Harvard Business School Press, Boston, MA.

Milton, D.G." The Compleat Entrepreneur." *Entrepreneurship Theory and Practice* **13**(1989): 9– 19.

Miner, J.B., 1997. *A Psychological Typology of Successful Entrepreneurs.* Quorum Books, Westport, CT.

Morris, M.H., Lewis, P.S., 1995. The determinants of entrepreneurial activity: implications for marketing. *European Journal of Marketing* 29 (7), 31–49.

Morris, M.H., Sexton, D.L., 1996. The concept of entrepreneurial intensity: implications for company performance. *Journal of Business Research* 36 (1), 5 – 13.

Penrose, E.T. (1963). *The Theory of the Growth of the Firm.* Oxford: Basil Blackwell.

Powell, T.C., Ma, H., 1996. Organizational resilience: extraordinary performance in hostile environments. Paper presented at the 16th Strategic Management Society Conference at Phoenix, Arizona.

Say, J. B., 1815, tr. 1967. *A catechism on political economy*(trans. By Jojn, Richter). New York:Augustus. M. Kelly.

Saylor, M., 1987.Home-basedenterprise development and craft mareting. *Proceedings of International Rural Entrepreneurship Symposium.* Economic Research Service, USDA, Knoxville, Tenesse, pp 57-70.

Schumpeter, J. A. 1949. *Economic Theory and Entrepreneurial History–Change and the Entrepreneur, Postulates and Patterns for Entrepreneurial History.* Harvard University Press: Cambridge, MA.

Schumpeter, J.A., 1934. *The Theory of Economic Development.* Harvard University Press, Cambridge, Ma.

Sexton, D. L., Bowman-Upton, N., 1991. *Entrepreneurship, Creativity and Growth,* Macmillan, New York.

Shane, S., Venkataraman, S., 2000. The promise of entrepreneurship as a field of research. *Academy of Management Review* 25 (1), 217– 226.

Shane, S.A., 1994. Are champions different from non-champions? *Journal of Business Venturing* 9, 397– 421.

Stearns, T.M., Hills, G.E., 1996. Entrepreneurship and new firm development: a definitional introduction. *Journal of Business Research* 36 (1), 1 – 5.

Tan, J., Tan, D., 2005. Environment–strategy coevolution and coalignment: a staged-model of Chinese SOEs under transition. *Strategic Management Journal* 26 (2), 141– 157.

Venkataraman, S., 1997. The distinctive domain of entrepreneurship research: an editor's perspective. In: Katz, J., Brockhaus, R. (Eds.), *Advances in Entrepreneurship, Firm Emergence, and Growth,* vol. 3. JAI Press, Greenwich, CT, pp. 119–138.

Vesper, K.H., 1990. *New Venture Strategies.* Prentice-Hall, Englewood Cliffs, NJ.

Zahra, S. and Covin, J. (1995) Contextual Influence on the Corporate Entrepreneurship-Performance Relationship. *Journal of Business Venturing,* 10: 43-58.

Zahra, Z.A., 1995. Corporate entrepreneurship and financial performance: the case of management leveraged buyouts. *Journal of Business Venturing.* 10 (3):225– 247.

Chapter 6

Dynamics of Entrepreneurship Development in Agriculture: Basics to Advances

A.K. Singh[1] and Lakhan Singh[2]

[1]*DDG (Extension), ICAR, New Delhi*
[2]*Principal Scientist (Agril Extension), Zonal Coordination Unit,*
Zone IV (ICAR), G.T. Road, Rawatpur, Kanpur – 208 002, U.P.

Entrepreneurs play an important role in developing and contributing to the economy of a nation. But entrepreneurship development is not much emphasized in developing countries. We see more entrepreneurs in comparatively more developed areas. There is problem of unemployed population, seeking wage earners' career and unaware about the wide open opportunities for entrepreneurial career. This is, by and large, because of the lack of education about entrepreneurship development. In most of the countries, entrepreneurship development has not found any place in the education curriculum.

There is plenty of food. People are poor. People are sick. To solve this problem, entrepreneurship qualities have to be developed among the rural youths in the rural areas in the field of agriculture and related fields. It can be possible through human resource development at field level through Krishi Vigyan Kendras working in whole country. A huge network of 557 KVKs has been developed, where rural youths may be trained for establishing their enterprises.

Need and Scope of Entrepreneurship Development

☆ Cereal growers continue to remain poor and poorer.

☆ Low purchasing power and low accessibility.

☆ Agriculture cannot absorb surplus agricultural labourers.

☆ Non-farm sector has greater scope.

☆ Raw agricultural produce vs. value added products.

☆ Demand driven market economy.

☆ Exploiting opportunities in sunrise sectors.

☆ Farm diversification for sustainable development.

☆ Private extension and contract farming options.

☆ Human resource utilization – labour intensive work.

☆ Horticulture and related enterprises.

☆ Precision agriculture for cost reduction on farms.

☆ Harnessing ICTs for Rural development.

☆ Gearing farmers to meet export potential.

Why Entrepreneurship?

☆ Challenging career options.

☆ Increasing national production.

☆ Dispersal of economic power.

☆ Weakening monopoly.

☆ Harnessing youth vigour.

Income generating activities, self employment and entrepreneurship development should have the following features:

☆ Innovative

☆ Risk management

☆ Urge for excellence

☆ Sustenance and growth

☆ Diversification

☆ Organizing resources

☆ Management

☆ Marketing

Characteristics of Entrepreneurs

There is need of few important characteristics in entrepreneur to execute his enterprise successfully. Some of the characteristics are as under:

☆ Innovative, creative and hard/smart working

☆ High achievement motivation

☆ Goal setting behaviour – SMART goals

☆ Specific, measurable, achievable, realistic and time bound goals

☆ Moderate risk taking – not high nor low

☆ Time conscious – Meeting deadlines on Time

☆ Time conscious – Seasonal and Timely

☆ Problem solving and not problem avoidance

☆ Action oriented – believes in action and see results

☆ Well aware of personal strengths

☆ Ability to perceive obstacles properly

☆ Market savvy and business acumen

☆ Pragmatic and realistic in their approach

Entrepreneurial Development Involves

☆ Stimulating the entrepreneurial motivation.

☆ Providing support to the potential entrepreneurs.

☆ Helping them to sustain and manage their enterprises.

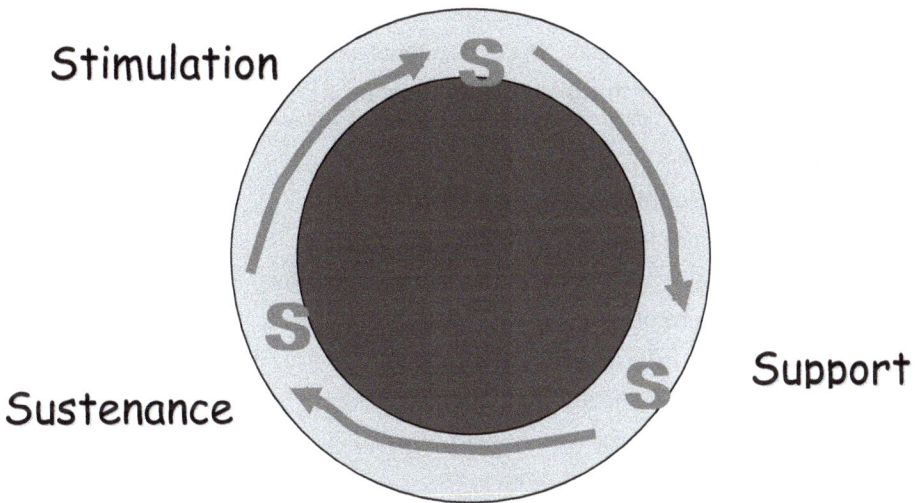

Stimulation Phase

☆ Stimulating environment

☆ Caste and traditional occupations

☆ Entrepreneurial Motivation Training (EMT)

☆ Achievement Motivation

☆ Need for achievement – personal and social

☆ Risk Taking behaviour

☆ Goal setting behaviour

☆ Identifying personal strengths and perception

☆ Achievement Planning Orientation

☆ Measuring and enhancing motivation

☆ Profit motive and economic motive

☆ Sustaining the initial enthusiasm and interest

Support Phase

☆ Business Plan for agri-business

☆ Identifying enterprises – Micro-screening

☆ Market survey and analysis

☆ Developing Project Proposal

☆ Feasibility studies and Financial Analysis

☆ Institutional support – training, banks, *etc.*

☆ Seeking loans from Banks

☆ Setting up or launching the enterprise

☆ Trial run and teething troubles

☆ Cash flows and managing costs

☆ Price fixing and marketing agreements

☆ Agreements for input supply and labour

Sustenance Phase

☆ Teething troubles and management

☆ Management issues – production, personnel

☆ Marketing issues – price fixing and competition

☆ Management of working capital and finances

☆ Running to full capacity and problems

☆ Capturing and sustaining market shares

☆ Diversification and growth of enterprise

☆ Production cycles and payments

☆ Dues recovery in marketing and supplies

☆ Loan recovery, repayments and discipline

☆ Financial health and treatment of sick units

☆ Cooperatives and group counseling

New Initiatives and Case Studies

☆ Poultry farming in Siddharthnagar district has been promoted by KVK. Shri Hidaytullah Khan, an entrepreneur started layer farming with the capital investment of Rs. 29,30,000. Total return being received is Rs. 67,55,000 with a net profit of Rs. 16,28,904. He is having 12000 birds. Net profit per bird for 12 months comes to about Rs. 136.

☆ About 100 Piggery farming units have been established by KVK, Gonda in Gonda district of Uttar Pradesh.

☆ Custom hiring services for resource conservation implements in eastern Uttar Pradesh.

☆ Apiary promoted by KVK, Aligarh in 14 districts of Uttar Pradesh.

☆ Madhya Pradesh – ITC e-chaoupals and rural super markets.

☆ Tomato ketchup - Contract farming in Punjab –Pepsi and HLL.

☆ Many more still not documented.

Beekeeping Promoted by KVK, Aligarh

Achievements in Bee Keeping in KVK, Aligarh

Year	No. of Courses Organized	No. of Trained Entrepreneurs	No. of Practicing Beekeepers	No. of Colonies
2000	6	148	15	660
2001	5	46	31	2617
2002	4	54	40	3502
2003	5	117	46	4515
2004	6	167	53	10458
Total	26	532	185	21752

Environmental Impact

☆ Enhanced production of mustard by 10-15 percent.

☆ Non use of any chemical in honey production.

☆ It promotes growing of nectar based crops/plants enriching environment.

☆ Honey production does not exhaust any existing natural resource base.

☆ Honey production does not affect existing cropping system rather the boxes are migrated as per availability of flora.

Promotion of Vermiculture by KVK, Unnao

Physical achievement of Vermiculture Unit

Total No. of units established	750
No. of Villages	38
No. of Trainees	965
No of beneficiaries	365

Entrepreneurship Development as an Alternative Extension Approach

☆ Creating more jobs rather more food

☆ Developing non-farm sector

☆ Absorbing the surplus rural labour

☆ Improving quality of life - Human Development

☆ Changing priorities of Extension – broad basing

  ☐ Health and hygiene, and sanitation,

  ☐ Education, Standard of living, and Happiness

☆ Making farmers quality - conscious

☆ Educating farmers on cost reduction in farming

☆ Increasing awareness about food safety

☆ Increasing awareness about Codex Alimentations

☆ Providing better food products to people

☆ Improving nutritional security of the future people

☆ Developing sunrise sectors – value addition.

Entrepreneurship Awareness

Arousal of motivation for seeking, sensing opportunities is must. Developing capacity for selecting, planning and setting of ventures assessing, locating and mobilizing resources play a greater role in smooth running of an enterprise. Acquiring ability to manage venture enterprise and its growth again promotes the entrepreneurship development.

Entrepreneurial Motivation Training

Now more emphasis may be given to promote entrepreneurship among rural youths in the context of globalization. This work can be performed by KVKs. Motivation for entrepreneurship development has been ignored earlier. For this, capable trainers may be created at KVKs through organizing well designed entrepreneurial motivation training programmes where more thrust may be given on practical aspects.

Value Added Extension Services to Farmers through Agri-clinics and Agri-business Centres

Agri-clinics are those which provide expert services and advice to farmers on cropping practices, technology dissemination, crop protection from pests and diseases, market trends and prices of various crops in the markets and also clinical services for animal health, *etc.*, which would enhance productivity of crops/animals. Agri-business centre are those which provide input supply, farm equipment on hire and other services.

The central sector scheme of agri-clinics and agri-business centres was launched on 9th April, 2002.

The Need of Agri-clinics and Agri-business Centres in Present Extension system

Thousands of agriculture and allied graduates passed out from universities are going unemployed. On the other hand, there is acute shortage of trained manpower in agriculture extension. Agri-clinics and Agri–business centres are managed by professionally qualified agriculture and allied graduates, located in rural areas providing value added extension services to the farmers. Thus, extension gaps can be reduced.

Unemployed agriculture and allied graduates are eligible under the scheme. A management graduate with experience can also be part of group project in which others are agriculture and allied graduates.

Facilities Provided Under the Scheme

Selected eligible graduates are provided with two months free residential training on agri-business in their respective states. Trainees are exposed to potential agriventures in their locality, market survey based detailed project reports are prepared individually by the graduates, submitted to banks through the training institutes. One year handholding support is provided to trainees in the field through the training institutes. Efforts are made to link them to other government schemes and agri-business activities of agri-business companies. Loan and subsidy provision is made to eligible trainees.

Contact for Admission

Eligible graduated have to contact the recognized training institutes in their respective states or write to Agri-Clinics and Agri-Business Centres Cell, MANAGE, Rajendranagar, Hyderabad-500 030, Andhra Pradesh along with attested copy of marks card, photograph, residential proof and non refundable application fee of Rs. 500/- drawn in the name of MANAGE, Hyderabad.

Applications are processed at the level of training institutes by the screening committee, eligible candidate are interviewed to assess their seriousness in self employment. Selected candidates have to deposit caution money of Rs. 1000/- with the training institute which is refundable to candidates after satisfactory completion of the training. There is no training fee to be paid by selected candidates. Entire cost of training and handholding is met by Govt. of India.

Training Module and Methodology

Training is conducted for a period of two months. Motivation, personality enhancement, basic principles and practices of business management, agri-business management, IT skills, potential agri-ventures, market survey, and project preparation are taught along with hands on experience in the chosen field. Market survey based detailed project is prepared by trainees with the help of experts and submitted to banks through the training institutes. One year handholding support is provided to trainees in the post training period through training institutes facilitating the establishment of Agri-clinics and Agri-business centres. Loan and subsidy provision is also available for eligible candidates.

Loan and Subsidy Provisions Under the Scheme

Loan is provided by Nationalized Banks, RRBs, and Co-operative Banks to the eligible trained candidates up to Rs. 10.00 lakhs per individual project and up to Rs. 50.00 lakhs for group project. 25 percent of the capital cost is subsidized through credit linked back ended subsidy and first two years total interest is waived off. Candidates trained after 01/04/2004 are eligible to avail subsidy benefit under the scheme. Up to Rs. 5.00 lakh loan, security is waived off.

Provisions under ATMA to Support Agri-clinics and Agri-business Centres

Provision of Rs. 2.5 lakh per ATMA is made to support Agri-clinics and Agri-business centres established in the district. Agri-clinics and Agri-business centres are used in implementation of ATMA activities for which service charges are provided through the provision.

Success Rate of the Scheme

Out of 13660 candidates trained, 4550 have established Agri-clinics and Agri-business centres Scheme in 36 categories of activities. The present success rate is 33.30 percent.

Distribution of Established Agri-blinics and Agri-business Centres in the Country

State wise and activity wise distribution of Agri-clinics and Agri-business centres in the country are as follow:

Impact of Agri-clinics and Agri-business Centres on Agriculture Extension

Agri-clinics and Agri-business centres supplement the efforts of public extension, provide specialized extension services and also generate self employment opportunity to unemployed agriculture graduates. Recently conducted MANAGE impact study reveals that an agripreneurs with 32 months of business experience is covering 3013 farmers spread over 38 villages. There was 28.8 per cent increase in the income of the farmer and 17.4 per cent increase in yield of the farmers due to service provided by agripreneurs. Monthly average earning of agripreneurs found to be Rs. 7950. Agripreneurs also created self employment opportunity for other four people in the rural areas.

Role of Agri-clinics and Agri-business Centre in Future Extension

Small and medium scale agri-clinics and agri-business centres scheme of today are agri-business companies of tomorrow. Such agri-ventures are managed by professionally qualified agriculture graduates who help in professionalizing the

	MANAGE		
	PROGRESS OF AGRICLINICS AND AGRIBUSINESS CENTRES SCHEME		
	Period From : 01-04-2002 To: 30-10-2015		
S.No.	Name of the State	No. of Applications Received	No. of Candidates Trained	No. of Agri-ventures established
1	Andhra Pradesh	883	825	306
2	Arunachal Pradesh	35	32	3
3	Assam	601	597	200
4	Bihar	3387	3282	1208
5	Chandigarh	3	3	1
6	Chattisgarh	638	527	250
7	Delhi	25	17	3
8	Goa	10	9	4
9	Gujarat	1326	1274	503
10	Haryana	601	552	202
11	Himachal Pradesh	423	418	108
12	Jammu and Kashmir	1332	1276	175
13	Jharkand	619	618	155
14	Karnataka	3105	3037	1244
15	Kerala	203	183	51
16	Madhya Pradesh	1354	1294	521
17	Maharashtra	10272	10189	4723
18	Manipur	416	413	126
19	Meghalaya	11	11	3
20	Mizoram	34	34	0
21	Nagaland	177	174	21
22	Odisha	549	506	106
23	Pondicherry	112	111	68
24	Punjab	534	531	197
25	Rajasthan	2769	2623	971
26	Sikkim	10	9	0
27	Telangana	1037	982	361
28	Tamil Nadu	5191	5116	2690
29	Tripura	4	2	1
30	Uttar Pradesh	9437	9171	4574
31	Uttarakhand	418	392	114
32	West Bengal	777	725	216
		46293	44933	19105

AGRI-CLINICS AND AGRI-BUSINESS CENTRES CELL
NATIONAL INSTITUTE OF AGRICULTURAL EXTENSION MANAGEMENT (MANAGE), HYDERABAD
ACTIVITY-WISE CATEGORISATION OF AGRI-VENTURES
From 01-04-2002 To 17-06-2016

Name of the Agri-venture	AP	ARN	ASS	BHR	CHA	CHD	DEL	GOA	GUJ	HP	HRN	JAM	JHA	KAR	KER	MAN	MEG	MP	MS	NAG	ORS	PON	PUN	RAJ	SKM	TG	TN	TRI	UP	UTC	WB	Total
01. Agri-Clinics	127		21	155	20				39	12	19	11	72	191	16	6		61	881		27	16	29	146		86	595		610	18	25	3,185
02. Agri-Clinics and Agribusiness Centres	67	2	30	494	63			1	253	25	97	49	26	426	7	25	2	169	1,422	3	20	17	32	163	1	87	606		2,611	26	47	6,772
03. Agro-Eco Tourism								1	1					1					7								1					11
04. Annual Feed Unit	1		2						1			2		1				1	16	1						1	5		9			47
05. Bio-fertilizer production and Marketing	3		1	3					4		1	2		15	1				30		2			9		9	9		12	2		102
06. Contract Farming	5				2				1	1				13		1		1								6	21		11			66
07. Cultivation of Medicinal Plants	1		1	16	3					4	2		2	14		1		3	6		2			31		5	10		12			112
08. Direct Mkt.	1		1	6	2				6	3		3	1	16	1	1		4	66		4			5		2	8		38	2	1	168
09. Farm Machinery Unit	8		3	11	8				32	16	9	8		28				46	207		3		3	14		8	89		196	7	1	713
10. Fisheries Development	4		4	104	34					1	3	2	7	12	14	10		3	40		5			2		4	43	1	29	1	18	349
11. Floriculture	3				1				3	5				21					33				4	3		11	14		8		3	108
12. Horticulture Clinic	2			6	3					7	2		2	23		2		5	62		1					3	26		11	1		170
13. Landscaping + Nursery	11		1	1				2	1	1	1	4		15	3			1	46		4		1				24		1			113
14. Nursery			5	21	3				13	8	7	11	2	45	3	8		8	164		6		11	15		10	91		53	2		506
15. Organic Production/ Food Chain					3				4				1	20		2		1	15		1		5	8		2	13		10			88
16. Pesticides Production and Marketing				1					4				1	10				2	14				2				1		2			40
17. Value Addition	4		2	38	6				5					10		2		8	120	3	10		3	10		1	22		25	3	2	276
18. Fishery clinic				4								1	1						3								6					15
19. Seed Processing and Marketing	13		1	28	10		1		19	3	12			56		1		13	23	1	3	2	26	14		31	31		39	3	7	335
20. Soil Testing Laboratory	2		1	23	1				6		1			12				2	16		1		1			1	25		5		3	102
21. Tissue Culture Unit	2													11				1	8								2					28
22. Vegetable Production and Marketing	1		1	1	21		1		3	9	7	2	8	7	1	2		7	33		2	13		6		7	21		56	3	35	245
23. Vermicomposting / Organic manure	7		5	125	3				6	3	3	2	6	39	2	2		6	82		1	17	16	71		12	43		53	6	1	495
24. Veterinary Clinics	2		91	66	11	1	1		17	2	6	23	16	11		28	1	1	323	7	2		12	83		6	152		22	3	10	875
25. Crop Production	3		1	3						1	1			27				11	41		11		2	49		2	13		11	2	2	197
26. Dairy/Poultry/Piggary/Goatary	50	1	35	107	56				130	8	18	24	13	253	3	18		219	1,593	6	4	12	55	385		65	1,041		1,216	54	77	5,443
27. Rural Godown				5										2				2	8					1			3		28			49
28. Production & Marketing of Bio-Control Agents					1									4					4								2		8			18
29. Agriculture Journalism									1					3					3							2			2			16
30. Sericulture				1										7		19			13		6		1				7				1	49
31. Mushroom Cultivation				16					1		2		2	2					3		1						39		5	2	5	99
32. Apiary				8	2						14	3	2	2													2		65	1		101
Total	321	3	206	1,246	253	1	3	4	557	108	205	176	163	1,302	51	128	3	576	5,283	21	106	77	203	1,024	1	363	2,964	1	5,148	140	256	20,893

agriculture extension. Agripreneurs bring innovations and entrepreneurship to rural areas resulting in employment generation in villages. Private investment is expected to increase in agriculture sector.

Chapter 7

Agripreneurship Development in India

Ram Bahal

Former Principal Scientist, Division of Agricultural Extension, IARI, New Delhi – 110 012

India is having 329 million hectares of land area of which 143 million hectares is under cultivation. Despite all the natural advantages, India's productivity of food grains per hectare is no more than three-fourths of the world average. Out of 29 states and 6 union territories, only five states in India, namely Himachal Pradesh, Punjab, Haryana, Uttar Pradesh and Madhya Pradesh – produce more grain than their populations can consume. The combined population of the five states is less than one- third of the total of the country. Remaining two third of the population lives in 24 states and six union territories which are still food-deficit. This requires transport of lakhs of tonnes of food grain, involving high costs and pilferage. The effort should have been, to make all the states self-sufficient with respect to food grains. The growth rate of grain production during the ninth plan has been less than the population growth rate. Per capita availability of grain and per capita calorie intake, which were less than the minimum required for adequate nutrition, have further declined. According to Human Development Report 2003, the percentage of the undernourished in India, has reached 24. The environmental status is though not alarming in comparison to developed countries, it gives an early warning to take appropriate precautionary measures. The main reason, for poor performance of the farm sector has been the adverse terms of trade policies for agriculturists in addition to the mismanagement of natural resources and lack of trained agriculture managers. Agriculture today is diversified in so many areas and sub-areas which are not the same what it was two decades ago. It has shifted from subsistence to commercial activity, but faces enormous challenges including trained agripreneurs. Despite so many agricultural research, teaching and extension organizations, Indian agriculture is still treated as traditional or conventional.

It has placed unparallel demands on the capabilities and competence of human resources responsible for linking production with consumption vis a vis economic-environment equilibrium. Existing system of education, due to many limitations is not in a position to cater the rising demand for agri-business managers. Therefore, there is great need of agripreneurship development in the country. Considering the scope of the topic, an attempt is made to delineate the agripreneurship, strengths to promote agripreneurship, supports, incentives and promoter of agripreneurship, key challenges for HRD skills, status of present day education, and institutions supporting agripreneurship in this chapter.

Diversity of Indian Agriculture

Agriculture is producing food, feed, fuel, fiber and other goods by the systematic growing/harvesting of plants, animals and other life forms. "Agriculture" may commonly refer to the study of the practice of agriculture (also, "agronomy" or "agricultural science). Agriculture encompasses many subjects, including aquaculture, cultivation, animal husbandry, and horticulture. Each of these subjects can be further partitioned: for example, cultivation includes both organic farming and intensive farming, and animal husbandry includes ranching, herding, and intensive pig farming. Agricultural products include fodder, (starch, sugar, alcohols and resins), fibers (cotton, wool, hemp, silk and flax), fuels (methane from biomass, ethanol, biodiesel), cut flowers, ornamental and nursery plants, tropical fish and birds for the pet trade, and both legal and illegal drugs (biopharmaceuticals, tobacco, marijuana, opium, cocaine).[3] It has been mentioned by many western authors that despite developments in the filed of agriculture research, Indian agriculture is still traditional. It may be true due to the fact that agriculture in India is performed by villagers, who won generally small and marginal size farms, poorly equipped, using traditional equipments and also the traditional practices.

One of the salient features of traditional farming systems throughout the developing world is their high degree of biodiversity. These traditional farming systems have emerged over centuries of cultural and biological evolution and represent accumulated experiences of indigenous farmers interacting with the environment without access to external inputs, capital, or modern scientific knowledge (Chang, 1977; Grigg, 1974). Using inventive self-reliance, experiential knowledge, and locally available resources, traditional farmers have often developed farming systems with sustained yields (Harwood, 1979). [1]

How it is Called Traditional Agriculture?

Definition: Traditional agriculture is an indigenous form of farming, result of the co-evolution of local social and environmental systems and that exhibit a high level of ecological rationale expressed through the intensive use of local knowledge and natural resources, including the management of agro biodiversity in the form of diversified agricultural systems.[1] The term traditional is usually associated with primitive agricultural systems or pre-industrial peasant agriculture. Traditional farming usually is based on practices that have been passed down for many generations.[2]

Problem of Changing Traditional Agriculture

The real problem is that although we may need to change, it is hard to actually do it. To try something new is just too hard for some to face. And even if you decide to do something new–say for a current "traditional" family farmer to try alternative crops and direct marketing–it is still very hard to actually do it. It is hard to market yourself and your farm if you have never been a salesman. It is hard to think in terms of square feet when you have been working with hundreds of acres. But usually, if you persevere and work hard, the change will be well worth it ".[12]

It is presumed that unless agriculture educated young people are engaged in agriculture with appropriate training and incentive the Indian agriculture will remain traditional. According to Krishi Vigyan Kendra (PIRENS) Babhaleshwar (India) there are 11,900 graduates from agriculture and allied sectors passing out from agricultural universities in India while there are 2405 graduates in agriculture and allied fields passing out every year from Maharashtra state. However, only 2000 are able to get employment in Government/Private sectors. Thus there remains a vast pool of around 9900 graduates in the country who can support and boost agricultural production process if viable business opportunities are provided to them. On the other hand there exists an agricultural extension system which is fast shrinking resulting in wide extension gaps between those who require improved technologies and those who generate them. Thus, in the aftermath of this wide extension gap and the vast pool of unemployed agriculture pass outs remains to be tapped to provide support to the extension system and ultimately to help improve agricultural productivity. The traditional agriculture may be transformed to commercial agriculture through agripreneurship development.

What is Agripreneurship?

Definition: Agripreneurship is defined as generally, sustainable, community-orientated, directly marketed agriculture. Sustainable agriculture denotes a holistic, systems-oriented approach to farming that focuses on the interrelationships of social, economic, and environmental processes.

What is Entrepreneurship?

1. Entrepreneurship is a process of change comprising the following three behavioral components:
   ☆ The identification, evaluation and exploitation of an opportunity.
   ☆ The management of a new or transformed organization so as to facilitate production and consumption of new goods and services.
   ☆ The creation of value through successful exploitation of new idea (*i.e.* innovation).
2. Entrepreneurship is the dynamic process of creating incremental wealth. This wealth created by individuals who assume the major risks in terms of equity, time, and/or career commitment of providing value for some product or service. The product or service itself may or may not be new

or unique but value must somehow be infused by the entrepreneur by securing and allocating the necessary skills and resources (Robert Ronstand).

Why Agripreneurship Development?

There are so many areas and sub-areas in agriculture. Within each area, enormous number of commodities are available. Based on the needs, agro-climatic conditions and available resources, agripreneurs can adopt some commodities and flourish on it. On other hand, there are uncounted jobless agriculture graduates looking for jobs. The agriculture production is much less in comparison to many other countries. Therefore, it becomes essential to train these jobless agriculture graduates agri-business management and also provide finance to develop their own business in agriculture. Agripreneurship is required because:

1. Increasing demand for organic/quality food both in India as well as abroad. Market growth of around 15-25 per cent per year.

2. Competitive advantages for many primary production activities in agriculture. Rain-fed farming, tropical fruits and vegetables, livestock, animal husbandry, aquaculture, wild craft, *etc.* are produced through real low cost production methods.

3. Private sector is willing to enter into agri-business at all levels of operations. Changing consumer demand and retail revolution has opened the doors for investment by private sector in agri-businesses like Reliance, Bharati, Pantaloon, Carrefour, *etc.*

4. To reduce mall nutrition: The women and children of the country are malnourished. As Pt. Jawaharlal Nehru said that "You can tell the condition of a nation by looking at the status of its women.

Strengths for Promoting Agripreneurship in India

India is an agriculture based economy having 329 million hectares of land area of which 143 million hectares is under cultivation. India's population is 1027 million (531.3 million male and 495.7 million female) as on 1st March 2001(*Jain and Goria, 2006*). There are about 103 million farm families spread over 127 agro-climatic zones of the country with a variety of crops and animal production systems. There are 29 states, 6 union territories, 602 districts and 5600 blocks in the country. The country have 41 State Agricultural Universities, 5 Deemed to be Universities, 47 Central Institutes, 31 NRCs, 5 National Beureaux, 558 KVKs, 120 Zonal Agriculture Research Stations, 10 Trainers Training Centers and 44 Agriculture Technology Information Centers. There are about 30,000 scientists/researchers working in all theses organizations. On other hand there is a vast pool of around 9900 agriculture graduates in the country who can support and boost agricultural production process if viable business opportunities are provided to them. Therefore, the country has a great strength to promote agripreneurship.

Supports and Incentives for Agripreneurship and Agri-business Development

There are a number of Government and Non-government agencies supporting agripreneurship development through public-private partnership. Some of them are listed as follows:

1. The Ministry of Agriculture, Government of India has launched a unique programme to tap the expertise available in the large pool of agriculture graduates. Irrespective of them being fresh out of colleges or not, or whether currently employed or not, can setup Agriclinic or Agri-business Centre to offer paid professional services for enhancement of agricultural production and income of farmers" - *The Ministry of Agriculture, Government of India.*

2. "Small Farmers Agri-business Consortium (SFAC), New Delhi, the implementing agency is providing free start-up training to graduates in Agriculture, or any subject allied to Agriculture like Horticulture, Sericulture, Veterinary Sciences, Forestry, Dairy, Poultry Farming, Fisheries, *etc*. Once Agri-entrepreneurs setup their agriclinics and agri-business centres, SFAC will also support in establishing linkages for sustainability" - *Small Farmers Agri-business Consortium.*

3. "National Institute of Agricultural Extension Management (NIAEM), Hyderabad, is coordinating two months training programme for agri-entrepreneurs through selected institutes across the country. The course comprises of entrepreneurship and business management, as well as skill improvement modules in the chosen areas of activity by the agri-entrepreneurs" - *MANAGE, Hyderabad.*

4. "National Bank for Agriculture and Rural Development (NABARD), vide circular dated 23/07/2001, loans will be provided by banks for setting up agri-business centres. An individual can avail loan for the outer ceiling project cost of Rs. 10.00 lakhs and for joint/group projects, the ceiling is Rs. 50.00 lakhs, pro-rata" - *NABARD.*

5. The Krishi Vigyan Kendra (PIRENS) Babhaleshwar has been designated as a recognized training center by MANAGE, Hyderabad after taking cognizance of its innovative extension methodology and unique professionalism in providing vocational training for rural youths.

Characteristics of Entrepreneurs

The major characteristics of entrepreneurs that have been listed below by many commentators are useful for even agripreneurship.

☆ *Self confident and multi-skilled*: The person who can 'make the product, market it and count the money, but above all they have the confidence that lets them move comfortably through unsheltered waters'.

☆ *Confident*: In the face of difficulties and discouraging circumstances.

☆ *Innovative skills*: Not an 'inventor' in the traditional sense but one who is able to carve out a new niche in the market place, often invisible to others.

☆ *Results-orientated*: To make be successful requires the drive that only comes from setting goals and targets and getting pleasure from achieving them.

☆ *A risk-taker*: To succeed means taking measured risks. Often the successful entrepreneur exhibits an incremental approach to risk taking, at each stage exposing him/herself to only a limited, measured amount of personal risk and moving from one stage to another as each decision is proved.

☆ *Total commitment*: Hard work, energy and single-mindedness are essential elements in the entrepreneurial profile.

Why Agriculture to Agri-business?

The Ministry of Agriculture, Government of India, in association with NABARD has launched a unique programme to take better methods of farming to each and every farmer across the country. This programme aims to tap the expertise available in the large pool of Agriculture Graduates. Irrespective of the time of graduation or the status of employment, one can set up one's own Agriclinic or Agri-business center and offer extension services to innumerable farmers. Agri-business Centres would provide paid services for enhancement of agriculture production and income of farmers. Centers would need to advise farmers on crop selection, best farm practices, post-harvest value-added options, key agricultural information [including perhaps even Internet-based weather forecast], price trends, market news, *etc.* Initiated by Small Farmers' Agri-business Consortium [SFAC] and co-ordinate by MANAGE, the training programme is being offered free for selected candidates for a period of two months and is funded by the Ministry of Agriculture. The Agriclinic was inaugurated at SASTRA in the month of October 2002. SASTRA's Agriclinic has produced 50 per cent success stories as against the National average of 20 per cent.[13]

Key Issues: Agriculture to Agri-business

The following issues are favorable to promote agripreneurship and agri-business:

☆ **Policy issues:** From absolute control and management of agriculture by Government, today it is being opened to public-private partnerships.

☆ **Production technology issues:** From input/s oriented technology development for increasing production, today practice based value addition is being promoted.

☆ **Quality and certification issues:** Demands from consumer/s for better quality has forced Government/s to establish regulatory mechanisms for quality certification.

☆ **Logistics and supply chain issues:** Modern retail formats requires efficient and dedicated supply chain and management facilities.

☆ **Human resource issues:** Lack of appropriately trained human resource is today considered as the biggest constraint in conversion of agriculture to agri-business or agripreneurship.

Challenges for HR Skill Development in Transforming Agriculture to Agri-business

It has been experienced that there is great rush for getting admission in coaching centrs of Medical, Engineering and Civil Services in our country. If student get admission and succeeded to get a medical degree, he may either get suitable job or start his own clinic. Similarly engineering degree holder may either get some job or starts his own industry. The civil services are always more prestigious. But being agriculture based country; there is no rush for agriculture education as in the case of Medical and Engineering. Agriculture graduates are generally unemployed and unable to start their own enterprise due to lack of proper training and education in agri-business. Therefore, agriculture institution managers need to modify their teaching plans towards job oriented courses specially agripreneurship development and agri-business. It has been realized by other authors that:

☆ Agriculture to agri-business potential has neither been recognized by the academic/s nor is being promoted as a career making opportunity among practitioners.

☆ Region-location specific agri-business opportunities needs to be identified and accordingly suitable human resource training programs should be launched.

☆ Learning processes for human resource to be engaged in agri-business enterprises should be designed to suit the candidates as per the job requirements.

☆ The focus to upgrade the skills of conventional practitioners of agriculture into agri-business human resource will provide the missing links.

Status of Present Day Education in Agriculture

☆ Education is based on scientific developments carried out during 1950s to till date. Mostly candidates are trained in input based technologies and extension works.

☆ Focus on external inputs for increasing productivity. The skills for enabling value addition in agriculture is not imparted.

☆ Focus on government jobs (extension and research). Managerial skills are not included in the curriculum/s.

☆ Absence of value addition and supply chain management subjects. Even trained candidates do not learn the skills to manage agri-business enterprises.

☆ Accessible to careerists. For very long candidates joining these courses have mainly been attracted because of government jobs with total exclusion of practitioners wanting to create their own agri-business enterprises.

Keys to Success in Agri-business

There are no easy answers to agri-business success. Farmers who have started successful agricultural ventures recommend the following points may be considered:

☆ Choose something that you love to do. But it has good market and scope.

☆ Create a high quality produce.

☆ Start small and grow naturally. Don't try to become milliner overnight.

☆ Make decisions based on good records.

☆ Produce based on customers needs.

☆ Establish a loyal customer base and don't fool them.

☆ Provide more than just a food, produce or service. Provide an experience.

☆ Get the whole family or partner's involvement.

☆ Keep informed.

☆ Plan for the future. Set goals for your business and establish a plan of action to achieve them.

Strategy and Support to Promote Agripreneurship Development[5]

Agri Clinic and Agri business Centers: These are centers started by trained agriculture graduates which provide expert services and advice to farmers on cropping practices, technology dissemination, crop protection service, Market trends and prices of crops in markets and provide clinical services for plant and animal health which would enhance the productivity of crops and animals. Agri business centers are those which provide inputs supply farm equipment on hire and such services. Agricultural graduates may choose from a range of potential areas. The scheme is open to agricultural graduates in subjects allied to agriculture like horticulture, sericulture, forestry, veterinary science, animal husbandry, dairy, poultry, Pisciculture, Agri engineering, food technology and other allied activities. Under the scheme, the unemployed agricultural graduates are provided start up training at designated training centers. The successful candidates can later apply for start up loans for any specified venture. The project can be taken up by trained agri graduates either individually or on jointly/group basis. The outer ceiling for cost of project by an individual would be Rs. 10 lakh and for project by a group would be Rs. 50 lakh. The group may be of 5 of which one could be a management development and management. Projects up to Rs. 10,00, no margin money but projects over 10,000, 15-25 per cent of project cost. There is no Collateral security for a loan unto Rs. 5.0 lakhs as per the latest RBI Circular for the Agricultural graduates who have successfully completed the Agriclinics and Agri-business Training. The period of loan will vary between 5-10 years depending on the activities. The repayment period may include a grace period which is discretionary extending up to a maximum of two years.

Training Support

As an integral part of the programme, specialized training is being provided to agricultural graduates interested in setting up such center. Training programmes are being launched in recognized training institutes through out the country to prepare the professionals in entrepreneurship and agri business ventures. The training is free for selected candidates for a period of two months in selected training institute.

Institutions Promoting Entrepreneurship and Agripreneurship

Realizing the importance of agripreneurship development, a number of government and non-government organizations started promoting agripreneurship and educating agriculture graduates. Some of them are listed below:

1. National Science and Technology Entrepreneurship Development Board

The National Science and Technology Entrepreneurship Development Board (NSTEDB), established in 1982 by the Government of India under the aegis of Department of Science and Technology, is an institutional mechanism to help promote knowledge driven and technology intensive enterprises. The Board, having representations from socio-economic and scientific Ministries/Departments, aims to convert "job-seekers" into "job-generators" through Science and Technology (S and T) interventions.[6]

Objectives

1. To promote and develop high-end entrepreneurship for S and T manpower as well as self-employment by utilising S and T infrastructure and by using S and T methods.
2. To facilitate and conduct various informational services relating to promotion of entrepreneurship.
3. To network agencies of the support system, academic institutions and Research and Development (R and D) organisations to foster entrepreneurship and self-employing using S and T with special focus on backward areas as well.
4. To act as a policy advisory body with regard to entrepreneurship.[9]

2. The Entrepreneurship Development Institute of India

The Entrepreneurship Development Institute of India (EDI), an autonomous body and not-for-profit institution, set up in 1983, is sponsored by apex financial institutions, namely the Industrial Development Bank of India (IDBI), IFCI Ltd. ICICI Ltd and State Bank of India (SBI). The Institute is registered under the Societies Registration Act 1860 and the Public Trust Act 1950. The Government of Gujarat pledged twenty-three acres of land on which stands the majestic and sprawling EDI campus.[10]

EDI aims at:

☆ Creating a multiplier effect on opportunities for self-employment,

☆ Augmenting the supply of competent entrepreneurs through training,

☆ Augmenting the supply of entrepreneur trainer-motivators,

☆ Participating in institution building efforts,

☆ Inculcating the spirit of 'Entrepreneurship' in youth,

☆ Promoting micro enterprises at rural level,

☆ Developing and disseminating new knowledge and insights in entrepreneurial theory and practice through research,

☆ Facilitating corporate excellence through creating entrepreneurs (entrepreneurial managers),

☆ Improving managerial capabilities of small scale industries,

☆ Sensitizing the support system to facilitate potential and existing entrepreneurs establish and manage their enterprises,

☆ Collaborating with similar organisations in India and other developing countries to accomplish the above objectives.

☆ The institute's basic strategy to realise its mission has been to concentrate on some broad areas to achieve its objectives. The areas so selected have to satisfy the twin criteria of social relevance and the institute's capability.

3. Wadhwani Centre for Entrepreneurship Development

The Wadhwani Centre for Entrepreneurship Development (WCED) fosters and supports this entrepreneurial spirit. The vision of the WCED is to be a world renowned entrepreneurship centre by nurturing the spirit of enterprise in students and building a network to promote their efforts in the business environment. The WCED also offers entrepreneurs opportunities to learn and to interact with the right sources to build global competitiveness. The Centre has taken up the task of supporting aspiring entrepreneurs throughout the cycle, from conceptualization to grounding of their projects.

☆ WCED takes a holistic approach, covering training, research, mentoring and networking activities.

☆ The Centre provides opportunities for real time learning through interactions with some of the world's best business leaders.

☆ The WCED is funded by the Wadhwani Foundation, a not-for-profit organisation set up by Romesh Wadhwani, the founder, Chairman, and CEO of Symphony Technology Group.[11]

4. Morarka Foundation

Not for profit voluntary organization was set up by Mr. Kamal M. Morarka in 1993 to create wealth for the nation through innovations in resource management and capacity building of the people. From being a successful grass-root level implementing agency for 10 years, today it is also being recognized as a leading resource organization offering solutions for sustainable agriculture in the world.[8]

☆ The Foundation is today also recognized as Training cum Resource Centre and is rated as a highly successful member of "Network of Agri-clinics and Agri-business Centres by Agricultural Graduates Scheme" of the Department of Agri. and Coop., Ministry of Agriculture being implemented by SFAC, New Delhi and MANAGE, Hyderabad.

☆ In last five years Morarka Foundation has assisted over 500 youths to become agripreneurs in India. Under Agri-Clinic and Agri-Business Centre promotion scheme of Ministry of Agriculture about 150 youths trained for two months nearly 70 percent of them have taken up entrepreneurial activities.

5. MANAGE

The National Institute of Agricultural Extension Management, popularly known as MANAGE, is an apex national institute set up in 1987 as an autonomous society under the Ministry of Agriculture, Government of India. MANAGE is the Indian response to the challenges of management in a rapidly growing agricultural sector. As a management institute, MANAGE has a mandate to assist the State Governments, the Government of India and other public sector organizations in effective management of their agricultural extension and other agricultural management systems. MANAGE is a nodal institute for conducting International Programmes and organizing study visits for foreign delegates in the above areas.[7]

MANAGE offers its services in 5 streams. They are:

1. Consultancy
2. Management Training
3. Management Education
4. Management Research
5. Information and documentation services
6. National Institute for Entrepreneurship and small Business Development (NIESBUD)[14]

The National Institute for Entrepreneurship and small Business Development (NIESBUD) was established in 1983 by the Ministry of Industry (now Ministry of Small Scale Industries), Govt. of India, as an apex body for coordinating and overseeing the activities of various institutions/agencies engaged in Entrepreneurship Development Particularly in the area of small industry and small business. The Institute which is registered as a society under Govt. of India Societies Act (XXI of 1860) started functioning from 6th july, 1983. The policy, direction and guidance to the Institute is provided by its Governing Council whose Chairman is the Minister of SSI. The Executive Committee consisting of Secretary (Small Scale Industry and ARI) as its Chairman and Executive Director of the Institute as its Member Secretary executes the policies and Decisions of the Governing Council through its whole-time Executive Director.

Objectives

1. To evolve standardised materials and processes for selection, training, support and sustenance of entrepreneurs, potential and existing.
2. To help/support and affiliate institutions/organisations in carrying out training and other entrepreneurship development related activities.

3. To serve as an apex national level resource institute for accelerating the process of entrepreneurship development ensuring its impact across the country and among all strata of the society.

4. To provide vital information and support to trainers, promoters and entrepreneurs by organising research and documentation relevant to entrepreneurship development.

5. To train trainers, promoters and consultants in various areas of entrepreneurship development.

6. To provide national/international forums for interaction and exchange of experiences helpful for policy formulation and modification at various levels.

7. To offer consultancy nationally/internationally for promotion of entrepreneurship and small business development.

8. To share internationally experience and expertise in entrepreneurship development.

9. To share experience and expertise in entrepreneurship development across National frontiers.

Conclusion

It is true that India is seventh largest country in area and having 143 million hectares under cultivation. But food shortage in many states, malnutrition, and unemployment is increasing day by day. The real farmers are generally small and marginal in land holding and resource poor. They are still using traditional agricultural technologies. In order to dovetail the situation country needs a great number of agri-entrepreneurs. Fortunately, the Ministry of Agriculture, Government of India, in association with NABARD has launched a unique programme to take better methods of farming to each and every farmer across the country. In addition, Small Farmers Agri-business Consortium, Morarka Foundation, National Institute of Agricultural Extension Management, Hyderabad and some other public and private agencies started helping agricultural graduates by training them and providing sufficient fund. All out effort is to be made to develop agripreneurs to take advantages of the system and develop modern agriculture. In order to achieve this goal, some basic changes may be required in research, teaching and extension organizations in the country. For examples all State Agricultural Universities and Central Research Institutes need to start a department of agripreneurship development or agri-business management and impart training for the agriculture graduates. Then they will not feel shy to work in the fields and manage agriculture as a commercial enterprise. Unless these efforts are made practical, self-sufficiency in food grain and job security for agricultural graduates will be un-achievable task in new millennium.

References

Jain, S. P. and Goria, S. (2006). Indian National Digital Library for Agriculture) INDLA): A proposal for sharing of e-resources. The fifth International

Conference of the Asian Federation for Information Technology in Agriculture. p. 103-10.

Chang, J.H. 1977. Tropical agriculture: crop diversity and crop yields. *Econ. Geogr.* 53: 241-254.

Grigg, D.B. 1974. *The Agricultural Systems of the World: an evolutionary approach.* Cambridge University Press, Cambridge.

Harwood, R.R. 1979. Small Farm Development – Understanding and Improving Farming Systems in the Humid Tropics. Westview Press, Boulder.

[1]http://www.cnr.berkeley.edu/~christos/articles/traditional_ag.html

[2]http://www.tropag-fieldtrip.cornell.edu/tradag/default.html

[3]http://en.wikipedia.org/wiki/Agriculture

[4]http://www.gisdevelopment.net/application/agriculture/overview/mi04115a.htm

[5]http://www.kvk.pravara.com

[6]http://www.nstedb.com/

[7]http://www.manage.gov.in/About/about.htm

[8]http://www.morarkango.com/agripreneurship/ac_abc.php

[9]http://www.nstedb.com/

[10]http://www.ediindia.org/

[11]http://www.isb.edu/wadhwani/

[12] http://www.smallfarmtoday.com/Ridge/2000/JanFeb.asp

[13]http://www.sastra.edu/aboutus/agri-business.asp

[14]http://niesbud.nic.in/objectives.htm

Management Issues in Entrepreneurship Development

Chapter 8

Enterprise Resource Management

Ashok Mangaraj

Brain Stain,
Behind Fire Station, Baramunda
Bhubaneshwar – 751003

Enterprise is a system of operations converting/transforming resources into outputs in a meaningful manner. It has to be well thought out/planned, properly implemented and evaluated generating utility and profit in a sustainable manner. It should be fascinating to the entrepreneur and within his control. The resource pool has to be studied to its source for maintaining the dynamics through the cycle of the enterprise. A resource can be anything that has identity. *e.g.*, human beings, corporations, and bound books in a library, an idea, skill, information and knowledge can also be considered resources. For an entrepreneur in farm and farming sector it can be:

☆ Financial resources

☆ Physical/Geographical resources

☆ Goals and outcomes

☆ Human resources

☆ Policies and legislation

☆ State academic and technical standards

☆ Youth and adult services

☆ ITKs and Success stories *etc*.

Evaluating Your Resources

Before committing to a new enterprise, there are always fundamental questions that ought to be addressed. These may be practical (What are the business/

management skills of those involved?), organizational (Does everyone involved agree on how the business should be run?), or philosophical (Does everyone involved know, understand, and agree on the objectives, both short- and long-term?).

Marketing

☆ Where am I going to sell the products?

☆ Who is the customer?

☆ What is the size of the potential customer base?

☆ Where do the customers live, and how will their location influence my selling to them?

☆ What are the customers' needs and desires?

☆ Am I going to sell directly to consumers?

☆ Am I going to wholesale to the commodity market?

☆ What are the seasonal price fluctuations I can expect?

☆ What are the quality standards that I must meet?

☆ How many hours will it take to research direct markets?

☆ Are there legal or food-safety considerations?

Personal

☆ Do I have time to devote to this new enterprise?

☆ Does the workload correspond with the time of year I want to work?

☆ Will the new enterprise complement my current enterprises?

☆ Do I have written objectives describing the desired outcome?

☆ Do I have the skills and experience necessary to do this?

☆ Do I like to supervise people?

☆ Have I managed a business before?

☆ Do I have enough personal energy to do this?

☆ Can I count on my family members for support?

☆ Do I care what the neighbors think about my new enterprise?

☆ Why do I want this enterprise?

After you have determined that the enterprise is something you really want to do, consider these additional questions (for land-based enterprises):

Land

☆ What is the water drainage like?

☆ Are the soils suitable?

☆ What is the seasonal rainfall pattern?

☆ What will happen to my enterprises during a flood or drought?

☆ Are these plants or animals adapted to this climatic region?

☆ Are there water resources available for irrigation or for watering livestock?

☆ Do I want concurrent uses for the land such as wildlife conservation, fishing, or hunting?

Buildings and Machinery

☆ Do I have adequate facilities?

☆ What additional machinery will I need?

☆ Can I rent or borrow machinery or storage facilities?

Labor Needs

☆ How much labor will be required?

☆ What is the source of labor?

☆ How much will it cost?

☆ Is seasonal labor available?

☆ Will I need housing for my workers?

☆ Does this enterprise use existing labor in off-seasons?

Why Enterprise Resource Management?

For most farmers, the path to sustainability includes practices that go beyond environmentally conscious production and good stewardship of resources. It includes the development of products that take advantage of a farm's strengths, tapping into markets that bring a better financial return, and developing strong business and entrepreneurial skills to keep the farm poised for success.

Increasing your Competitive Advantage

The value of an ERP solution tailored to your industry becomes obvious in every area: reduced energy and material consumption, less tied-up capital, faster response to customer requirements, safer and more stable processes.

What is Enterprise Resource Management?

ERM (enterprise resource management) describes the system that lets an enterprise manager access to its network resources efficiently. ERM (enterprise resource management) also describes the skill that manages all of a firm's assets and resources, including such basic applications as general ledger, accounts payable and receivable, as well as manufacturing, inventory, and human resources. It is all about being able to make better decisions quicker and more effectively. Today, competitiveness means making the most of all resources in your selection of enterprise, input management and output throw along with a sharp eye on the farm byproducts: Knowledge, time, technology, money, customer relationships – all the way down to your farm's most valuable asset, human resources. IT-based management solutions are the key to doing this. They can consolidate information to assist in decision-making, support management processes and accelerate flows.

They enable efficient planning and control of all purchasing, value creation and sales and marketing processes.

Farm ERP (Enterprise Resource Planning)

A complete resource-planning for managing multiple enterprises in farms, useful for optimization of resources, input management, HR management, and financial management within it. Farm ERP gives complete resource planning for exports and domestic marketing of Agricultural produce. It manages the information regarding day-to-day work activity on Farm.

Features of Farm ERP

Information of farm, farm business and allied business can be maintained under the various sections like:

☆ Personal Information: Personnel information, family information *etc.*

☆ Land Information: Land, plots, land specifications *etc.*

☆ Irrigation information: Irrigation facilities, available water *etc.*

☆ Crop information: Crop grown and work done respect to crop *etc.*

☆ Human Resources information: Human resource info associated with farm.

Advantages of Farm ERP

Due to globalization and mechanization in agriculture, agriculture exports is increasing; presently many countries in the world are implementing food safety and quality standards for agricultural production. While exporting, it is necessary to submit some reports of production history or management practices.

☆ Software is useful to the exporters and growers to maintain their records.

☆ Crop planning for next season and year, gives planning of assets, requirement of inputs and human resources.

☆ Successful management of agri-business, demonstration or research plots.

☆ Reports generation for GAP, Codex and other standards are made easy.

☆ Better control over stock and inventory costs.

☆ Proof of good practices adopted in the farm.

☆ In line with Food safety and Food quality initiatives.

What is a "Perfect Farm Enterprise"?

☆ It is an enterprise that makes a bunch of money for the farm operator?

☆ It is an enterprise that makes the farm family operators comfortable and farming enjoyable?

☆ It should be a combination of both?

Evolution of ERP

Material Requirements Planning (MRP)

What material needed?

When and how purchased?

How managed?

Maximize efficiency of physical, financial assets

MRP 1960's

☆ Master production schedule

☆ Material requirements planning

☆ Capacity requirements planning

MRPII 1970's

☆ Sales and operations planning

☆ Simulation

☆ Forecasting

ERP 1990's

☆ Sales and distribution

☆ Material management

☆ Farm maintenance

☆ Quality management

☆ Financial accounting

☆ Controlling

☆ Investment management

☆ Human resource management

Enterprise Resource Planning (ERP)

Align Operations with Customer Demand

For entrepreneurs that produce and package products through recipe and formula-based processes, enterprise resource planning (ERP eliminates the risks and overhead associated with customizing and maintaining solutions to address several challenges. The system helps process entrepreneurs to manage manufacturing operations with dynamic forecasting and scheduling, formula-based production and yield management, quality control, inventory management, complex product costing, and streamlined regulatory compliance.

Financials

Financials Management applications enable you to efficiently manage and optimize your business operations across multiple components, divisions, and sites throughout your supply chain. Financials module is a central source of financial information that leverages data from across your enterprise and helps you manage your critical financial processes – from planning through decision-making. Enables to create vendor and customer diaries containing information to help manage the business more efficiently. Diary entries can be set with future progression dates to remind users to follow up on various tasks. Helps focus in key areas such as the chart of accounts, currencies, fiscal calendars, balance types, and accounting periods to facilitate local business requirements. Reduces or eliminates guesswork and re-keying by sharing data such as payment terms, bank codes, and customer and supplier information across all applications. Creates detailed and accurate financial reports, such as profit and loss statements, that are organized by product or job/activity.

Competitive Advantages

☆ Automates and streamlines your accounting closing cycles.

☆ Lowers your per-production costs with more efficient operations across finance.

☆ Delivers key financial results to farm managers regularly.

☆ Grows your farm operations in multiple currencies and languages.

☆ Improves interactions with suppliers and customers.

☆ Eliminates the need for spreadsheets for reporting, budgeting, and costing.

☆ Eliminates surprises with increased budgetary control over expenditures.

☆ Improves management analysis of your business operations with integrated costing capabilities.

☆ Provides insights and improved decision-making across your extended enterprise.

Production/Manufacturing Management

Production Management applications help you make informed, real-time production decisions, ensures quality, increase farm output, contain costs and improve marketing performance. With applications designed to address market demand and need-based production operations, you can consistently produce and supply quality products. Integrated with Financial Management applications, you have the added benefits of comprehensive management of your order-to-cash, production planning-to-execution and procure-to-pay processes. Solutions include customer preference and inventory management, study of market demand, product costing, quality maintenance, materials planning, and production.

Production-related activities are tracked which determines actual cost and variances.

☆ Production Specifications

☆ Quality Control checks

☆ Need-based and Fixed-time Production

☆ Flexible Units of Measure

- ☆ Input/Output-based Process Specifications
- ☆ Batch Scaling
- ☆ Location-Specific Specifications
- ☆ Product Costing
- ☆ Production
- ☆ Material Inputs
- ☆ Recording Machine and Labor Time
- ☆ Recording Process Variables – Lot and Batch Record
- ☆ Recording QC Test Results
- ☆ Recording Production Outputs
- ☆ Recording Job Status and Performance
- ☆ Yield Reporting
- ☆ Job Costing

Quality Management and Regulatory Compliance

Quality control (QC) and Quality assurance (QA) are integral parts of the process; quality can be planned and managed just like any other process. This gives you precise control over all materials and products, including co-products, by-products and end-products. Throughout your entire process, from planning to production and distribution, ERP enables you to be more diligent in selection of land and other inputs, ordering of materials, storage of inputs and machineries, requirement of labor and time, time bound operations, quality and quarantine measures, harvesting and output management and yield optimization. Additionally, using ERP, you can define and measure customer quality specifications at critical steps in the process. You can maintain the quality of your processes and products within strict tolerances to minimize delays and associated costs, and ultimately better satisfy the expectations of your customers.

Integrated Regulatory Management

In addition, Integrated Regulatory Management capabilities help you streamline your compliance efforts in today's environment of increasingly stringent local, national, and international standards.

Materials Management and Inventory Control

The Materials Management and Inventory Control function provides the ability to manage and view materials and products across the process. Manage materials, intermediate goods, or finished goods on the basis of defined characteristics such as shelf-life, pH, moisture content, and potency level. Alternatively, manage on text values like color or grade, among others. Inventory Control comprises materials management and warehouse management.

Data Collection

The success of your business depends on how well you compete in your market.

As an entrepreneur, this means quality production, faster time to market, quicker inventory turns and reduced production and transportation costs. To achieve these goals, you need real-time data that accurately depicts your production-manufacturing process. With this information, you can identify tactical problems that must be addressed immediately, as well as trends that may require process improvements.

Examples of Streamlining Inventory Management with Data Collection

☆ Automate tracking of all material movements

☆ Immediately capture details when goods are received

☆ Gain real-time shop floor visibility

☆ Streamline cycle of operations

☆ Receive immediate error alerts

☆ Confirm dispatch details – reduce shipping errors

☆ Improve customer satisfaction

Key Benefits

☆ Gain insight into to hundreds of critical questions in time arming everyone in the enterprise with the critical information they need to reach important goals.

☆ Monitor progress and performance more easily whether it is as per the customer's needs and preferences.

☆ Avoid potential threats and capitalize on opportunities by giving decision-makers the necessary tools to quickly identify key trends.

☆ Increase transparency and accountability by enabling all decision-makers–from family to front-line employees–to easily track and measure their own performance against farm objectives.

☆ Eliminate guesswork stemming from not having the right information at the right time.

☆ Improve responsiveness through dynamic alerts and notifications, as well as through timely and accurate reports.

☆ Act with confidence, knowing your decisions are based on accurate, timely data.

Starting a Small Business

Introduction

Starting a small business offers rewards and challenges that attract thousands of new entrepreneurs each year. Owning a business for many entrepreneurs is the dream that allows them the opportunity to be their own boss, earn money for themselves, work flexible hours, work at home, and work at something they enjoy. However, with the rewards, come the challenges and more than 50 percent of all new businesses fail in the first five years of existence. Most businessmen and bankers agree that the main reasons for the high failure rate is the lack of experience, business and financial planning, and management skills. Those taking on a new business venture must begin by developing sound business and marketing plans to guide them through the initial start-up years. They must possess and display strong management skills and have the ability to convince potential lenders to finance their business venture. They will need to research the type of business they intend to start, the feasibility of starting the business, the best locations to ensure a successful outcome, and more.

Questions to Ask Yourself Before Starting a Business

1. Is there a true need for the products or services your enterprise will offer in the marketplace?
2. Are you just anticipating a need in order to justify the great idea you have?
3. How do your perspective customers describe the product or service benefits, by size, dimensions, current needs *etc.*?
4. Where did the need originate for your product or service?
5. How long will the demand last for your product or services?
6. Is the product service friendly?
7. Is the demand going to last long enough for the initial investment?
8. Will your product or service have to be redesigned to fit into an ever changing marketplace?
9. Who are your customers, and can they afford your product or service?
10. Who will be the first ten people you will sell your product or service to?
11. How much time will pass between purchases by a typical consumer?
12. What sales methods and techniques will be necessary to market your product?
13. Can your business keep up if the demand rate goes up for your product or service?
14. How much will your sales methods cost your business?
15. Who and where is your competition?
16. Do you know and understand your competition?
17. What is your competition doing right now?
18. How will your business respond to the competition such as price, warranties, and customer service?
19. Can you afford to be competitive and beat out your competitor?
20. How much time and money must be spent selling the product or service before it is purchased and paid for?
21. How will your product demand change as the price moves up or down through time?

Financial Questions for Potential Entrepreneurs

1. What real estate, machinery, equipment, technology, and systems will be required to support the enterprise now and in the future?
2. Are the above mentioned at a price you can afford?
3. How will you pay for them (mentioned above)?
4. Is your proposed per-item selling price at each market level realistic?
5. Does your intended pricing accommodate the price tolerances of your intended consumers?
6. What kind of per item "gross profit margin" (*e.g.* the difference between the cost of the item and its sales price) does your pricing allow?
7. Is the margin between the selling price to your buyer (*e.g.* the wholesaler or retailer) and the price to the ultimate consumer sufficient to appropriately compensate the external distribution channels you'll require?
8. Will everyone in the distribution chain earn a return sufficient to keep them involved?
9. What is your pricing strategy?
10. Is your pricing strategy flexible enough to counter competition?
11. What is your break-even point in terms of the volume of items that must be sold?
12. How many items must be sold, in what period of time, to get you a cash flow break-even point?
13. How much cash will have to be invested in your venture to achieve a level of operation that produces break-even sales levels?
14. Will you receive reasonable return on investments?
15. Do you have enough cash available to get you to that break-even point and provide you with reasonable standard of living in the process?
16. How much time will pass, on average, between the point at which you sell an item and the point at which you collect the cash from sale?
17. What will you do if purchasers don't pay as requested?
18. How will this affect your projected cash flow?
19. What percent of your personal liquid net worth and credit capacity will be obligated at the point at which you start to make sales?
20. What will your contingent liabilities be at that point?
21. How do you expect to fund operating expansion and to acquire needed capital equipment?
22. Will the nature of your business appeal to investors and lenders?
23. What will it cost you in terms of interest expense, equity give-up, covenants, restrictions, etc.?

24. What will be the internal rte of return generated by the business (*e.g.* the interest rate that is required to present-value both cash inflows and cash outflows to an equal amount or a "zero-sum gain")?

25. What financial or personal give-ups will you experience?

26. Is this the best use of your time?

27. How will you eventually exit from this business?

28. If someone is selling a business consider this, why are they selling?

29. What evidence of past performance can they give you?

Marketing Tips Every Entrepreneur Should Know

1. The most important order you ever get from a customer is the second order.

2. In direct mailing, spend 10 percent of your budget on testing.

3. Understanding and adapting to consumer motivation and behavior is not an option, but an absolute necessity for competitive survival.

4. A well-designed catalog mailed to a qualified response list will probably bring a 2 per cent response.

5. Know the power of repetition. Be sure that your message is consistent.

6. The two most common mistakes companies make in using the phone is failing to track results and tracking the wrong thing.

7. Marketing activities should be designed to increase profits, not just sales.

8. It cost five times as much to sell to a new customer than to an existing customer.

9. Selling what your customers need instead of what they want can lead to failure.

10. Don't think that product superiority, technology, innovation or company size will sell itself.

11. Don't neglect or ignore your current customers while pursuing new ones.

12. People don't buy products, they buy the benefits and solutions they believe the products provide.

13. The average business never hears from 96 per cent of its dissatisfied customers.

14. Fifty percent of those customers who complain would do business with the company again if their complaints were handled satisfactorily.

15. It is estimated that customers are twice as likely to talk about their bad experiences as their good ones.

16. Marketing is everyone's business, regardless of title or position in the organization.

17. Exaggerated claims can produce inflated expectations that the product or service cannot live up to, thereby resulting in dissatisfied customers.

18. Get to know your prime customers–the 20 per cent of product users account for 80 per cent of the total consumption of that product class.

19. Telephone-generated leads are likely to close four to six times greater than mail-generated leads.

20. The two ways to sustain superior performance are to take exceptional care of customers through superior service and quality, and to constantly innovate.

21. The Rule of Thirds: For a given catalog, one third of the merchandise will sell well, one third will sell OK, and one third will bomb.

22. Properly designated and used, research efforts can significantly lower the risks of marketing failure.

23. There are three foundations for marketing decisions: 1) research, 2) experience, and 3) intuition. None is complete without the other.

24. Marketers should spend 25 per cent of their time in the field, learning from and listening to their customers.

25. About 90 per cent of all product introductions ultimately either fail or fall short of potential.

26. Personal influence (word-of-mouth) has a more decisive role in influencing behavior than advertising and other marketer-dominated sources.

27. The more information that can be given visually, through observation or actual product demonstration, the greater the impact on awareness and stimulation of interest.

28. Gaining the consumer's attention is possibly the most formidable challenge a marketer faces.

29. The consumer's perception of price is usually more important than actual price.

30. The customer is not dependent on you; you are dependent on the customer.

31. If you don't stay in contact with your customer, someone else will attract that customer's business.

32. Plan your business operations for the convenience of the customer. Make it easy to do business with you.

33. If customers call with questions, your catalog isn't doing its job.

34. Forty percent of people who receive free gifts can remember the name of the advertiser as long as six months later.

35. The two biggest mistakes in marketing are spending too much and spending too little.

36. Remember that small shares of gigantic markets can be abundant and profitable.

37. Identify your competitive advantages, and then focus your marketing effort on them.

38. Always remember that it's best to under-promise and over-deliver.

39. It's hard to manage something that you can't measure.

40. Develop back-end marketing strategies; ways to resell to customers through updates, new versions, complimentary products.

41. If you can't be first in a category, try selling up a new category to be first in.

42. Marketing is not a battle of products; it's a battle of perceptions.

43. Find ways to turn current customers into a sales force.

44. The credibility and persuasiveness of your marketing efforts increases in direct proportion to the amount of specific information you provide.

45. Treat sales transactions not as an end, but as a beginning of a relationship.

ERM - Enterprise Resource Manager enables an organization to:

☆ Manage and guide production - sales activities by direct or indirect reps, agents and partners

☆ Provide complete customer support and professional services tracking with Service Desk

☆ Forecast revenue with accuracy and predictability, even near end of quarter

☆ Generate timely product/service quotations

☆ Record Purchase Orders and initiate the provisioning process

☆ Manage inventory with multiple warehouse locations for shipping and receiving

☆ Issue accurate Invoices on time for products/services actually delivered

☆ Keep accurate Order Backlogs and Deferred Revenues

☆ Calculate personalized sales commissions

☆ Track, assign and manage Fixed Assets

☆ Approve procurements and payments and print checks to vendors

☆ Monitor enterprise health with up-to-date Business Activity Monitoring (BAM)

☆ Eliminate human errors throughout the cycle of operations

Steps in Developing a Co-owned Agricultural Business

1. Define the Opportunity
2. Identify Organizers
3. Set Goals for the Business
4. Conduct a Preliminary Assessment of the Project
5. Initiate a Business Plan
6. Conduct a Detailed Feasibility Study and Revise Business Plan
7. Organize the Business Entity

8. Capitalize the Business
9. Implement the Plan

Diminishing government subsidies combined with globalization and increased competition are causing farmers to consider additional and alternative business strategies to increase profit. Common strategies have been to produce a specialized product (such as certified seed), vertically integrate into processing agricultural commodities, produce commodities that are new to the region. The resources needed to construct and operate a new agriculture business can be extensive, often requiring more capital, human resources, and risk exposure than one individual can provide or sustain.

> ☆ An alternative is for farmers to pool their resources of capital, commodities, knowledge, and management skills to develop a business that captures economies of scale few farmers could build individually. A common question is how to shape an idea into a successful business; that is, how to organize investors, assemble needed resources, construct the facility, and initiate operation in response to a market opportunity.

1. Define the Opportunity

Developing a new business often starts with an individual who recognizes an opportunity and contacts others who also may be interested in exploring the possibility. As the members of the group begin their discussions, they will share their perceptions about the opportunity and the industry. For example as they look to the future, the group members will likely consider their expectations about:

Market potential–Will consumers be willing to buy the product? In what form, when and where are consumers likely to want the product? Will they have enough income to afford the product? How might the market change over time?

Production practices–Are the inputs and services necessary to operate the business available? What technology can be used to produce or process the product? What resources do the organizers have available? What resources are available in the community? How might production practices change over time?

External factors–Which government programs, policies, and regulations will apply to the business? Which competitor offers a similar product or service? Is there access to information? How might the competition change over time?

Members of the group also may find it beneficial to describe their reasons for pursuing this business opportunity. Following such discussions, the group can decide whether the opportunity justifies proceeding with the project.

Discussions at this stage of the planning process are only preliminary. Organizers should expect to revisit these topics several times and gather additional information and analyze the issues more thoroughly each time. An extensive understanding of current and long-term market potential, production practices, and competition is critical in developing and revising the business plan during subsequent steps of the planning process.

2. Identify Organizers

The next step is for the members of the group to look at themselves to determine their level of interest in the project. How much time is each member willing and able to commit to the project? Does the group have enough time to make the project successful? How much cash are they willing to commit? What is their level of understanding about the potential business? An adequate interest and understanding, as well as a willingness to commit time and cash, are needed to start the project.

Depending on the size of the group, one or two individuals may be identified as the project champions to keep the project moving by arranging meetings, assuring that other members of the group are informed, making contacts with knowledgeable individuals, gathering information, promoting the idea of developing the business, and assuring that the planning process continues. The project champion may be the member of the group with the most interest and time to commit to the project, rather than the individual with the initial idea or the most available capital. An alternative to designating a project champion is to develop an alliance with an established group in the industry. Likewise, several individuals (*a steering committee*) could be identified to assist the project champion by reviewing progress, providing ideas, and assisting when necessary. These persons may be only temporary leaders who start the project. In many situations, they will be replaced by a permanent management team as the business develops. Additional issues the group may want to address include:

☆ Setting parameters for the project to define which ideas will be investigated; for example, will the group investigate establishing a feedlot for finishing cattle, or a feedlot for back grounding feeders? Will the group focus on irrigated vegetable production for sale to wholesalers, or will the group package and distribute the produce for retail sale to consumers?

☆ Establishing a time line for each step of the process. Depending on the size and complexity of the project, developing a new business could take as little as nine to 12 months, or as long as three to five years.

☆ Determining the necessary level of commitment and progress to justify moving to the next step.

☆ Identifying sources of "seed money" to study the project in addition to the time and cash organizers are willing to invest to study the opportunity, are there other sources of seed money, and who will apply for that assistance? How will organizers be compensated for their investment of time and seed money if the business becomes operational? What are the disclosure requirements? Are there any limits on how this seed money can be used? Organizers may want to document their agreement.

3. Set Goals for the Business

The next step for the group is to set goals for the project; for example, is the goal for the project to generate profit for investors, to provide employment opportunities in the community, or both? If locating the business in another community or region will enhance profit, will the organizers choose greater profit or local economic growth? In many cases, major operations and markets can be successfully established outside the region to capture the markets. Profits are then "harvested" and returned to the investor's community where they can be used for other economic activities. Goals are necessary. They are the basis for making future decisions. The group may ask itself, "From among our alternatives, which one will cause us to most effectively and efficiently reach our goal?" Goals also provide benchmarks for measuring progress.

It may be helpful to have each member of the group review (and share to the extent they want to) his or her career and personal goals. From these goals, each individual can specify some goals for the new business. "What is it that each one of us wants to get out of this business? Is it a return on investment? Is it a market opportunity for a commodity we already produce? Is it an opportunity to produce a different product or commodity? Is it an employment opportunity for one or more of the investors? Is it a strategy for managing our risk exposure? What is our vision for the business? What activities and what scale of operation do we envision for the business?" The answer to these questions will not be the same for each individual. Instead, the group will likely work toward developing a business that fulfills various goals of the group members.

Open and honest communication is a necessity for a co-owned business. Willingness to discuss goals may be an early indicator of the level of trust among group members, and how members of the group will interact when difficult decisions need to be made in the future. The process of setting goals for the business also may reveal who remains involved in the project. Some members may withdraw from the group at this time if they recognize that the business will pursue goals that are incompatible with their personal or career objectives. In addition, this step may be an opportunity to review the timetable for the project, as well as the tasks of the project champion and steering committee.

Setting goals also should include developing an initial mission statement for the business; that is, one or two sentences which summarize the business' goals and strategies for fulfilling them.

4. Conduct a Preliminary Assessment of the Project

The next step of the process is to conduct an informal, relatively low-cost (in terms of cash and time), preliminary assessment of the project. In addition to gathering information and surveying the interests of potential investors (but without creating much publicity and speculation), the group will want to consider issues related to production, processing, distribution, marketing, capitalization, and management. Gathering information from persons knowledgeable with the industry can be invaluable during this step. More specifically, the organizers will likely address the following issues:

☆ **Scope of the business**: What commodity, product, or service will the business produce? Will it produce a commodity, process a commodity produced by the investors, process a commodity produced by non-investors, perform one step in the processing sequence and rely on others to complete the process, or provide a service (such as marketing for a fixed percentage or fee) without ever taking title to the commodity?

☆ **Market for the commodity or product**: What is the market for the product? Who will be the customers, what do they want to buy, where would they want to buy, how much are they willing to pay, and how will the business be paid? Will the business sell on the wholesale market or to retail consumers? This step involves contacting potential buyers to ascertain their reactions and interests. Are there retail outlets interested in establishing a relationship with the business? What are the potential local, regional, domestic, and international markets? How might these markets change over time? Are they likely to expand or contract? What distribution facilities or network is available for marketing the product? The result of this initial market assessment may cause the group to revise the scope of the business.

☆ **Competition**: Which other businesses currently produce, or are likely to produce, a similar product or service? What appears to be their strengths and weaknesses? What strategies are they apparently using and how effective do those strategies appear to be? Are these firms likely to alter their production capacity or modify their product? Is there any opportunity to cooperate with an existing business? Is there any opportunity to form an alliance or joint venture with an existing business?

☆ **Production technology**: What are the current and potential technologies that the business can use? Is the technology proven? What inputs are needed? How much labor would be required? How much capital would be required? Are there any specific sitting needs? Visiting comparable firms or meeting with representatives of businesses that offer the needed equipment, technology, and expertise is often an effective means of acquiring information. However, careful management is needed to control the time and cash committed to such visits.

☆ **Availability of inputs**: What inputs would the business need and what might be the availability of those inputs, now and in the future? Answering this question often requires visits with potential suppliers to ascertain their reactions and interests. For a business that will process agricultural commodities, this step may include a survey of producers to determine their interest in establishing a working relationship with the business. The survey also may request information about the amount of equity farmers would be willing to invest in the business, especially if the business may be organized as a cooperative. Additional issues to address may include who else may be interested in buying these inputs, and should the business enter into a long-term contract with the suppliers.

☆ **Availability of workers**: Is there an adequate number of workers available in the region with the skills necessary to meet the needs of the business? Does the community offer the services, educational opportunities, and other amenities necessary to attract an adequate workforce? Is the necessary level of management expertise available? The group may find that it will need to hire an experienced manager from another region of the country. The group also may need to hire other professionals (*e.g.* production specialist) on a consulting basis. These individuals do not need to be hired or retained at this time, but an understanding of their availability can be beneficial as the planning effort proceeds.

☆ **Location and community infrastructure**: Is proximity to inputs and markets critical for this business to be profitable? What is the perishability and transportability of the inputs and products? What types of roads, trucking and rail services, airports, communication systems, water resources, and waste disposal are needed for the business? What locations might meet these needs? What people and resources are available in these communities? What will be the criteria for selecting a location?

☆ **Size of operation**: What are the alternative sizes of operation for the business? What size is needed to match the expected scope of the business? What size of operation will best fit the expected market for the product? Which size offers the best economies of scale? What is the minimum size necessary to earn an acceptable rate of return? At what size are dis-economies of scale likely to be experienced? What size of operation are the communities and permitting agencies willing to accept?

☆ **Capital needs**: This issue is perhaps the most crucial to success. How much investment is required to develop, construct and begin operating the desired business? How much capital will be needed to continue operating the business in the future? An engineering firm may be able to help project construction costs. Is the necessary capital available? What rate of return would investors expect to receive? What interest rate would lenders likely charge? How quickly would the lenders expect to be repaid? Are there any opportunities for receiving financial assistance from federal, state or local governments?

☆ **Organizational structure**: Who will be allowed to invest in or become a member of the business; that is, who will be the investors? Do the investors not only have rights but also obligations? For example, an investor may have the right to share in the profits but also an obligation to provide a specified amount of raw product at a specified grade and at a specified time. What will be the minimum and maximum investment? Will members be producers, investors, or both? These decisions should reflect the group's goals. Will the business be organized as a corporation, limited liability company, limited partnership, or cooperative? Businesses intending to engage in farming activities will need to review the state's corporate farming laws to assure they are in compliance. If intending to organize as a cooperative, the group will need to decide whether a closed or an open membership cooperative will be proposed to investors.

☆ **Government policies**: Which local, state and federal government policies will impact the operation of the business? These policies may encompass a broad range of issues, such as labor law, environment regulations, taxation, economic development incentives, food safety concerns, and trade policies. What federal, state, and local permits are necessary?

☆ **Risk Exposure**: What are the risks associated with initiating and operating the proposed business including the uncertainty of market, technology, production, finance, government and management? What opportunities are available to manage these risks? If the business primarily will provide a service for its investors, such as marketing organization, will the business acquire ownership of the product or will the farmers retain ownership?

The preliminary assessment is not the last time organizers will consider these issues. They can expect to revisit, refine, and revise their vision as additional information becomes available and decisions are made during the planning process.

Decision

Does the preliminary information justify continuing to study the project by completing a formal feasibility assessment? Has an adequate general description of the project been developed so a detailed study can be conducted? Are the organizers still willing to work together? Will they support the business after it begins operating? These questions likely will be discussed and answered at a meeting of the organizers. If the responses are positive, the group is ready to move onto the next steps.

5. Initiate a Business Plan

The information already gathered and the results of the preliminary assessment can be the basis for an initial formal business plan; that is, a general description of how the business will operate and market its product. The purpose of a business plan is to record decisions as well as organize information to be shared with investors, lenders, and regulators as part of disclosures, loan applications, and permit applications. However, the business plan will most likely be revised several times as new information becomes available and the business continues to be developed. This also may be the time to review and refine the mission statement.

Components of a Business Plan

☆ Overview of the Business and Industry

☆ Description of the Business, and its Commodity, Product or Service

☆ Production Strategy and Technology

☆ Market Analysis

☆ Site Selection

☆ Personnel Needs

☆ Ownership and Management

☆ Capitalization

☆ Future Opportunities

☆ Critical Concerns

☆ Financial Projections and Other Supporting Information

6. Conduct a Detailed Feasibility Study and Revise Business Plan

At the same time that a business plan is being prepared, the group will want to complete a formal feasibility study to determine the feasibility (cash flow) and profitability of the business. A feasibility study will include thorough financial and market analyses for the new business, including detailed revenue and cost projections. It should also analyze the impact changing conditions may have on the business' profitability (sensitivity analysis).

The feasibility study should be done by an unbiased company, person, or group and reviewed by an independent source. Many groups retain highly specialized professional firms or individuals (such as a consulting firm, an accountant, an economist, or a marketing specialist) to assist in the market and financial analyses. However, decisions about the business must be made by the organizers, even though a consultant, for example, has been retained to conduct the study or assemble the plan.

To help cover the cost of a feasibility study, some groups seek financial support from local and state economic development organizations. Other groups rely on the organizers and potential investors to provide some or all of the resources needed to complete the feasibility study. One strategy a few groups have used is to require individuals to provide "seed money" in order to be "qualified" to invest at a later time.

Preparing a business plan and feasibility study often causes organizers to reconsider some issues (such as scope of the business or size of the operation) as well as resolve additional questions such as:

☆ Will the new venture be limited to conducting business only with investors, or will non-investors also be allowed to conduct business with the new firm? This question assumes that the new business relates to the investors' other businesses; for example, a business that processes or markets agricultural commodities the investing farmers will produce.

☆ Do potential suppliers need to adopt new management practices in their farm operations to assure there will be a supply of commodities at the time, in the quantity, and of the quality needed by the business? Including such information in the business plan would help producers decide whether to invest in the new business.

☆ What method will be used to value the inputs that will be supplied by the investing farmers? For example, will the commodity be priced by an outside order buyer or by a purchasing agent employed by the business?

☆ How will the business market its product? A key component of the feasibility study is the market study. What is the current market structure and how might it change in the future? How might the number of customers, level of competition, and market environment evolve over time? Will the product be marketed to a single buyer based on preferred supplier arrangements, will the product be marketed to whoever provides the highest bid when the product is ready for sale, or will the business market directly to consumers? The detailed feasibility study is an opportunity to *review and possibly revise the business' marketing strategies*. Some groups may find it helpful to retain a *marketing specialist* at this point in the planning process. This is a critical component of any business; no business can succeed without a market for its product.

☆ What will be the par value of equity stock shares? What will be the investors' obligations to supply inputs or agricultural commodities to the business? What will be the business' obligation to purchase the investors' commodities?

☆ How will the business be operated? An initial description of facilities and operating strategies is often developed at this point in the planning process.

☆ What contingency plans can be developed in preparation for unexpected events both problems and opportunities?

☆ Is the group ready to hire a manager at this time? Some firms, especially larger ones, begin the process by hiring a general manager or chief executive officer (CEO) as the first member of the permanent management team. This individual would take the place of the project champion and lead the effort of continuing to develop the business plan. Again, initial investors may be expected to provide much of the cash needed to pay this cost.

☆ How will ownership in the business transfer among investors? Will the business control who can acquire an ownership interest?

Market opportunities, revenue projections, distribution strategies, production practices and technology, operating costs, financing alternatives, alternative business structures, and human resource policies/strategies (such as training programs and a wage, benefit and incentive package) are all part of analyzing the project's feasibility. With this information, organizers can decide whether to proceed by comparing the business' projected performance to the group's goals.

Decision

Does the project appear feasible? Will it generate cash when and in the amounts needed to meet expenses? Will the business earn an acceptable rate of return for the investors? Will the business generate a rate of return comparable to other opportunities (including production agriculture or purchasing stock in food companies)? Will the business likely fulfill the goals set by the organizers? If the project appears to be reasonable, the next steps involve developing the formal business entity, capitalizing the business, and implementing the plan.

7. Organize the Business Entity

As the business plan and feasibility study are being completed, the organizers will want to develop a formal business structure. One question will be how to be organized as a partnership, limited liability company, corporation, cooperative, or some other form of business. The goals of the group, such as who will be the owner, who will control the business, what risk exposure are the investors willing to accept, tax considerations, and financing arrangements, are issues to consider in selecting a business structure. Competent professional advice can be invaluable in making this decision.

An attorney skilled in formation of agricultural businesses should be employed to help set up the prospective firm and organize a board of directors (if one is required), develop the business' information packages for potential investors, attend to needed legal requirements associated with promoting the business proposal, and obtain sitting permits, construction contracts, and trademark registration. A financial advisor often is employed to assure that business reality is reflected in the information packages provided to potential investors. An accountant should be hired to establish an accounting system for the business.

The group is now ready, based on the business plan, to develop the information and disclosure documents (such as a prospectus) for potential investors and lenders. These documents must meet federal and state law filing and disclosure requirements, but may not contain the level of detail included in a thorough business plan.

☆ Retain professionals necessary to complete the project, such as a construction engineer or architect.

☆ Preliminary construction plans could be started. This also may be the time to negotiate an option to buy or lease a site for the business, to begin the process of acquiring necessary government permits, and to initiate discussions with possible construction contractors. If not already hired, a permanent general manager or CEO should be recruited and in place by this point in the process to assist with the remaining steps of the developing the business. However in some cases, the organizers postpone hiring this person until the necessary capital has been raised during the capitalization phase.

8. Capitalize the Business

The business will likely use a combination of equity capital provided by investors and debt capital provided by lenders based on interest rates, risk, alternative investment opportunities, lending requirements, and investors' interest in the project.

Solicit equity investors. Formal documentation for the stock offering and loan applications will need to be completed; some of this will have already been developed as part of the business plan. Potential equity investors in the business must be identified. In the case of a cooperative, a series of information and sign-up meetings would be held to discuss the proposal and to sell shares. Sometimes it may also be necessary to provide information to the investor's lender. Provision for returning money invested if the membership target is not met within a specified time period must be clearly indicated.

Related Questions may Include

☆ How often will the investment opportunity be extended? Just once, or will investors have several opportunities to become involved?

☆ What will be the deadline for investing in the business? If the business will have a relatively small number of investors, will the time period for deciding whether to invest be extended for selected individuals?

☆ Will investors be required to pay the full amount immediately, or will they be expected to contribute only a portion of the investment with a commitment to provide the remainder at a later time?

☆ Will investing in the business also *obligate* the investor to do business with the firm (such as delivery a specified quantity of commodities to be processed or marketed)?

☆ How many investors does the group want to work with; what is the targeted size? Is there a minimum or maximum desired number of investors?

☆ Is there a minimum or maximum amount of investment per investor?

Is there a minimum or maximum desired level of total investment? Will there be enough equity to sustain the business during a period of low profitability? Will investors be expected to contribute additional capital to sustain the business if necessary?

Decision

Is the amount of capital that investors are willing to contribute adequate to justify proceeding with the project? Does the group want to continue developing the business, or does it need to revise or abandon the proposal?

Now that they share ownership of a new business, the investors will likely meet (a stock or shareholders meeting), and if the number of investors justifies, elect a board of directors. On the other hand, if there is a small number of investors, the entire group could function as a board of directors. The directors would assume the role of the steering committee.

Negotiate for debt capital. As commitments from investors reach specified levels, discussions can proceed with potential lenders regarding loan amounts, covenants, maturity, and price. Lenders will likely specify a level of equity for the individual investors before they will consider the loan application.

Organizers also often investigate state and local *economic development assistance* alternatives that could be used in capitalizing the new business. The assistance may be in the form of debt, operating subsidies, or equity. Important questions to ask include what are the disclosure requirements and will there be limits on how these funds can be used.

Based on the level of interest among investors, lenders, and government, the organizers may want to review and possibly revise the business plan.

Decision

The success of capitalizing the business will determine whether the project moves forward to the next phase is rethought, or abandoned. If the project organizers have been successful in obtaining the necessary equity capital, and if lenders have given approval to the needed loans, the project is ready to proceed with constructing, purchasing, leasing or contracting to acquire needed facilities. Alternatively, if either equity capital or debt capital has not been secured, project organizers must re-evaluate the project proposal and adapt it to fit available capital, or postpone or abandon the project.

Questions Investors Should Consider

Farmers and ranchers interested in forming an agricultural business also should ask at least five questions regarding the impact of that investment on their own farm and ranch businesses.

☆ What are the potential returns from the business investment?

☆ What risks are the businesses exposed to?

☆ How will business investment influence the investor's farm/ranch operation? Will the business risk in the farm/ranch increase or decrease? Will the farm or ranch business be jeopardized if the investment in the agricultural business be lost?

☆ How will your lender view the investment?

☆ How will investment in the agricultural business impact your personal or business goals?

9. Implement the Plan

The last step is implementing the plan, which includes constructing the facility and beginning operation.

Acquire Facilities

☆ *Location:* A site for the proposed business will need to be selected and acquired.

☆ *Construction Design.* An engineer, along with the CEO or production committee, will study selected sites and develop a final design for facilities.

☆ *Negotiate the Construction or Refurbishing Contract:* Will this include a bidding process?

☆ *Government Permits:* The business will need to apply for and acquire necessary federal, state, and local government construction and operating permits.

☆ *Business Contracts:* The business' management will want to negotiate necessary contracts with input suppliers and product buyers. A contract will not be needed for all suppliers and buyers; the primary focus may be on those firms whose long-term commitment is critical to the success of the business. For some businesses, negotiating an agreement to use technology controlled by other firms is critical. Developing market opportunities and a distribution network is vital for all businesses. Negotiations also are an opportunity for the group to help suppliers understand how the suppliers can meet the needs of the business. In the short-term, construction contracts need to be finalized. Purchasing various insurance policies also protects the new business against loss from some risks.

Begin Operation

☆ *Hire the workforce:* Recruiting, hiring and training employees would be one part of initiating the business operation.

☆ *Prepare to Operate:* The CEO or management team will need to develop delivery schedules for inputs and inform investors and other suppliers to prepare for starting operations. Likewise, buyers of the business' product need to be informed that the production process is about to begin.

☆ Test production or processing facility and make necessary refinements.

☆ Move to full operation, including continuous monitoring of market opportunities, production practices, and the business' financial performance.

Summary

Developing a new business is an involved process, and one that may be complicated by the novelty of the product being produced and marketed or the number of co-owners. Although steps can be suggested, they will vary depending on the situation. Generally, developing a new co-owned business involves defining the opportunity, identifying the organizers, setting goals for the business, determining the market potential for the commodity, product, or service, conducting a preliminary assessment of the project, developing a business plan, conducting a detailed feasibility study, organizing the business entity, capitalizing the business, and implementing the plan. Continual reassessment and deciding whether to proceed are part of developing a successful new business. Careful analysis, thorough planning and matching the market pace can reduce errors and enhance the profitability and success of a new business

Choosing an 'Alternative' Enterprise

There are many kinds of enterprises that can be profitable in a rural area:

1. Nontraditional crops, livestock, and other farm products.
2. Service, recreation, tourism, food processing, forest/woodlot, and other enterprises based on farm and natural resources.
3. Unconventional production systems such as organic farming and aquaculture.
4. Direct marketing and other entrepreneurial marketing strategies.

When considering alternative enterprises, you should look first at your farm's underutilized resources and your area's market opportunities. Underutilized resources might include unused buildings, or manure that could be sold as fertilizer. New market opportunities may arise as a result of changing demographics in your area–there may be an increase in immigrant families who want specialty foods, or of affluent business people that commute to a metropolitan area.

One very important change in national demographics is the number of people who have become dissociated from the land. In an effort to re-establish that bond, young consumers are often eager to support small farms and they're willing to put their money where their mouth is. The huge increase in the number of farmers' markets around the country not only means that consumers are interested in fresh produce, it also reflects their desire to have a different kind of food shopping experience. An extension of these encounters is a farm visit, -'Agri-tourism'. Consumers like to feel that they are helping to keep small, family farms alive. This kind of experience requires farmers to learn new skills: how to deal with the public, the ability to assess unique opportunities on the farm, and the vision to produce a feeling as well as a product.

Criteria for recommending these enterprises are:

☆ Low initial start-up cost relative to the ability to generate income

☆ High gross profit margin

☆ Relatively low maintenance requirements

☆ High cash flow relative to expenses

☆ History of high success rates among new enterprises

☆ High demand, low supply in the current marketplace

☆ High product distinctiveness

☆ Relatively size-neutral profit potential

Increasing the Effectiveness of Assistance for Farm Entrepreneurs

In April 2005, in Southern United States, a study was conducted to review and compile a list of strategic activities that would increase the effectiveness of assistance provided to farmers that would increase the effectiveness of sustainable enterprise and got the following feedback:

Suggested Research and Education

The following are suggested activities focused around five areas that were felt to be some of the most important to increase the effectiveness of sustainable enterprise development assistance to farmers.

Research and Education on Sustainable/Regional Food Systems

1. Identify the key components and principles of a sustainable food system. How do they connect and relate to make up the system?

2. Create an educational dialogue on sustainable food systems and their viability, their impact on the community, and opportunities for the farmer. What components are in place? What components need to be developed and how?

3. Facilitate the development of strategies (or "blueprints") for building more sustainable food systems in each state. This would include some type of assessment of current components and infrastructure.

4. Specifically describe the opportunities for farmers in the system, and help assistance providers understand how to provide assistance in a way that takes advantage of the benefits of these emerging systems.

5. Identify needed next steps for research and for successfully implementing sustainable/regional food systems in our region.

Enhance Networking, Collaboration, Cooperation, and Partnerships

1. Map where sustainable enterprise development resources are allocated.

2. Facilitate the coordination of resources through enhanced networking, collaborations, and partnerships.

3. Help farmers have input on development of resources and become more equal partners in assistance programs.

Create a Marketing Information Service for Direct Marketers

1. Develop a system of on-going research on markets and price information at local levels. Especially look at market analysis on consumer demand for sustainably produced products.

2. Assess, consolidate and deliver this information to farmers. Develop a system for this information to be easily accessed by farmers.

3. Conduct research on gaps in marketing information.

Enhance Enterprise Information and Entrepreneurship

1. Compile and circulate successful enterprise models that could be used by farmer entrepreneurs. Focus on components or principles of the enterprises that can be applied to other enterprises or other locations.

2. Provide training for assistance providers on how to analyze new enterprises, how to find and use enterprise models and budgets, and how to create enterprise budgets.

3. Facilitate entrepreneurship training for assistance providers, using expertise that is already in our region. Find a way to institutionalize this within organizations and agencies that assist farmers.

Research Farmer Coops and Facilitate Development of New Collaborations

1. Conduct research on farmer cooperatives (or other marketing collaborations) to determine what assistance is successful, what assistance is not successful, and why.

2. Establish a staffed program for facilitating development of collaborative marketing and other sustainable enterprise activities among farmer groups.

One Plan Production Scheduler

One Plan specifically addresses the Finite Scheduling needs of demand driven process and consumer products manufacturers with a strong emphasis on ease of use, visibility and collaborative planning.

Key Benefits of Ross One Plan

The key benefits of Ross One Plan include:

☆ Improved visibility and decision support - provides clear visibility of your schedules and commitments, with excellent decision support facilities to help you consistently make the best scheduling decisions.

☆ Improved response to change - dynamic alert capabilities help you identify and respond rapidly to changing demand and resource/materials availability in the optimum way.

☆ Improved teamwork, collaboration and performance management - forward looking analytics and reporting tools with inbuilt workflow capabilities allow factory management to set and monitor performance targets. Schedules, targets, analytical results and reports are easily shared across the organization, maximizing communication between different stakeholders.

☆ Improved operational efficiency - A powerful scheduling engine with flexible sequencing logic allows delivering significant improvements in operational efficiency, including reduced changeovers, increased capacity utilization, reduced inventory levels and reduced labor costs. The results are higher customer service levels at lower costs.

☆ Rapid deployment - Implementation timescales are typically about 10 weeks per factory, and the ease of use of the system means only 1 day of end user training is required.

☆ Rapid ROI with low cost of ownership - No hardware, software, IT support or capital outlay required.

Integrating Farmer-driven, Value-added Enterprises into Sustainable Agriculture Systems

From August 1998 to April 2003 Southern SAWG managed a project entitled "Integrating Farmer-driven, Value-added Enterprises into Sustainable Agricultural Systems." This project was partially funded by the Southern Region SARE Program.

Ten Most Important Lessons Learned

1. The development of better services to help farmers, farmer organizations and farm-based entrepreneurs hatch new businesses or expand current businesses that add value to sustainably produced farm products is greatly needed.

2. Even though the focus of the project was on adding value, farmers need assistance in a whole systems, whole farm manner. Assistance needs to be

delivered in a whole systems way or at least coordinated by someone who helps the farmer keep whole farm goals in mind. For instance, assistance solely on processing and market development will not be adequate for a farm that also has enterprise barriers related to production, business management, or even planning.

3. Providing assistance in a whole systems manner must also include an aspect of personal counseling or coaching. Often thinking patterns and emotional patterns of the farm manager are barriers just as difficult to overcome as lack of technical know-how. Some of the most important assistance a consultant can provide is more interpersonal than technical.

4. Assistance is much more useful if it is delivered in a "one stop" system. Most farmers don't have the time or skills to hunt and peck through numerous resources, hoping to find information on value-added enterprise development one kernel at a time.

5. When farmers are matched with consultants who can talk through problems with them and help them create a plan, the farmers are more motivated to take calculated business risks and create value-added operations.

6. The most valuable sources of information are producer entrepreneurs who have gone through several trials and errors already, and created successful enterprises. Yet these producers may not have the time or teaching skills to train others. Assistance providers can take advantage of these practicing experts by facilitating a visit to their farms, facilitating a discussion between the experienced entrepreneurs and farmers needing assistance, paying the producer entrepreneurs for their time, and following up the visit with further analysis and information.

7. Assistance must be tailored to each client, along with numerous opportunities and channels for clients to provide feedback on the assistance, the quality of help being provided, and reassessment of future goals.

8. Consultants or organizations that facilitate value-added enterprise development need to know how to access other appropriate experts, including business planners, accountants, bankers, legal experts, and marketing professionals. If some sort of database or network of possible consultants and organizations could be developed and maintained, it would provide an important tool for service providers.

9. Organizations and agencies that focus on sustainable agriculture need more expertise in small business management and entrepreneurship, and organizations that focus on small business management and entrepreneurship need more expertise in on-farm applications. There is a need for networking and collaboration between organizations with good business and entrepreneurship information, and organizations that work with family farmers and the sustainable agriculture community.

10. Prospective farmer clients, consultants, facilitators, organizational directors, and evaluators all need to be involved in planning, implementing, and evaluating program efforts. Dedicated participation will help ensure that the services are farmer-friendly, organizationally sound, and contain appropriate training procedures.

Areas Needing Additional Study

1. The keys to successful farm-based, value-added enterprises that we developed should be explored with other entrepreneurs around the country, refined by additional research, and updated as appropriate.

2. There is a need for information or models that will help a farmer entrepreneur decide when it is advantageous to add value to an item themselves, when it is better to partner with someone else to add value, and when it is better to not add value. Research comparing management, labor, investment, and net return under each condition is needed to help farmers make more informed decisions.

3. We need to create innovative ways to deliver training and services–including providing expert consultants–to farmers in an individualized way that is also cost effective.

4. There is a need to create more effective ways to have farmer entrepreneurs' network with one another. We need to find ways to overcome barriers of long distance, lack of time, and the reticence to disclose proprietary information.

5. A system of compiling and sharing names of local experts who can act as consultants to farmer entrepreneurs is needed.

Chapter 9
Eventuality Management

Ashok Magaraj

Brain Stain,
Behind Fire Station, Baramunda
Bhubaneshwar – 751003

Risk

What is Risk?

Risk is any uncertainty about a future event that threatens your enterprise/ entrepreneur's ability to accomplish its mission. Viable threats endanger your enterprise's core assets and thereby limit your ability to provide critical services. Although your "fund balance" may be minimal, and equipment may be second generation, your agency has vital assets at risk. Some of the most significant risks which entrepreneurs face are highly subjective. These include:

☆ Credit risk

☆ Operations risk

☆ Fraud risk

☆ Market risk, especially in inefficient markets

Risk is a personal experience, not only because it is subjective, but because it is individuals who suffer the consequences of risk. Although we may speak of organizations taking risk, in actuality, organizations are merely conduits for risk. Ultimately, all risks which flow through an organization accrue to individuals' stock holders, creditors, employees, customers, board members, *etc.* Enterprise risk management is about people. It is about how they think and interact. It is about what they know and what they don't know. It is about their strengths and weaknesses. It is about the complex combination of factors which shape the decisions we make. One of the fundamental challenges of enterprise risk management is the fact that individuals who take risks on behalf of an organization are not always the same people who suffer the ultimate consequences of those risks. Finally, the notion that expected excess returns are proportional to risks taken is simply wrong. These are some of the significant risks which enterprise risk management seeks to address-

Damage to Property: Every organization owns some property even if it is antiquated office furniture and computers. The damage or destruction of the nonprofit's property could impair the organization's ability to continue operations. A fire, tornado, flood, wind, explosion, vandalism, theft or electrical malfunctions can cause damage to the organization's property. The need to abandon or temporarily vacate unusable offices would severely impact the operations of most if not all nonprofits. The organization could face substantial costs to locate and establish temporary or new offices. A property loss has both operational and financial consequences.

Employment risks: Arise from the existence of extensive laws regulating the employer-employee relationship. The laws apply to all aspects of the employment relationship – including the hiring, supervision, and termination of employees. Non-compliance can result in an employment claim and significant fees, fines or settlements.

Fraud: Every nonprofit is vulnerable to fraud. The theft or misappropriation of funds can have severe consequences. A single major theft of funds or equipment could jeopardize seriously a nonprofit's viability. The organization may suffer a cash flow crunch, loss of donor confidence, reduction in services and a loss of jobs. The public reporting of a loss could be devastating as media reports affect donations, the availability of volunteers and public goodwill.

Legal Requirements: In addition to laws regulating the operation of nonprofits, many organizations must follow operational rules and regulations. If an organization fails to comply with these regulations, the regulatory authority may impose a fine, suspend its operating license or permanently close the operation. To manage the risks of legal compliance, every organization must research and keep current on the rules, regulations and statutes that apply to its operations. Noncompliance can both initiate governmental action and create the basis for a liability action against the organization.

Misuse of Funds: An excellent way to monitor an enterprise's progress is through its use of funds. Many farms receive gifts or funding with restrictions or limitations on its use. The improper use of these funds can cause the funder to withdraw the money, require repayment of the expended funds, and refuse to provide future funding.

A similar risk is the use of funds for purposes other than serving the enterprise's mission. Funds inappropriately expended can lead to the loss of the entrepreneur's tax exempt status or other legal actions. As pressures continue to mount for nonprofits to meet social needs, it is often easy to lose sight of the farm's mission.

Tax Liabilities: An organization must pay the appropriate employment taxes and income. Failure to pay these taxes will lead to large fines. A nonprofit may also be responsible for charging and remitting sales tax on items sold. Also, unrelated business income is becoming a significant concern as nonprofits seek creative ways to raise funds. Every nonprofit is responsible for knowing and paying its tax liabilities.

Fundraising: The financial risks for fundraising are two-fold and extend beyond the theft of the money raised. First, an entrepreneur must protect itself from unscrupulous fundraising. Many entrepreneurs have discovered fictitious groups raising funds on their behalf. However, the organization never receives any of the money. An entrepreneur may also suffer losses stemming from injuries at a fundraising event staged by the fictitious group. The entrepreneur should respond quickly whenever it discovers someone using its name and logo without authorization. The second issue concerns the selection and use of sponsors and cause-related marketing partners. An entrepreneur may spend hours and many dollars to negotiate a sponsorship arrangement only to later discover a flaw with the new partner. The potential damage to an organization's reputation and goodwill could have a lasting impact

The following are some of the wide range of uncertainties, which threaten a business/enterprise:

☆ **Natural Perils**: Damage by fire, windstorm, flood *etc* is probably some of the immediate apparent threats to the business of a group. This may lead to scarcity, increased prices for surviving stocks and for those produced when operations are resumed, and therefore to an increase in earnings.

☆ **Loss of Personnel**: Death or injury to employees, compensations, accidents, time that has to be spent for assessing this losses, reports, proceedings, loss of production due to it, cost of recruiting someone else for that post *etc.*

☆ **Labour risk**: Availability of suitable labor may be fundamental to its success. Vulnerability to the effects of staff dissatisfaction may also be crucial and if measures are introduced which might increase unrest.

☆ **Liability risk**: Every business faces the possibility that a single event could involve it in crippling liabilities to third parties.

☆ **Technical risk**: The introduction of new process carries risks, which could have either positive or negative results.

☆ **Marketing risk**: The launch of any new product involves risk of rejection. There is always a possibility that a change in needs, attitudes, tastes or fashion may render it obsolete.

☆ **Political and social risk**: Introduction or abolition of grants or local incentives may have either a favorable or an adverse effect on business. Such changes may reflect a development in general public opinion or be the result of a campaign by a sectional pressure group, which may be able to influence the way the group business is carried on, even in advance of legislation.

☆ **Environmental risk**: The constraints upon permissible contamination of air, water, and land are becoming severe all the time, and legal penalties can be costly for a group.

What is Risk Management?

Risk management increases the probability of attaining goals successfully and fulfilling a mission by anticipating challenges and planning alternatives around possible barriers. Before you can take positive steps to reduce your risk exposures, you have to know what risks you face. You must analyze and identify the areas within your farm day to day operations that hold the potential for causing losses. Risk Management (at its simplest) is a "process for managing the risks that you can identify – and insuring those you can't manage." It uses accepted managerial techniques in order to preserve the assets of the organization or entity. The Risk Management process is comprised of two separate, but equally important components, risk control and risk financing.

Risk control involves identifying the organization s risk exposures, examining the various alternatives available to either eliminate those risks that can be eliminated or mitigate the effects of those that can not be eliminated, selecting the best alternative or combination of alternatives to deal with each risk exposure, implementing the chosen techniques, and monitoring the process for the purpose of altering or improving the program based on the observed results. Risk financing is the method or methods by which an organization chooses to pay for those losses that result from the various risk exposures the organization faces.

Frequency vs. Severity

By looking at past occurrences, we frequently can predict future events. By examining how frequently a loss causing situation has occurred in the past, we can predict with a certain degree of accuracy how often it will occur in the future. By reviewing the cost, or severity, of prior losses we can make an informed decision as to which conditions deserve priority attention and how best to finance the larger losses.

Risk Management Tools

The primary functions for the risk management tools are to assist in the assessment of risks, to assure that risk assessments address all pertinent aspects of the program and to provide specific means of overcoming the underlying bases for the risks. The WBS, SOW and Proposal are recommended as structures for assessing risks. Make-or-buy decisions, development tests and engineering analyses are, of course, means of mitigating the risks by overcoming inexperience and/or a lack of knowledge of specific issues. Risk management as a shared or centralized activity must accomplish the following job or tasks to be performed or carried out in time for better management:

☆ Identity concerns

☆ Identify risks and risk owners

☆ Evaluate the risks as to likelihood and consequences

☆ Assess the options for accommodating the risks

☆ Prioritize the risk management efforts

☆ Develop risk management plans

☆ Authorize the implementation of the risk management plans

☆ Track the risk management efforts and manage accordingly

Risk Control Techniques

Once your risk exposures are identified, the next step is to choose the technique, or combination of techniques, best suited to effectively eliminate or control the exposure. There are five basic risk control techniques. Some can produce the desired results in and of themselves; others work best when used in combination, dependent on the particular exposure being dealt with. These five basic techniques are defined here:

Risk Avoidance

Voluntarily choosing to no longer participate in the activity that creates or causes the loss. If you no longer provide the service or perform the function that created the loss exposure in the first place, you are no longer faced with the exposure. Examples of risk avoidance would include not using old seeds after three years, not growing susceptible varieties. For law enforcement, risk avoidance is not always an option. There are some things we just have to do, but it is nonetheless a desirable technique where its implementation does not significantly interfere with the delivery of vital and necessary police services.

Prevention of Losses

Prevention involves measures or activities undertaken before a loss occurs, in an attempt to prevent the loss causing event from happening, or to render its impact less significant. Examples of preventive measures are the creation and implementation of sound policies that provide appropriate guidance to frontline workers, continuous and on-going in-service training, and patrol cars equipped with prisoner screens and the issuance of latex gloves for the prevention of infection. The primary objective of loss prevention is to reduce the frequency with which the loss causing event occurs.

Reduction of Losses

Reduction techniques can be implemented either before or after a particular loss occurs, in an attempt to reduce the amount of the loss or damages that may result. Having a trial plot test first, keeping perishable products in cold storage, Sprinkler systems, fire extinguishers, soft body armor, and vehicle safety belts are examples of reduction measures. These activities are intended to minimize the potential severity of loss. They do not prevent the loss causing event from occurring but can only reduce or control the severity of loss occurring.

Segregation of Resources

This technique actually consists of two separate elements duplication and separation, both having substantially the same goal: to segregate the agency's resources so that no one event can significantly impair the overall operation of the organization. Basically, segregation involves not placing all your eggs in one basket. Duplication involves the use or creation of spares or backups, to be used only in the event the primary or original item is damaged or destroyed. Examples of duplication are tape backups of computerized data, spare patrol cars left in reserve, or an Emergency Operations Center (EOC) housed in another location that is only utilized if a power failure or natural disaster renders the primary communications facility inoperable. These items for the most part sit, unused, until after a loss occurs. Duplication efforts are intended to reduce the severity of potential losses because the department can still function, although possibly with less efficiency.

Separation is similar to duplication in those facilities, operations or items are duplicated in other locations. The difference is that these facilities or items are used on a daily basis. Examples of separation would be the creation of precincts or mini-stations that provide full service on a daily basis, and can serve to take up the slack in the event the primary location becomes disabled or inoperable. Separation also targets potential severity, but because of its daily usage, can actually increase the frequency of losses.

Transfer of Risk

Transfer techniques are used to transfer, or move, the risk from one party to another. The most common examples of transfer strategies are the use of waiver forms, hold harmless agreements, insurance policies, and contracting with others for services such as prisoner transports or lodging. Ideally, to receive maximum benefit from transfer arrangements, the organization strives to transfer both legal and financial responsibilities for an incurred loss, although this is not always possible.

Objectives of Risk Management

Can be classified into two groups:

Pre-loss Objectives

The most important include:

☆ **Economy:** this involves an analysis of safety program expenses, insurance premiums, and the costs associated with the different techniques for handling losses.

☆ **Reduction of anxiety:** certain loss exposures can cost greater worry and fear for the risk manager, key executives, and stockholders than other exposures.

☆ **Meeting externally imposed obligations:** this means the firm must meet certain obligations imposed on it by outsiders.

Post-loss Objectives

☆ Survival of the firm

☆ Continue operating

☆ Stability of earning

☆ Continued growth of the firm.

☆ Goal of social responsibility is to minimize the impact that a loss has on other persons and on society.

Risk has always been a part of agriculture. But, farming in India is a ball game that has changed dramatically over the past few years. Increasingly, farmers and ranchers are learning that it is now a game with new rules, new stakes, and most of all, new risks

The nation's most successful farmers are now looking at a deliberate and knowledgeable approach to risk management as a vital part of their game plan. For them, risk management means farming with confidence in a rapidly changing world. It is the ability to deal with risks that come with new, attractive farming opportunities.

Farming has Five Primary Sources of Risk

☆ Production

☆ Marketing

☆ Finance

☆ Legal

☆ Human resources

Major Sources of Production Risks

☆ Weather

☆ Pests

☆ Diseases

☆ Machinery efficiency

☆ Quality of inputs

☆ Interaction of technology with other farm and management characteristics

Risk Management Strategies to Lower Production Risks

☆ Enterprise diversification

☆ Crop insurance

☆ Contract production

    ❐ Poultry

    ❐ Livestock

☆ New technologies

    ❐ Genetically altered seed

    ❐ Precision farming

Reducing Marketing Risks

☆ Develop a marketing plan

    ❐ The goals and objectives of your business should drive the marketing plan.

- ☐ Know your production costs.
  - ▲ Use enterprise budgets
  - ▲ Keep good records
  - ▲ Determine break even prices
- ☆ Exhibit marketing plan discipline
- ☆ Have a contingency plan
  - ☐ Prices may be lower
  - ☐ Yields may not be as good as expected
- ☆ Learn and use marketing tools
  - ☐ Storage
  - ☐ Cash sale

Financial Risk has Three Basic Components

- ☆ The cost and availability of debt capital
- ☆ The ability to meet cash flow needs in a timely manner
- ☆ The ability to maintain and grow equity

Lowering Financial Risks

- ☆ Maintain good financial records
- ☆ Keep an eye on interest rates
- ☆ Ensure liquidity and adequate cash flow
- ☆ Buy crop, liability, and personal insurance
- ☆ Control family living expenses

Legal Issues Associated with Agriculture

- ☆ Structural issues
  - ☐ Sole proprietorship
  - ☐ Partnerships
  - ☐ Corporations
- ☆ Contract arrangements
  - ☐ Leases, notes, mortgages
  - ☐ Government farm programs
- ☆ Statutory obligations
  - ☐ Tax reporting
  - ☐ Safety requirements
  - ☐ Use of pesticides

☆ Tort liability
- ❏ Injury tort
- ❏ Toxic tort

☆ Environmental liability

Human Resource Issues

☆ Managing people
- ❏ Hiring, job descriptions
- ❏ Orientation and training
- ❏ Employer/employee interaction
- ❏ Performance appraisal
- ❏ Compensation
- ❏ Discipline
- ❏ Wills and estate planning

Remember

☆ Use a comprehensive strategy of production, marketing and financial responses to risk, rather than separate ones

☆ Structure the business to minimize exposure

☆ All responses to risk have a cost
- ❏ *i.e.,* insurance premiums, a higher price foregone through forward contracting

☆ Make changes as the business grows

Change

The Three C's of Change

☆ Customers take charge

☆ Competition intensifies

☆ Change becomes constant

Change is a continuous challenge as customer expectations change and can sometimes change faster than organizations can respond.

TQM Initiatives Rest on Three Quality Principles

☆ Customer focus

☆ Process improvement

☆ Total involvement

Nine Recommendations for Change Initiatives

1. Consider all cost-reduction alternatives

2. Front-end-load any sizable layoffs
3. Spread the pain
4. Leverage human development
5. Create-and reward-shared wins
6. Let everyone know what the score is
7. Beware of fetishes
8. Identify and integrate multiple perspectives
9. Take time out to reflect

Types of Change

Project Change

☆ Variation

☆ Foreseen Uncertainty

☆ Unforeseen Uncertainty

☆ Chaos

Organizational Change

☆ Technology-led change

☆ Performance improvement

☆ Transformational change

☆ Organizational-led restructuring

☆ Culture and behavior-led transformation

Strategies for Overcoming Resistance to Change

☆ Education and communication

☆ Participation and involvement

☆ Facilitation and support

☆ Negotiation and agreement

☆ Manipulation and co-optation

☆ Explicit and implicit coercion

Encouraging Change

☆ To convince people to change they must be engaged in the decisions - *empowered*

☆ *"Collaborative not Authoritative"*

☆ People have good ideas - use them

☆ Theory Y shows that they really care

☆ They want the esteem - engage them

☆ *Make them the force for change*

Effective Ways of Managing Change

1. **Information**: With inadequate information rumors spread which could be more damaging? Proper and timely information builds confidence and credibility. One should be prepared to answer legitimate questions.

2. **Involvement**: Allowing all those who will be effected with the change to participate in the discussion regarding change might bring in greater acceptance. This reduces WE-THEY syndrome.

3. **Support and reassurance**: if the persons know that their seniors are there to support them in case something goes wrong with the change it gives confidence.

4. **Guidance**: Specially in the technological field where change can be anticipated to provide adequate guidance or training to adopt the new role will be quite helpful.

5. **Presence and proximity**: Even after introducing the change there might be necessity for modification. Availability of the supervisor who is willing to listen and discuss could help.

6. **Talking about it**: One always feels that old is gold, though it will not be true. Yet to talk about the past together allows emotional stream to be let off. Once steam is out, there will be greater acceptance of the present reality.

7. **Clarification**: All established media of communication upward, downward and sideways will help in removing lot of misunderstanding. If this could be done by somebody, who is fairly senior, its reliability could be more.

8. **Respect for values and dignity**: Some are observed with their value system. To deny or suppress it could make one more adamant. Accepting it and yet try to point out that value of incumbent also without total compromising may be helpful. Personal dignity is to be maintained.

9. **Hope**: Change in broad way with great expectation and this hope of something achieving better has to be sustained.

10. **Listening**: No change is permanent. Hence one should listen to the grape –vine or the option leaders to determine whether some modifications or changes are necessary on the change already made.

Change is inevitable. We can neither wish it away nor eliminate it. We have managed change in a positive way so that the organization and hence the people grow.

One needs to CHANGE

Initiate CHANGE

Effect CHANGE

Sustain CHANGE

Managing Conflict

☆ **Conflict**

    ❏ Incompatible behaviors that make another person less effective

☆ Dealing with the Two Faces of Conflict

    ❏ **Competitive conflict:** parties are pursuing directly opposite (win-lose) goals.

    ❏ **Cooperative conflict:** a mutually reinforcing experience (win-win) that serves the best interests of both parties.

☆ Conflict Triggers

    ❏ **Conflict trigger:** Any factor that increases the chances of conflict.

        ▲ Types of triggers

        ▲ Ambiguous or overlapping jurisdictions.

        ▲ Competition for scarce resources.

        ▲ Communication breakdowns.

        ▲ Time pressure.

        ▲ Unreasonable standards, rule, policies, or procedures.

        ▲ Personality clashes.

        ▲ Status differentials.

        ▲ Unrealized expectations.

Years of conflict have disrupted livelihoods, resulting in the loss of assets such as fishing boats and equipment, and household possessions, and have limited access to the natural resource base. Conflicts also restrict mobility, affecting people's capacity for earning a living and curbed access to services and markets. The mobility restrictions and high overall poverty levels have encouraged overexploitation of the natural resource base and environmentally destructive practices.

Potential Impacts

It is argued that any development project set in a conflict-prone region may reduce, create or intensify violent conflicts by: changing access to individual or collective resources, especially nonrenewable ones; creating or exacerbating socioeconomic tensions; changing the material basis of economic sustenance or food security; or changing the content of, or control over, existing political, economic or social systems.

Conflict sensitivity is the ability of an organisation to:

☆ Understand the context in which it operates;

☆ Understand the interaction between its intervention and the context; and

☆ Act upon the understanding of this interaction, in order to avoid negative impacts and maximise positive impacts.

Any intervention by organisations working in, on, or around conflict, or in regions of unstable peace will have an impact on the context. Thus, all actors can benefit from implementing conflict sensitivity in their operations to ensure that, at a minimum, their programming does no harm and that, ideally, they have a positive impact on the conflict.

Conflict Analysis is a central component of the conflict-sensitive development approach. Conflict analysis is a systematic study of the profile, causes, actors, and dynamics of conflict. It provides a foundation to inform conflict-sensitive programming, particularly in terms of understanding the interaction between intervention and context.

Entrepreneurial Capacity

Aims to identify whether an entrepreneur has appropriate human and organizational capital in place to minimize negative and maximize positive impacts on the conflict dynamics of the environment(s) where the specific enterprise is to be implemented. Human capital includes staff and partner skills, knowledge and experience. Organizational capital includes departments, structures, financial resources, organizational culture and learning.

1. Does the Entrepreneur Have?

☆ Experience and expertise of the particular enterprise area

☆ Conflict management skills (negotiations, mediation *etc.*)

☆ Conflict/context analysis skills

☆ Understanding and solid knowledge of the local context

2. Does the Entrepreneur Have?

☆ Institutional memory in particular project/geographic area

☆ Regular presence/representation in the project location

☆ Effective internal/external information sharing mechanisms

☆ Effective M and E system (including monitoring and evaluation of context dynamics)

☆ Clear and appropriate internal division of responsibilities in connection with this project

☆ Effective internal conflict management mechanisms

☆ Ability to assess and consider socio-economic-political trends in the global context (broader than project geographic area)

Steps in Conflict Resolution

☆ Involved parties must acknowledge that there is a conflict

☆ Put out the fire, calm the emotions of those involved

☆ Create an atmosphere in which there is a desire to solve the conflict

☆ Define the problems so that all parties should have a mutual understanding

☆ Identify potential alternatives for solving these problems and list them out

☆ Mutual discussions on the available alternatives

☆ Select mutually the acceptable alternative

☆ Establish a formal agreement

☆ Implement the selected alternative and arrange for actual monitoring of the implementation

Disaster/Crisis

Management Initiative: Mission and Goal

☆ Minimize loss of life and property

   ❏ Prepare

   ❏ Mitigate

   ❏ Respond

   ❏ Recover

☆ Provide consolidated source of disaster-related information and services (www.disasterhelp.gov)

☆ Establish information interoperability infrastructure

☆ Leverage existing assets

☆ Streamline disaster-related processes

☆ "Bottom up": responder/stakeholder driven requirements

Consequences

☆ Resources are Dispersed across many Actors

☆ Knowledge about Resource Availability is Dispersed Across Multiple Actors

☆ Knowledge about Requirements is distributed across Multiple Actors

☆ Knowledge about Response Actions is Distributed

☆ Possibly Overlapping Tasks are done by Multiple Actors

Requirements for Collaboration Support

☆ Multi-Actor Participation

☆ Global in Scope

☆ Synchronous and Asynchronous

☆ Should accommodate a variety of Infrastructures

☆ Various levels of functionality and graceful degradation

☆ Should include distributed databases, and knowledge bases, expert locaters, multi-media interactions, desk-top sharing

Crisis Response Procedure

Continuing communities

- ☆ Standards of care and legal issues
- ☆ Education and training of stress management professionals
- ☆ Education and training for incident coordinators
- ☆ Family support services
- ☆ On-scene support services
- ☆ Peer counselling
- ☆ Re-mobilisation for large-scale disasters
- ☆ Coordination of supporters in a crisis
- ☆ Sources of funding and other material resources
- ☆ Critical incident stress de-briefings

Crisis and Post-incident Communities (incident-specific)

- ☆ Victim support
- ☆ Support for relatives
- ☆ Support for supporters
- ☆ Sources of funding and other material resources, appeals *etc.*
- ☆ Communications with the media
- ☆ Peer counselling

"The main task is to take immediate care of accident victims and people involved in incidents to train supporters, to set quality standards for care and therapy measures, to provide scientific assistance as well as building up of a network in which effectively unifies the many centers and initiatives already existent in this field."

Activities

- ☆ Crisis and disaster response coordination for provision of support (psychological/emotional/material) for the people involved in major incident.
- ☆ Community of practice to support Teams
  - ❏ 'In background': preparation, planning, setting of standards, mutual support and knowledge and information exchange among professionals and 'permanent' members of the community, education and training of professionals and 'supporters'.
  - ❏ Post-crisis (incident-specific): resources, exchange of knowledge and information, support, provision of ongoing therapy.

Problem

Problem Analysis

This process is designed to help the manager to deal with a problem that already exists. Problem analysis is particularly helpful when a problem occurs without warning and without apparent cause. The process consists of several steps:

1. Define the Problem

Before a problem can be dealt with, it must be defined. All persons involved must be clear and in agreement about what is wrong. A deviation statement is being created, that is, a specific description of the malfunction or problem. The exact matter should be pin pointed.

2. Describe the Problem in Four Dimensions

The manager will need to know:

☆ **Identity**: what is being explained

☆ **Location**: where is it observed.

☆ **Timing**: when the problem occurs.

☆ **Magnitude**: how serious the problem is?

Consideration of these factors will help in bringing possible causes of the problem to light. It will also serve to discount other factors.

3. Reading the Clues

that is extracting the key information's. What is distinctive. The answers to is/could/be/but/is not list are made and are applied to each factor to make a comparison. Example – a car stalls. So whether it is due to non-maintenance, driven more frequently, driven in rough roads *etc*.

4 Testing for the most Probable Cause

After all the possible causes are identified then it is important to ask each one of them – if this is the true cause of the problem then how does it answer these clues. If after regular maintenance the car does not stall then we get sure that the problem was due to non-maintenance.

Potential Problem Analysis

Decision analysis and problem analysis are designed as strategies for coping with immediate situations and currently occurring problems whereas potential problem analysis, on the other hand is designed to assist managers in dealing with the future. Potential problem analysis is a method of predicting future problems and dealing with them before they have a chance to happen. The two questions asked during potential problem analysis are: "what can go wrong" and "what can we do about it now". Broken down further, potential problem analysis requires the identification of four aspects of the proposed course of action.

☆ Vulnerable areas of the proposed understanding

☆ Specific potential problems –within those vulnerable areas, especially those that pose enough of a potential problem to warrant taking preventive action.

☆ Likely causes- of the potential problems and courses of action to take to prevent these problems.

☆ Contingent actions – to be taken, should the first preventive action fails.

Tools to Assist in Processing Ideas for Problem-Solving

☆ Forced Field Analysis

☆ Problem Tree Analysis

☆ Problem-Opportunity Analysis

☆ Problem Space-Policy Space Analysis

☆ Brainstorming

☆ Managing Long Lists: Categorizing

☆ Managing Long Lists: Prioritizing

☆ Nominal Group Technique

☆ Fish jar and Role Plays

☆ Case Study Method

Decision Making

"Decision is not merely fact-finding or statistical analysis. It is not merely logic, creativity, nor intuition. It is a power process in which all these factors are combined. The mind pulls this way and that, hesitates, and finally swings over. In the end a man's basic instrument for decision-making is himself.

Challenges in Decision Making

☆ Technological changes

☆ The dilemma of roles

☆ Emergence of consumerism

☆ Management of man of today

☆ The aura of cynicism among managers

☆ Emergence of Officers' Association

As Peter Drucker (1960) Pointed Out

"The most common source of mistakes in management decision-making is the emphasis on finding the right answer rather than the right question. It is not enough to find the right answers; more important and more difficult is to make effective the course of action decided upon. Management is not concerned with knowledge for its own sake; it is concerned with performance".

Handling Stressful Situations

CLEAR ways to handle complaints or conflicting views.

☆ **Remain calm**: Try to reduce the anger and stress by maintaining a calm, friendly attitude while asking the person(s) to explain the situation. Maintaining composure sends the message that the problem can be resolved without conflict. It lowers defensiveness so that the issue can be discussed

☆ **Listen effectively**: Let the other person or persons explain they're point-of-view. Let them get rid of remaining anger. Hear all the details so that you can better understand what is happening and to convey the message that you are sincere in resolving the dispute. Don't interrupt. Listen for the details and the feelings involved.

☆ **Empathize**: You need to let the other person(s) know that you understand how the situation could arise and you don't blame anyone. Empathy reduces the negative feelings and brings rationality to the process.

☆ **Apologize**: Apologize to let the persons know that you are sorry that the problem arose or that something wasn't done before the conflict. The gesture models a measure of respect and let's others begin to open their minds to resolution or atonement. It indicates a clear desire to resolve the issue.

☆ **Remedy the situation**: Once calm is restored and the facts have been expressed, search for ways to deal with the issue. Ask those involved for suggestions. If an action can be taken immediately, do so. If not, decide on a strategy to deal with the issue. Be sure to follow-through and revisit the issue to see if it is resolved satisfactorily. If not, problem-solve again.

Steps in Decision-making

☆ Identifying and defining the problem
☆ Developing various alternatives
☆ Evaluating alternatives
☆ Selecting an alternative
☆ Implementing the alternative
☆ Evaluating both the actual decision and the decision-making process

Stress Management

Barriers create stress. Stress refers to pressure, emphasis, demand on energy, strain *etc*. Stress is a psychological response to the pressure of daily living – yet stress is vital to effective living. Certain amount of stress is necessary. Stress triggers and initiates action.

Behavior

☆ Impatient

☆ Speak humidly

☆ Walkout rapidly

☆ No time to spare

☆ Exhibits nervous gesture

☆ Gets upset and angry soon.

☆ Preferred only to listen

☆ Enjoys only winning.

Approaches to Coping Stress at Workplace

I. Change your habits and life style –start with yourself.

☆ Good nutritional habit

☆ Get fit and kip fit

☆ Play active games

☆ Raise your self-image

☆ Assertive behavior-very important.- wherever necessary say-yes/no.

☆ Effective management

☆ Regular relaxation habit.

☆ Regular exercise

II. Self Management

☆ Cooperative life style and supportive relationship at work.

☆ Share success and failure

☆ Learn to respect.

☆ Face stress and prepare yourself tuning of minds and positive thinking.

III. Organizational measure

☆ Belief in effective human resource utilization

☆ Belief in development of employees

☆ Provide meaningful or enriching jobs

☆ Adapt to change

☆ Promote positive attitude

☆ Education and awareness for fitness and health.

☆ Stress counseling

Chapter 10
Business Accounting: An Overview

Prasant Kumar

Professor,
Faculty of Commerce, Banaras Hindu University,
Varanasi, Uttar Pradesh

The knowledge of accounting is extremely helpful in any entrepreneurial endeavor. If you decide to become an entrepreneur, you need to know whether business earned profit or incurred loss during the period and how is the financial health of the business. You will have difficulty in making your dreams a reality if you are unable to understand and interpret accounting data for financial control and decision making. As a matter of fact, accounting is necessary not only for business activities but also for non-profit organizations like colleges, hospitals, clubs *etc.* Thus, a sound background in accounting will be extremely valuable to an entrepreneur in establishing and running a business activity successfully.

Running a business involves a large number of transactions relating to various items of purchase and sales frequently. Similarly, payments are made and received regularly. An entrepreneur cannot remember the exact details of each and every transactions taking place during a period. This necessitates the recording of transactions in a systematic and analytical manner so as to ascertain correctly the profit earned or loss incurred during the particular period.

Meaning of Accounting

Accounting is often called the language of business as it communicates or reports the performance and financial health of a business enterprise. The accounting is an art of recording, classifying and summarizing business transactions and events which can be measured in terms of money and analyzing and interpreting result there of and communicating in the form of financial statements *i.e.* Profit and Loss

Account and Balance Sheet so that intelligent decisions can be made about the concern. The process of accounting is depicted with the help of the following flow chart of accounting:

Flow Chart of Accounting

| Identification |

(of Transactions)

| Measurment |

(in terms of money)

| Recording |

Objectives of Accounting

1. To maintain systematic and complete records of business transactions as per accounting principles.
2. To ascertain profit or loss of the business.
3. To depict the financial position of the business.
4. To communicate accounting information to users.

Double Entry System

The whole accounting system is based on double entry system. The double entry system means every transaction has double effect *i.e.*, when we receive something then we give something else in return. For example, in case of purchase of goods in cash we receive goods and pay cash. Of the two accounts one account is debited while the other account is credited with an equal amount. Thus, on a particular date the total of all debits must be equal to total of all credits because every debit has corresponding credits.

Types of Accounts

In order to record each and every transaction of a business properly all the transactions have been classified in three types of account:-

1. Personal Accounts

These accounts record business dealings with natural person, artificial person and representative persons.

2. Real Accounts

These accounts relate to the tangible and intangible real assets.

3. Nominal Accounts

These accounts deal with expenses, incomes, losses and gains.

Rules of Debit and Credit

For proper recording of different transactions in different books of accounts, the knowledge of the rules of debit and credit is essential. When to debit and when to credit depends upon the type of accounts involved. The following table shows the rules of debit and credit for each account:

Type of Account	Debit	Credit
Personal Account	Receiver	Giver
Real Account	What comes in	What loses out
Nominal Account	Expenses and losses	Incomes and gains

Books of Account

In practices, the following books of account are maintained by each and every organization irrespective of their size and activities. These books are:

1. Cash Book
2. Purchase Book
3. Sales Book
4. Purchase return Book
5. Sales Return Book
6. Ledger
7. Journal Proper

1. Cash Book

Cash book is a book of original entry. This book is used to record all transaction relating to cash receipts and cash payments. This is divided into two parts. The left hand side is used to record cash receipts while right hand side is used for recording cash payments. Cash includes money in all its forms like coins, currency notes, cheques and bank drafts.

Types of Cash Book

The Cash Book may be of four types:

i. Single column Cash Book (Cash column of Amount only)
ii. Double column Cash Book (with cash and bank columns of amount)
iii. Triple column Cash Book (with Cash and Bank columns of Amount and one column for Discount)
iv. Petty Cash Book (for small and recurring cash transactions)

2. Purchase Book

The purchase book is ment for a chronological record of all merchandise purchased on credit only, similarly the credit purchases of capital goods (assets) are recorded in journal proper. The entry in the purchase book is made on the basis of invoice/bill received from the supplier. The following three steps are under taken:

i. An entry is made in purchase book with the help of invoice/bill

ii. A simultaneous entry is made on the credit side of the relevant suppliers account in the ledger.

iii. At periodical intervals may be weekly, fortnightly or at the end of the month, the periodical total of the purchase book is posted on the debit side of the purchase account in the ledger.

3. Sales Book

All credit sales of merchandise are recorded in sales books only. For each credit sales the firm prepares the invoice/bill showing full details of goods sold. The flowing three steps are undertaken:

i. An entry is made in the sales book with the help of carbon copy of the invoice prepared by the firm.

ii. A simultaneous entry is made on the debit side of the respective customers' account in the ledger.

iii. At periodical intervals may be weekly, fortnightly or at the end of the month, the periodical total of the sales book is posted on the credit side of the sales account in the ledger.

4. Purchase Return Book

Sometimes it becomes necessary to return to the supplier a part or whole of the goods bought on credit for a variety of reasons such as defective, damaged, goods not as per order *etc*. All such returns of goods are recorded in the purchase return book.

5. Sales Return Book

This book is used to record sales returns from customers of the goods sold on credit to them. The goods may be returned by the customers for variety of reasons defective goods, inferior quality, not up to the sample *etc*.

6. Ledger

The ledger is also the prime book of accounting system. The ledger is a book which contains all the accounts of real, personal and nominal nature of a business enterprise in a summarized and classified manner. The position of an account on a particular date can be ascertained only from the ledger because various transactions pertaining to a particular account of different dates will be recorded on one page. Every transaction recorded in different books is posted to the concerned accounts in the ledger and balanced as and when needed.

7. Journal Proper

Journal proper is a residuary book in which those transactions are recorded which can not be recoded in any other subsidiary book such as cash book, purchase book, sales book *etc*. The various entries recoded in this book are open entries, closing entries rectification entries, transfer entries and any other entries not finding a place in any other book of account.

Trial Balance

The trial balance is a statement showing the balances of all accounts in the ledger with a view to verify the arithmetical accuracy. If the total of debit side is equal to the total of credit side then it is presumed that the posting to the ledger in terms of debit and credit amount and balancing is correct. The statement helps in locating errors and rectifying them. It is the basis for preparing the financial statements. Generally the trial balance is prepared by the business firm at the end of every month.

Final Accounts

The main motive of doing the above mentioned process is to know the results in the form of profit or loss and financial soundness of the concerned. This objective can be achieved by the preparation of final accounts of the concerned. Usually profit and loss account and balance sheet are collectively known as final accounts.

1. **Final accounts of manufacturing organization**: The final accounts of manufacturing organization consists of the following statements:
   i. Manufacturing account
   ii. Trading account
   iii. Profit and loss account
   iv. Balance sheet
2. **Final accounts of trading organization**: The final accounts of trading organization consists of the following statements:
   i. Trading account
   ii. Profit and loss account
   iii. Balance sheet

Manufacturing Account

The enterprises engaged in manufacturing activities prepare manufacturing account. It is prepared to determine the cost of goods manufactured. The balance of this account *i.e.*, cost of goods manufactured is transferred to trading account. The manufacturing account reveal, the composition of manufacturing cost such as cost of material consumed, value of work in progress, manufacturing expenses *etc*. a general format of manufacturing account is given below.

Format

Manufacturing account of ——— for the period ending on ———

Particulars	Amount	Particulars	Amount
To opening work in progress		By sale of scrap	
To raw material consumed		By closing work in progress	
Opening stock		By trading account (cost of goods manufactured/ transferred)	
Add purchases Carriage award			
Loss close-up stock To wages To salary of workers manager To power electricity and water To fuel To depreciation on Plant and machinery Factory land and building			
To repairs Plant and machinery Factory land and building			
To insurance Plant and machinery Factory and land building			
To rent and taxes			
To royalty based on production			

Trading Account

Trading account is prepared at the end of accounting period. It is designed to show the gross profit or loss as a result of trading activities *i.e.* buying and selling goods and services. The preparation of trading account requires the passing of closing entries to transfer the balances accounts of all the concerned items from ledger to trading account. A general format is given below

Format

Manufacturing account of ——— for the period ending on ———

Particulars	Amount	Particulars	Amount
To opening stock		By sales Loss return	
To purchases Loss returns		By closing stock	
To direct expenses		By gross loss (Transferred to profit and loss Ac)	
To wages			
To carriage inward			

Particulars	Amount	Particulars	Amount
To freight inward			
To cartage inward			
To gross profit			
(Transferred to profit and loss)			

Profit and Loss Account

After preparing the trading account, the next step is to prepare profit and loss account. The profit and loss account is prepared to ascertain the net profit or loss of the business by matching income and expenses for a given accounting period. The balance of trading account (*i.e.* either gross profit or gross loss) is transferred to profit and loss account. The preparation of it requires that the balance of accounts of al concerned items of expenses and incomes are transferred from ledger by passing closing entries. The profit and loss account will show either net profit or net loss during a particular accounting period which is transferred to capital account in balance sheet. The general format is given below.

Format

Profit and Loss account of ——— for the year ended ———

Particulars	Amount	Particulars	Amount
To gross loss		By gross profit	
To salary (Salaries and wage)		By rent received	
To rent rates and taxes		By discount received	
To printing and stationery		By commission received	
To postage and telegram		By interest received	
To telephone, telex and fax		By dividend received	
To insurance premium		By income from investment	
To legal charges		By apprentice premium	
To trade expenses		By bad debts recovered	
To entertainment expenses		By interest on drawings	
To charity and donations			
To audit fees			
To office expenses			
To establishment expenses			
To livestock expenses			
To office lighting			
To license fees			
To bank charges			
To discount allowed			
To interest on capital			
To interest on loan			

Particulars	Amount	Particulars	Amount
To salesmen salary and commission			
To freight/carriage outward			
To packaging expenses			
To export duty			
To advertising and marketing expenses			
To traveling expense and conveyance			
To godown rent/warehouse charges			
To delivery van expenses			
To promotion expenses/ free samples			
To repairs and renewals			
To brokerages and commission			
To bad debts			
To depreciation			
To provision for bad debts			
To loss of goods by fire/theft			
To loss on sale of fixed assets			
To managers commission			
To net profit (Transferred to capital account)		By net Loss (Transferred to capital account)	

Note: The items included in the above profit and loss account are illustrative in nature and not exhaustive. Even though the exact number and name of items to be included will vary from firm to firm, the principles on the basis of which the profit and loss account is prepared, remains unchanged.

Balance Sheet

Having ascertained the operational results, the next step is to know the financial position of the business. For the purpose, the statement which is prepared is known as balance sheet. A balance sheet is a statement of assets and liabilities of the business enterprise at a given date. The right hand side of the balance sheet shows the nature and values of the assets while left hand side shows the nature and value of the liabilities. With the help of balance sheet, the financial solvency of the enterprise can be ascertained. An enterprise financial position is good if its assets value exceed its external liabilities. The general format is given below:

Format

Balance sheet of ——— as at ———

Particulars	Amount	Particulars	Amount
Capital		Fixed assets	
Opening balance		Goodwill	
Add: Net profit		Land	
(Less: Net loss)		Building	
Less Drawings		Plant and machinery	
		Furniture and fixtures	
Long term liabilities		Investments	
Loan		Current assets	
Current liabilities		Closing stock	
Income received in advance		Accured income	
Sundry creditors		Prepaid expenses	
Outstanding expenses		Sundry debts	
Bills payable		Bills receivable	
Bank overdraft		Cash at bank	
		Cash in hand	

Chapter 11
Prospects of Entrepreneurship: Promotion and Management

K. Ghadei[1], R. Panigrahi[2] and D.S. Bhati[3]

[1]*Lecturer, Extension Education, Institute of Agricultural Sciences, BHU, Varanasi*
[2]*SMS (Extn.), KVK, Kendrapara (OUAT), Orissa*
[3]*Assistant Professor(Extn), KVK, Ajmer, Rajsthan*

The Focus

Skills and blue collar jobs are not regarded as challenging or remunerative in Indian society despite our rich tradition in arts, crafts and architecture. We need to systematize the training for these sectors and give them the respect they deserve by initiating a major SKILLS MISSION to develop this sector. This will only lead to greater economic development of the country – opines our former president, Dr. APJ Abdul Kalam in his 'Mission India' publication.

Infact, job creators and not the job seekers have become the need of the hour. And Entrepreneurship is all about creating jobs through vocational promotion and employment generation. Unlike other countries, our vocationally trained population is very low at 5.06 per cent compared to Korea at 95.86 per cent. The main emphasis thus have to be on entrepreneurship education to educate individuals on entrepreneurial qualities and traits like risk bearing ability with confidence, ability to mobilize the resources, innovation and creativity, imagination, readiness, independent decision making, dreams for change to augment his business interest and make more earnings. National entrepreneurial policy on the other head should shift from growth-induced employment to employment-induced growth so that real entrepreneurs could come up in the form of answers to many economic and social problems faced by our vast nation.

Entrepreneur and Entrepreneurship

Entrepreneur is a person who takes the responsibility of initiating and establishing an enterprise or organization under conditions of uncertainty. He is hard-working, optimistic, tough minded and risk taker. He sets high targets or goals and tries to achieve those amidst odd situations. Joseph Schumpeter regarded "the entrepreneur as an innovator with potentialities of doing things in a new way." Entrepreneurship on the hand is a dynamic process with 'innovation' and 'risk bearing' as its two basic elements involving multiplicity of activities towards establishment of an enterprise. According to Jeffrey Timmons, (1995), "Entrepreneurship is the ability to create and build something from practically nothing. It is initiating, doing, achieving and building an enterprise or organization, rather than just watching, analyzing or describing one. It is the Knack for sensing an opportunity where others see chaos, contradiction and confusion. It is the ability to build a "founding team" to complement your own skills and talents. It is the know how to find, marshal and control resources (owned by others) and to make sure you don't run out of money when you need it most. Finally, it is the willingness to take calculated risks, both personal and financial, then to do everything possible to get the odds in your favour."

A scan of the definitions and introspection above hence, underlines the importance of entrepreneurship management to be backstopped by an Entrepreneurship Education base for lasting impact on employment and national economic development.

Entrepreneurship Education

Following the "catch them young" principle, it is essential to integrate entrepreneurship education into the higher education system. It would help circulate the necessary orientation among the upcoming employment seekers towards a future role of an able entrepreneur, employer and independent achiever. The text book learning being memorization dominated, fitting entrepreneurship education in an examination centered academics is a real challenge. Therefore, it is important to design a proper and effective course programme on entrepreneurship education keeping in mind the objectives and challenges it has to confirm ahead.

Key Points of an Effective Entrepreneurship Curriculum

☆ Motivational Chapters on basics of successful entrepreneurship at school levels.

☆ Branching out option to students to a full-fledged entrepreneurship course from higher secondary level.

☆ Emphasis should be on developing entrepreneurial competencies among students at college and university level to take up entrepreneurial carrier afterwards.

☆ Ticking course content to generate confidence and develop competencies among students to venture into the uncertain and risks world of business without any fear or hesitation.

☆ An effective Entrepreneurship Education Curriculum should be practical, implement able and appropriate including some of the following:

1. Nature of small business
2. Business opportunities
3. Global Markets
4. Business Plan
5. Entrepreneurial assistance
6. Market Location and Analysis
7. Business Finance and Pricing
8. Legal Issues
9. Human Resource and Business Management
10. Cost Benefit Analysis
11. Record Keeping
12. Credit Management
13. Risk Management

☆ Field exposure trips should be a part of the entrepreneurship curriculum to establish industry – institution linkage for a better learning experience.

Entrepreneurship Management

Only theoretical orientation does not suffice on the face of stiff challenge from field operation. Starting an enterprise is easy but running it successfully is a tough job. There are many instances where industries are being established with much zeal and enthusiasm only to become sick after one or two years. This is often the result of inadequate skill and expertise in operating an enterprise effectively and successfully. Sound Management of an enterprise is thus, of paramount importance which needs to be tackled at cognitive and community levels to reap the desired harvest.

Cognitive Entrepreneurship Traits

Mental make up is important before facing physical challenges. Some of the inherent characters essential with an entrepreneur are:

1. Self confident and multiskilled
2. Problem solver
3. Innovative thinking
4. Result oriented
5. Risk bearer
6. Total commitment
7. Hard work and single mindedness
8. Discipline
9. Optimism

Optimism refers to positive thinking with a hope of success forever. The A to Z of optimism according to Robert Schaller (2001) as applied to entrepreneurship management include:

Affirm that you can do it

Believe that you can achieve the goal

Commit yourself to a dream

Dare to try

Educate yourself

Find the Possibilities

Give and take attitude

Hope of holding on

Imagine the best outcome

Junk the junk of mind

Knock out depression

Laugh at your failure

Make it happen

Negotiate to compromise

Overlook and overcome follies

Persevere not to give up

Quit complaining

Recognize success or failure

Share the power and glory with others

Trade off for keep up

Unlock values for success

Visualize the dream before you

Work for better & better

X-ray your motive

Yield it all to God

Zip it up

Community Entrepreneurship Traits

Entrepreneurship is about taking initiative and organizing the men, money and material around to get the initiative fructified. Many more physical and financial efforts at community level (where the entrepreneur functions) are required over and above personal abilities in this regard. First of all he has to acquire the personnel management traits like:

☆ Planning the enterprise involves selecting objectives and strategies, policies, programmes and procedures for achieving them. Planning also includes decision making on production, pricing and marketing of products.

☆ Organising involves the establishment of infrastructural roles through determination of activities required to achieve the goals, grouping of activities, reporting and delegation of authority.

☆ Staffing demands defining the workforce, utilization of manpower, appointment, promotion and remuneration including downsizing and organizational unity.

☆ Leading the entrepreneurial unit interms of addressing the desire, attitude and behaviour of individual and groups amidst challenges towards opportunities.

☆ Controlling is the measuring and correcting the activities of staff to assure that events confirm to plan. It measures the performance against goals.

☆ Finance is most crucial that warrants the personal influence and rapport with the financial institutions to get it materialized. The entrepreneur has to show the required faith and credibility in this regard.

☆ Quality Control determines the future of one's aspirational climb up. Quality products create wide market thus making the entrepreneur's task of intervention easy amidst stiff challenges.

☆ Marketing linkage should be assiduously built up by making the products cost competitive, unique and indispensable before the consumers.

☆ Alternate opportunity refers to the next best option that could be explored or employed on the face of parallel enterprises and challenge to one's entrepreneurship ventures.

Conclusion

There is no thumb rule for a full proof entrepreneurial approach. It is about applying one's mind and soul to garner all possible opportunities towards self establishment, economic proficiency and employment generation. Studies reveal that 75 per cent of the unemployed graduates 'lack professionalism' towards entrepreneurship. Thus promotion of entrepreneurship proficiency could be taken at two levels *i.e.* entrepreneurial curriculum at academic level as well as skill training and exposure field trips at take off days to create a wholesome climate for the national prosperity through youth self employment.

References

Mehta, V.B. and Tilekar, S. N (2006). "Entrepreneurship Development in agriculture", *Varanasi News*, New Delhi.

Schaller, R. H. (2001). *Tough time never last but tough people do*, Orient Paperbacks, New Delhi.

Timmons, Jaffrey (1995). *New venture creation* (4th ed.), Bostan

Times City, (2007). "Job creators are the need of the house". 18th Nov. 2007, Sunday Times of India, Varanasi

Legal Issues in Entrepreneurship Development

Chapter 12

Labour Laws for Enterprise Establishment

Krishna Murali

Professor,
Faculty of Law, BHU, Varanasi, U.P.

Introduction

When we talk about labour laws for enterprise establishment, the first thing to talk about is that in our country, the workers working in the organized sector constitute only eight percent of work force and remaining 92 per cent are in the unorganized sector. The term unorganized labour has been defined as those workers who have not been able to organize themselves in pursuit of their common interests due to certain constraints like casual nature of employment, ignorance and illiteracy, small and scattered size of establishment *etc*. As per the latest survey carried out by National sample survey organization the total employment in both organized and unorganized sector in the country was of the order of 39.7 crore i.e, around 2.8 cr in the organized sector and balance 36.9 cr workers in the unorganized sector. Out of 36.9 cr workers in the unorganized sector, there are 23.7 cr workers employed in agricultural sector, about 1.7 cr in construction work, 4.1 cr in manufacturing, 3.7 cr in trade and 3.7 cr in transport, communication and services.

Specific Problems in Unorganized Sector

1. The workers in the unorganized sector are dispersed, unorganized and have poor bargaining power in general.
2. Suffer from seasonality of employment and lack of sustainable employment.
3. Low legislative protection due to scattered and dispersed nature of employment.

4. No formal employer employee relationship.

5. Inadequate welfare measures as it is difficult to reach them due to lack of organization and other factor.

The agricultural workers constitute by far the largest segment of the workers in the unorganized sector. In spite of the fact that these agricultural workers have such numerical strength they are extremely vulnerable to exploitation on account of low level of literacy, awareness, persistent social backwardness and absence of unionism and other forms of viable organization.

In order to take care of social security and welfare of unorganized sectors, two pronged strategy *i.e.*, legislative measures and implementation of welfare schemes and programmes have been followed so far. The legislative measures include:

Act	Year
The workmen's compensation Act	1923
The minimum wage Act	1948
The maternity benefit Act	1961
The contract labour (regulation and abolition) Act	1970
The bonded labour system (abolition) Act	1976
The inter state migrant workmen Act	1979
The building and other construction work Act	1996

Legislations Applicable to Agricultural Labour

A number of important legislation are applicable to plantation labour and have enabling provisions to extend them to agricultural labour. Some of such legislations are:

Act	Year
Payment of wages Act	1936
Maternity benefit Act	1961
Payment of gratuity Act	1972
Minimum wages Act	1948
The workmen's' compensation Act	1923
The equal remuneration Act	1976
Interstate migrant workmen Act	1979

Hence it was felt that existing labour laws have not be enable to protect the interest of agricultural labourers the proposal for enacting a comprehensive legislation for agricultural workers have been under consideration of the ministry of labour since 1975. However, as there was no consensus amongst the state governments, the matter could not be processed further.

Labour Laws in the Organized Sector

So far as the organized sector is concerned, it can be said that most of the benefits provided by the labour laws have gone to the organized sector. For an Entrepreneur intending to establish an enterprise, the law regulating labour-management relationship in the organized sector is given in the industrial disputes Act 1947. The trade union Act 1926 and the industrial employment (standing orders) Act 1946. While industrial disputes Act emphasizes on compulsory adjudication of industrial disputes the trade unions Act places exclusive emphasis on collective bargaining.

Industrial Disputes Act 1947

The preamble of the Act says that it is a law enacted for maintaining and promoting industrial peace and harmony. For an entrepreneur, it is very essential to know the various provisions of industrial Act.

The first thing to know in the Act is to know what is meant by the Industry. Industry is defined in the section II (J) of the Act, which says that industry means "any business, trade, undertaking, manufacturer, calling of employers and includes any calling, service, employment, handicraft, industrial occupation or avocation of workmen." The supreme court in the famous Bangalore water supply case laid down a "Triple Test" to decode this definition. According to the court, if there is an employer employee relationship, systematically or habitually cooperating with each other produce material goods or material services to the community at large, such an establishment can be called as an industry.

The next important thing to know is as to who is a "workman" defined in section II (S) of the Act. Section II (S) inter-alia defines a workman as a person who does manual skilled, skilled, unskilled, technical, clerical, supervisory or operational work.

The next important thing to know is what is ment by industrial dispute. Industrial dispute is defined in section II (K) inter alia, as a disputer or difference between employer and employees, employers and workman, workman and workman, relating to employment, non-employment, terms of employment or conditions of labour of any person. The supreme court interpreted the expression "any person" as a person in whose service conditions the workers as a class have got a direct and substantial interest. Industrial disputes are collective disputes, and sometimes even an individual disputes can become an industrial dispute, if it is supported by a trade union or a substantial number of workman. However, the legislature in the year 1965 has inserted new section 2-A converting certain individual disputes connected with or arising out of dismissal, discharge, retrenchment or any other termination to be deemed industrial dispute, notwithstanding that they are not supported by a trade union or a substantial number of workmen.

To resolve and settle the industrial disputes, the Act has provided for the mechanism of conciliation, arbitration and adjudication. One thing that is common to all these three mechanisms they involve the intervention of third party to settle or adjudicate the industrial disputes. The conspicuous thing about arbitration is the fact that parties choose their own arbitration unlike conciliation and adjudication where the authority to resolve the dispute is constituted by the appropriate government. In India the most prevalent method to settle dispute is through adjudication.

Strike and lockout are instrument of economic coercion under the Act in collective bargaining. Section 22 and 23 of that deal with the power of government to regulate strike and lockout. Apart from this under section 10 (3) and 10-A (4-A) of that, the government has the power to prohibit the continuance of strike and lockout by taking recourses to machinery provided under the Act to maintain and promote industrial peace and harmony. The government has very important role and its role is all pervasive. It has got the administrative power under section 10 of the Act to make the reference of industrial dispute to the authority constituted under the Act.

Apart from this the Act defines layout, retrenchment and closure which form the crux of the Act. Layoff is defined in section 2 (KKK) interalia as the failure, refusal or inability of an employer to provide work to a workman due to shortage of coal, power, raw materials, accumulation of stocks, natural calamity or any other connected reason. Retrenchment is defined in section II (OO) inter-alia as termination by the employer of the services of a workman for any reason whatsoever, as a punishment inflicted by way of disciplinary Action. Closure is denied in section II (CC) as permanent closing of the place of employment or any part thereof. The law obligates every employer to provide to every employee who has put in one year of continuous service, to layoff compensation retrenchment compensation or closure compensation whatsoever the case may be. The value of layoff compensation is 50 per cent of basic wages and dearness allowance, the value of retrenchment and closure compensation is 15 days of average pay for every completed year of service. The provision of layoff in chapter V-A are applicable to only a factory, firms or plantation employing 50 or more workmen while the provisions relating to retrenchment and closure in chapter V-A are applicable to all the industries. The provisions with regard to layoff, retrenchment and closure in chapter V-B are applicable only to a factory, mine or plantation employing 100 or more workmen. The most important thing in chapter V-B is that the employer has to secure the prior permission of the appropriate government.

Apart from this the Act also provides the security of employment to the workmen stipulating that no workman can be dismissed from service without conducting an enquiry complying with the principle of natural justice.

The Trade Unions Act – 1926

The trade unions permit the persons employed in the industry or trade to form a trade union for an effective collective bargaining. It also provides for registration of trade unions. Though registration is not mandatory under the Act. It also provides the registered trade unions to maintain "general fund" and "political fund" for efficient discharge of its duties. It also provides for civil and criminal immunities to registered trade unions to facilitate effective collective bargaining.

The Industrial Employment (Standing Order) Act – 1946

This Act provides for compulsory certification of standing orders with the primary objective of brining about uniformity and stability in the working conditions to promote industrial peace and harmony.

Social Security

Apart from this the state had passed many welfare labour legislations to provide social security to its working class. These legislations are:

Act	Year
The workman compensation Act	1923
Payment of wages Act	1946
Minimum wages Act	1948
The employee state insurance Act	1948
The factories Act	1948
The employees provident fund Act	1952
The payment of bonus Act	1965
The payment of gratuity Act	1972

Conclusion

Every entrepreneur wishing to start an enterprise must be conversant with these labour laws to become a successful entrepreneur. Instead of evading these laws to make profit the entrepreneur must try to implement these laws whole heartedly so that he/she can win the love and affection of working class and that in the long run would enable him/her more enduring profits.

Chapter 13
Forms and Formation of Enterprise: Legal Aspects

Manoj Kumar Padhy

Associate Professor, Law School,
Banaras Hindu University, Varanasi, U.P.

Introduction

The liberalization of the Indian economy and world trade as well as rising consumer prosperity have thrown up new opportunities for entrepreneurship development and opened up new avenues for growth. Demand for processed and convenience agro-product is increasing constantly because of urbanization, changing life style and habits of the people. Accordingly the Indian consumers are being offered newer high quality products made by using the latest technology. The need for entrepreneurship development can also be viewed from another angle, as increasing population and reducing opportunities in rural and urban areas have given rise to un-employment among educated youths.

Every entrepreneur aims at starting a business and building it into a successful enterprise. The entrepreneur innovates and combines resources in the form of 4Ms (man, material, machine and money) and brings them together to make the business venture profitable. The whole process of starting and successfully continuing an enterprise depends upon a good business plan. A good business plan is the key to setting up a successful business. A good business plan including other decisions comprises choosing a form of an enterprise and while choosing a form of an enterprise, the knowledge of the legal requirements for the formation of an enterprise is going to be a decisive or material factor.

With this background, this paper aims to provide information to the prospective entrepreneurs about the legal aspects of formation of different forms of enterprise so as to enable them to arrive at a conclusion as to selection of a suitable form of enterprise.

Meaning of the Term "Enterprise"

Literally speaking an "enterprise" is an economic institution engaged in the production and/or distribution of goods and services to earn profits and acquire wealth. The Micro[1], Small[2] and Medium[3] Enterprises Development Act, 2006 defines an "enterprise" as an industrial undertaking or business concern or any establishment by whichever name called, engaged in the manufacture or production of goods, in any manner, pertaining to any industry specified in the first schedule to the Industrial (Development and Regulation) Act, 1951 or engaged in providing or rendering of any service or services[4]. According to Black's Law Dictionary an enterprise is an organization or venture, esp. for business purposes. According to Chamber's Twentieth Century Dictionary an "enterprise" is nothing but is an undertaking.

Forms and Formation of Enterprises

Every business enterprise adopts one form of business organization or other. The forms of enterprise are the following:

A. Sole proprietorship

B. Joint Hindu Family Business

C. Partnership

D. Co-operative Organization

E. Public Limited Company, and

F. Private Limited Company.

A. Sole Proprietorship

Sole proprietorship is the oldest and most common form of enterprise, where the business is owned, managed and controlled by a single individual called the sole proprietor. However, while carrying out business, the sole proprietor can take the help of his/her family members. Other persons may also be appointed as

---

1. Where the investment in Plant and Machinery <= 25 Lakhs (in the case of enterprises engaged in the manufacture or production of goods) or where the investment in Equipments <= 10 lakhs (in the case of enterprises engaged in providing or rendering of services). Investment in plant and machinery excludes cost of pollution control, research development and industrial safety devices etc.

2. Where the investment in plant and machinery > Rs. 25 lakhs but <= Rs. 5 crore (Enterprises manufacturing and producing goods) or where the investment in Equipments > Rs. 10 lakhs but <= 2 Crores (Enterprises Providing or rendering services). Investment in plant and machinery excludes cost of pollution control, research development and industrial safety devices etc.

3. Where the investment in Plant and Machinery > Rs. 5 crore but <= 10 crores (Enterprises manufacturing and producing goods) or where the investment in Equipments > Rs. 2 crores but <= Rs. 5 crores (Enterprises providing or rendering services). Investment in plants and machinery excludes cost of pollution control, research development and industrial safety devices etc.

4. Section 2(e) of the Micro, Small and Medium Enterprises Development Act, 2006.

assistants or manager but, legally the business and the individual are the same. Its salient features are:

(i) Its formation and closure do not require elaborate legal formalities. Even agreement and registration of firm is not essential. So, it is easier to form and close a sole proprietorship firm[5].

(ii) The ownership vests on sole proprietor who is responsible for supplying whole capital and he is only entitled for the benefits or profits and liable for losses of the firm.

(iii) The liability of the proprietor in the case of losses is unlimited (extends beyond the capital invested).

(iv) The proprietor has absolute control over the firm and he takes all the business decisions, so, there is flexible operation.

Advantages	Disadvantages
☆ Ease of formation and closure	☆ Limited capital
☆ Flexible operation	☆ Unlimited liability
☆ Owner alone enjoys profits	☆ Lack of continuity
☆ Suitable for small scale business	☆ Limited managerial skills

B. Joint Hindu Family Business

It refers to a business which is owned by the members of a joint Hindu family[6]. This form of enterprise exists under Hindu law and governed by the Hindu Succession Act, 1986. The ownership vests in the head of the family who is known as "Karta". The business is owned only by male members of the Hindu joint family called "co-partners". Its salient features are:

(i) Membership is created by birth but not out of contract[7]. Only three successive generations in the male line constitutes the membership of this organization.

(ii) Registration is not necessary.

(iii) Management vests in 'Karta', whose authority is unquestioned in managing the joint Hindu family business.

(iv) The Karta has unlimited liability while the liability of the other members are limited to the value of their individual interests in the joint family.

(v) It has a perpetual life.

---

5.   The owner may be required to obtain a license form the local administration.

6.   For Hindu Law, See generally, Paras Diwan, *Modern Hindu Law*, Allahabad Law Agency (2002).

7.   In *State Bank of India* v. *Ghamandi*, 1969, SC 1330, it was held that the joint Hindu family business does not arise out of contract between members, but it comes into existence by operation of law.

Advantages	Disadvantages
☆ Continuity of existence	☆ Unlimited liability of Karta
☆ Ease of formation	☆ Limited capital
	☆ Limited Managerial skills

C. Partnership

Partnership is defined as relation between two or more persons who have agreed to share profits of a business carried on by all of them or any of them acting for all[8]. Partnership is controlled and regulated under the Partnership Act, 1932[9]. Its salient features are:

(i) The formation is free from cumbersome legal formalities.

(ii) Registration[10] is not compulsory; however, if the firm is not registered then a partner cannot file a suit either against the firm or any other partner for enforcing his rights arising out of contract. Similarly, an unregistered firm can neither file a suit against any third party nor against any partner[11].

(iii) The minimum number of partners must be two, while the maximum number can be 20 (for banking business 10)

(iv) The rights of ownership and control are jointly held by the partners[12].

(v) Liability of the partners is unlimited.

(vi) No partner can transfer his/her interest in the firm to any person[13].

(vii) The firm comes to an end on retirement, lunacy, bankruptcy or death of any partner.

Advantages	Disadvantages
☆ Ease of formation	☆ Limited duration
☆ Greater capital pool	☆ Non transferability of interest
☆ Pooling of managerial skills	☆ Unlimited liability
☆ Risk shared by partners	

D. Co-operative Organization

It is a voluntary association of ten or more members for mutual benefit and its

---

8.   Section 4 of The Partnership Act, 1932. Persons who have entered into partnership with one another are called individually "Partners" and collectively "a firm".

9.   For partnership see generally, S. Chand, *Business Law*, S. Chand and Company Ltd.; M.C. Shukla, *A Manual of Mercantile Law*, S. Chand and Company Ltd.

10.   Sec. 58 of the Partnership Act, 1932.

11.   Sec. 69 of the Partnership Act, 1932.

12.   Section 13 of the Partnership Act, 1932.

13.   Sec. 29 of the Partnership Act, 1932.

aims are accomplished through self help and collective effort[14]. Co-operatives differ from any other forms of business organization in the sense that its primary motive is service to the members rather than making profits[15]. It is controlled and regulated under the Co-operative Societies Act, 1912. Its salient features are:

   (i) It is a voluntary organization (free to join or leave)

   (ii) Minimum number of member is 10, but no upper limit.

   (iii) Registration is compulsory with the Registrar of co-operatives societies.

   (iv) It is a body corporate independent of its members.

   (v) Liability of every member is limited.

Advantages	Disadvantages
☆ Ease of formation	☆ Limited capital in comparison to company
☆ Greater amount of capital	☆ Inefficient management
☆ Continuity of existence	
☆ Limited liability	

E. Public Limited Company

In a practical way, a company means a company of certain persons registered under the Companies Act.[16] It is a voluntary association of members which is incorporated. It is controlled and regulated under the Indian Companies Act, 1956. Its salient features are:

   (i) Separate legal existence[17].

   (ii) Minimum seven members but there is no upper limit.

   (iii) Compulsory registration[18].

   (iv) Capital is constituted by the sale of its shares.

   (v) Minimum paid up capital is Rs. 5 Lakhs.

---

14. Co-operative societies may be classified into following categories: (i) consumers' co-operative; (ii) producers' cooperative; (iii) cooperative marketing societies; (iv) cooperative credit societies; (v) co-operative farming societies; and (vi) cooperative housing societies.

15. A cooperative society has much in common with Partnership. Yet there are differences between the two types of organisation. In a partnership mutual benefit is restricted to partners only, but it in a co-operative society it extends to its members as also the public.

16. For company law see generally, KS Anantharaman, *Lectures on Company Law and MRTP*, Wadhwa and Company, Nagpur, 1998; Avtar Singh, *Company Law*, Eastern Book Company, Lucknow, 2001.

17. Sec. 34 of the Companies Act, 1956.

18. To obtain the registration of a company an application has to be filed with the registrar of companies. The application must be accompanied by the following documents (Sec. 33 of the Companies Act, 1956):

   (a) Memorandum of Association; (b) Article of Association (If necessary). (c) The agreement , if any .

(vi) Shares are freely transferable.

(vii) Limited liability

(viii) Ownership is separate from management.

(ix) Formation of a company, powers and responsibilities of the directors and managers, raising of capital, holding company meetings, maintenance of audit of company accounts, powers of inspection and investigation of company affairs, reconstruction and amalgamation and winding up of a company *etc.* are dealt under the Companies Act, 1956.

Advantages	Disadvantages
☆ Continuity of existence	☆ Cumbersome legal procedure
☆ Greater capital pool	☆ Strict regulation
☆ Efficient and democratic management	
☆ Limited liability	

F. Private Limited Company

It is a voluntary association of not less than two and not more than 50 members. It is a very suitable device for carrying on the business of family and small scale concerns. Formation and regulation are dealt under the Indian Companies Act, 1956. Its salient features are:

(i) Compulsory registration but less cumbersome to organize in comparison to public limited Company for following reasons:

(a) No need to file prospectus[19] with the registrar.

(b) No need to obtain certificate of commencement of business.

(c) No need to hold statutory meeting.

(ii) Liability of members is limited.

(iii) Shares are not freely transferable between members.

(iv) Minimum paid up capital is Rs. 1 lakh.

Advantages	Disadvantages
☆ Continuity of existence	☆ Shares are not freely transferable
☆ Greater capital pool	☆ General public cannot be allowed to subscribe to its shares
☆ Less cumbersome to organize	
☆ Limited liability	

---

19. Prospectus is defined by Section 2 (36) of the Companies Act, 1956. "A prospectus means any document described or issued as prospectus and includes any notice, circular, advertisement or other document inviting deposits from the public or inviting offers from the public for the subscription or purchase of any shares in or debentures of a body corporate.

Concluding Observation

The success and growth of an enterprise depends largely on the choice of a suitable form of organization. The aforesaid discussion may enable a person to choose most suitable form of organization for a business venture. The selection of a suitable form of enterprise should be done after analyzing the following factors, namely: nature of business, mode of operation, financial requirements, nature of control, nature of liability, stability and legal necessities *etc*. At the same time before starting an enterprise the other factors such as location of enterprise, market position, production analysis and government policy should also to be taken into consideration.

Information and Communication Technologies (ICT) in Entrepreneurship Development

Chapter 14

Information and Communication Technologies (ICTs) in Entrepreneurship Development: An Emerging Paradigm

Meera Shaik N.

Directorate of Rice Research,
Rajendranagar, Hyderabad – 500 030

According to Peter Drucker an entrepreneur searches for change, responds to it and exploits opportunities. Innovation is a specific tool of an entrepreneur hence an effective entrepreneur converts a source into a resource. One way of converting sources into resources is harnessing information and communication technologies. The paper tries to explore the possibilities of harnessing ICTs in entrepreneurship development.

The world's young people now number more than they ever have. About half of the globe's population of 6.2 billion people is under the age of 25. In this backdrop no country can afford to ignore information and communication technologies (ICT) as an employment generator, whatever their stage of development. The paper describes as how ICTs have helped to create job opportunities for young people in India, in recent past. An attempt is made to give minor details of the cases from developing world which is considered appropriate for Indian conditions.

ICT can be broadly defined as a set of activities that facilitate, by electronic means, the capturing, storage, processing, transmission, and display of information. The term information and communication technologies (ICT) are used to encompass the production of both computer hardware and software as well as the means of transferring the information in digital form. It also includes low cost forms of communication such as radios. The more profound effect of ICT is likely to be in improving the efficiency and reach of the mainstream production of goods and services, in both the public and private sectors of the economy.

The entrepreneurship in the context of ICTs is discussed under the three broad heads.

1. ICTs in creating Entrepreneurial Opportunities
2. Developing Insights from the Select Cases
3. ICTs in developing and Sustaining Entrepreneurial Capabilities

Entrepreneurship development of yester years was not same as today. The concept is being changed rapidly. As the basics of entrepreneurship are going to be same, the way that is being carried out is changing fast. ICTs are playing important role in enhancing the efficiency of entrepreneurial activities on one hand and are creating new opportunities on other hand. These entrepreneurial opportunities are discussed under "ICTs in creating Entrepreneurial Opportunities".

In the context of interwoven nature of entrepreneurship with the rural development and macro economics, it is important to bring forth some cases as how these two concepts will go hand in hand. Harnessing ICTs in the rural development have been well documented (Shaik 2003), but the inherent strength with respect to the entrepreneurial backdrop was often ignored. Hence there is a need to closely examine ICTs, rural development and entrepreneurship. An attempt is made to discuss about the select cases where all these dimensions are well addressed. This is covered under the head "Developing Insights from the Select Cases". All the stages of entrepreneurship development from development to sustenance stage are influenced by information, communication and networking. Future entrepreneurial capabilities will depend largely on these three dimensions and ICTs are going to play an important role in bringing about efficiency in all these dimensions. Three cases are described under "ICTs in developing and Sustaining Entrepreneurial Capabilities" will give insights as how few initiatives were started in India and elsewhere.

ICTs in Creating Entrepreneurial Opportunities

Of late, young people are using ICT as a launching pad for initiating a range of entrepreneurial activities. With ICTs it is possible to explore low-income generation opportunities, involving telephony and the use of mobile phones, role of young people as information intermediaries, e-commerce and establishment of telecasters. Many of such paradigm shifts have been observed for last one decade and have potential to be generalized for over all entrepreneurship development.

Selling Telephone-Based Services

The worldwide expansion of mobile phone networks and the growth in the number of mobile phone subscribers has been phenomenal in recent years.

The availability of mobile phone networks in India opens up many opportunities for young people. One common option is to purchase a mobile phone through a micro credit program and to earn income by providing low cost phone calls to others. This is a very common income generating opportunity for rural youth in India. One example, among others is setting up of *Tata Indicom* PCOs in villages.

To further develop some insights as how mobile telephony entrepreneurship could be institutionalized, a case of Grameen Village Pay Phone program (VPP) is discussed here. Grameen Bank is a pioneer of small loans to the poor. The Village Pay Phone program makes it possible for a Grameen borrower to buy a mobile phone, and then to make the telephone available for others in the village to pay for phone calls, to send short message services (SMS) and to enable villagers to receive incoming calls. Grameen Telecom charges Grameen borrowers a wholesale airtime rate. Grameen Village Pay Phones operates in more than 2,000 villages in Bangladesh and an average of 100 additional villages is being connected each month. A typical pay phone owner can earn up to four times the average per capita income in Bangladesh. The phones are used for a variety of purposes. Farmers use them to find out where they can get the best prices for their crops, and relief workers are able to better coordinate disaster response measures. Villagers are also able to use the phones to communicate with local government officials.

Grameen Telecom is itself is a good example of entrepreneurial activity supported by partnerships with international agencies, international companies and other funding sources. The potential of Grameen Telecom as an income generator has been acknowledged by an international agencies and could be viable model for countries like India.

Tech-Mode 'Information Intermediaries'

India is an experimental country for the world to pilot test thousands of ICT projects aimed at rural development. Based on the success of several ICT projects, government of India has taken up several e-strategies to harness the power of ICTs for societal transformation. In many of such initiatives content, connectivity and capacity building have been identified as pre-requisites. In this back drop it is observed that the widespread use of English on the Internet has created the need for local content and applications to enable non-English rural Indians to make effective use of it. For the poor in particular, the vast amount of information on the Internet requires an intermediary to identify what is relevant and then interpret it in the light of the local context. Young people are well placed to perform this role of 'information intermediary'. A study conducted by Shaik N.Meera (2004) revealed that ICT projects to serve resource-poor farmers require qualified and well-motivated staff to serve as an interface with computer systems. This is a good opportunity for rural youth in near future. Ministry of Information Technology and National Institute of Smart Governance, Alliance 2007, and many more initiatives aim to establish Village Knowledge Centres (VKC) with a motto of every village a VKC. These Knowledge centres are going to be run by several young entrepreneurs who are imparted training by several institutes such as Jameshedji Tata National Virtual Academy.

Another option is for young people to use their skills in information technology to develop simple web sites/web pages/local content in local languages. For example, Swaminathan Foundation has set up Village Knowledge Centres, with special websites to provide a variety of locally relevant content. Another example is Warana Nagar rural network project, in Maharashtra. The district has 70 villages and is known for the strength of its cooperative societies. Villagers are using 'facilitation booths' to access agricultural, medical and educational information on the Internet.

E-commerce-Based Entrepreneurship in Remote Communities

Other low-income generating opportunities available to young people in remote locations is initiating e-commerce based entrepreneurship development. The Los Angeles-based Greenstar Foundation is setting up self-contained, solar-powered community centres in remote communities in India. Each centre offers an Internet connection, health facilities, including telemedicine, a classroom complete with distance learning equipment, and a business centre, through which traditional cultural products can be sold via the Internet. Traditional art, music, photography, legends, and storytelling in small villages can be recorded and brought to global markets through the Internet.

The projects are deliberately targeting areas without electricity. The approach is to use this market mechanism to sell cultural products in digital formats to pay for the hardware and connections needed and to produce ongoing revenue without the need for external funding. The projects are the product of public-private collaborations between governments, local ICT companies and international funding sources.

To generate income through e-commerce, Greenstar is encouraging a team of artists and teachers to record elements of rural Indian culture, working closely with the people of each village. The result is a powerful, unique collection of 'digital culture' – a gallery of music, artwork, photographs, video, poetry and other arts, which can be distributed in high-resolution digital form throughout the world, instantly and efficiently. The revenues from digital culture are used to fund basic needs of each village for its future, as decided by the people themselves.

Bridging the Gap between the Digital Economy and the Informal Sector

Another paradigm is the use of ICT to help bridge the gap between young people's opportunities for self-employment in the informal economy. For example, the Foundation of Occupational Development in India, which operates eleven telecentres, has also established a website called *India Shop* to provide a market outlet for indigenous crafts people. As a result, an isolated community is able to fetch much higher prices from international customers than from retailers in nearby cities. Reference has also been made to how communities in remote locations can make use of self-contained, solar-powered ICT centres to sell, among other things, traditional cultural products such as art, music, photography, legends and storytelling via the Internet. This is being done on a pilot basis in remote communities in India.

Another example of the use of ICT to help bridge the gap between employment for young people in the informal sector and the mainstream economy is India's Self Employed Women's Association (SEWA). Its 2,20,000 members are women who earn a living through their own labour or through small businesses. SEWA has been one of the first organizations in India to realise the potential for harnessing ICT to help women in the informal sector. It has sought to develop the organization's capacity to use computers by conducting awareness programs and imparting basic computer skills to its team leaders, 'barefoot' managers and members of its various member associations. Many of SEWA's member organizations have launched their own websites to sell their products in the global virtual market place. Since the entire membership of SEWA consists of poor self-employed women, giving its members access to software in the 'language of daily use' is of great importance. Hence, efforts are being made to develop software to enable grass-roots workers and members to make the best use of the tools provided by ICT. Recently, SEWA has started using telecommunications as a tool for capacity building among the rural population. SEWA uses a combination of landline and satellite communication to conduct educational programs on community development by distance learning. The community development themes covered in the education programs delivered include: organizing; leadership building; forestry; water conservation; health education; child development, the Panchayati Raj System and financial services.

Tele-centre Initiatives

Middle-income entrepreneurial opportunities can also be identified involving the use of ICT in the service sector. 'Mini telecentres' usually offer a single phone line (possibly mobile phone) with a three-in-one scanner/printer/copier, a fax machine and a PC with a printer, Internet access and a call meter. A 'telecentre' offers a number of phone lines, a call management system, fax machine, photocopier, several PCs with a printer, Internet access and perhaps a scanner. Finally, a 'full service telecentre' offers many phone lines and multimedia PCs with Internet access. Other equipment can include a high-volume black and white and/or colour printer, a scanner, a digital camera, a video camera, a TV, an overhead projector, a photocopier, a laminator, meeting rooms, and a video conferencing room.

India has seen a rapid growth in 'cyber kiosks' or 'telekiosks' which can provide access to business support services for underprivileged groups. These 'Internet kiosks' are often upgraded STD (Subscriber Trunk Dialing) booths that are common in India. These are small street shops, offering access to public phones for long distance calls. They number about 300,000 and have generated more than 600,000 jobs.

Promoting Public-Private Partnerships to Generate ICT-Related Employment

Use of public-private partnerships to create ICT-related employment opportunities for young people is another untapped area. Public-private partnerships refer to collaborative arrangements between governments and private enterprises or the NGO sector to generate employment or to deliver better services. One benefit of public-private partnerships by governments is to leverage additional investment to build public infrastructure or to deliver public services using private providers. Public-private partnerships can help leverage ICT-related employment because Governments need to attract not only investment funds but also the knowledge and expertise required to operate complex ICT facilities.

Young people have the opportunity to gain employment through the growth in remote processing facilities that are located outside the high-income countries. These provide a range of services from help lines, technical support, and handling reservations and sales to data conversion including voice to data transcription. Other remote processing includes payroll accounting to internal auditing and credit appraisals. High-end remote processing includes creating digitized maps of townships, utilities, roads, and other facilities. It is claimed that back office functions likely to grow in importance are settling insurance claims and summarizing legal documents, such as witness depositions.

Call Centres

A related source of ICT-generated employment for young people is through Call centres. These offer telephone-based services from a central office to customers in a variety of business sectors. Call centres handle telephone calls, fax, e-mail and other types of customer contact – in live and automated formats. Many young people in India have found work in call centres.

Developing Insights from the Select Cases

Many a times, promoting sustainable livelihoods of the rural people using ICTs is linked to the entrepreneurial development of the rural youth. At times ICTs may help rural people (irrespective whether youth or not) to change the old ways of doing the business and in becoming entrepreneurs of their own. These cases must be looked at with a view as how some ICT initiatives are creating entrepreneurial opportunities along with serving to bring about sustainable development of rural areas.

Enterprise Management Using ICTs: Case of Ascent

Athani, in the State of Karnataka is the heart land of Kolhapuri sandals and home to over 400 such families of artisans with a rich legacy. Footwear craft is their only livelihood. Prior to year 2000 most worked as low wage-bonded labour in footwear 'factories' owned by dominant traders. Their life and craft were demeaned - they lived on the very edge. Prior to 1998 skill training and technology up gradation intervention was targeted at men, apart from training some tools were provided which found their way to markets rather than application. It would not be wrong to state that the intervention was not gender inclusive, women were not reckoned as artisans even though it was an established practice that men fashion the hardy soles and women craft the intricate uppers with a role division. ASCENT initiated Project Enterprise (Jan 1999 - Dec 2002) sponsored by UNDP with technology support from CLRI (Central Leather Research Institute) and infrastructure by Government of Karnataka, the project objective was ensuring right price for the handcrafted footwear and transforming artisans to entrepreneurs, particularly women. The core focus was economic development and the core principal was - build, operate and transfer. The business front end and social back end needed constant balancing using an equality, equity and inclusive approach.

The outcome of this intense joint effort is ToeHold Artisans Collaborative (TAC). From a business perspective TAC has been an overwhelming success. The operation has achieved robust revenue growth in recent years and is achieving healthy profits, which are redistributing amongst artisans and self help groups (SHGs) of women in Athani. TAC is now a prominent player in the international market for ethnic footwear supplying international clientele in UK, Italy, Japan and Australia.

The artisans formed their collaborative called the ToeHold Artisans' Collaborative in October 2000 with their own brand 'ToeHold'. The brand is targeted at the niche segment or the class markets rather than mass markets for optimum profitability. The footwear is positioned as 'fashion accessory' and not just as craft. 150 women organized into eleven women Self Help Groups jointly exercise ownership of the collective. A Common Facility Center and Raw Material Bank with a Design Studio are set up at Athani (in two Vishwa sheds). The artisans –women and men, receive joint training in design development, entrepreneurial skills and leadership and soft skills. Artisans' direct exposure to international markets has improved their understanding of the international customer and the demands in terms of quality, delivery commitments and design. Each family now acts as a micro enterprise where the woman and man are 'co-preneurs'. The financial stakes are with women but men are equal partners to all other activities and inputs.

A vast collection of contemporary new designs has been developed. ToeHold Artisans' Collaborative exports to very competitive mainstream international fashion markets. The ToeHold artisans went on to participate in the India International Leather Fair, Chennai, Delhi Shoe Fair and in the GDS International Shoe Fair in Dusseldorf, Germany and International Leather Fair in Shanghai. By now, they were more receptive to new ideas and ready to understand the concepts of costing and pricing, and so far, these teams from Athani executed export orders worth nearly *USD* 225,000.

This unique model of enterprise, owned and governed by the artisans through women's Self Help Groups has taken the humble *'Kolhapuri'* to 'couture' status in the trendiest mainstream international fashion markets. This can act as a learning model for similar efforts in other sectors as well. This also throws light on the importance of artisan clustering and the ways in which some 2000 already existing artisan clusters in India can be helped with successful interventions from the state and civil society organisations. The model where SHGs manage the enterprise is an idea with much applicability and can be replicated for many kinds of enterprises.

Social Entrepreneurship Model: Case of Akshaya

The highpoint of the Kerala Model of development in the social sector is attributed to the creation of access points. Kerala has government owned access points in all the core sectors of development. A primary school for 1000 families, ration shop as the part of public distribution system, 'Anganwadi' for prenatal care, primary health centres for medical care *etc*. All the above mentioned access points were established owned and maintained by government. Setting up 5000 ICT centres, managing them, and to dynamically update the situations will be very difficult for the State Government primarily because of the constraints in the resources. It was therefore decided to start the centres with the help of local entrepreneurs. The huge unemployment percentage of the educated youths in the State also prompted the State Government to take this decision.

The concept of social entrepreneurship was brought in here, because of two reasons; one ICT for Development in the rural areas is a new concept, people have not started using ICT in their lives, except for the fact that some services like railway reservation are used without realizing it as an intervention by Technology. But to prompt the people to use Internet instead of telephone calls or online payment system to pay their bills is not an easy job, because of the hesitation to use technology. A Social entrepreneur, who is a native of that place, commands some social esteem, is well versed in social activities, badly in need of a job, ready to experiment, would be able to define the information, communication, and education needs of the common people around him, and to covert the need to a service with the help of government. Entrepreneur in Akshaya is clearly understood the fact that service delivery through his centre is an essential component for creating his business and maintaining the customer base. Many training programmes were organized for Akshaya entrepreneurs.

Akshaya Centre is owned by a single entrepreneur. There can be partnership firms, trusts, companies that start Akshaya Centres. But Local Bodies sign the agreement only with a single person easy operation of the arrangement. Akshaya entrepreneurs have to sign an agreement with government where he undertakes to start and run the centre for a period of three years. They are committed to do all the projects initiated by government during this time span. He has to sign a code of conduct to work as a social and economic catalyst by demonstrating the high spirits of democracy and perform the duties of a better citizen. Also he cannot indulge in any kind of immoral or anti social activities. Government can any time cancel the centre, if it violates the guidelines.

If the Akshaya Centre functions from a rented premise, Akshaya Entrepreneur has to sign a three year contract with the Building owner to maintain the centre at the same location for a minimum of three years. If the building is his own, he has to give an undertaking on Rs.50 Stamp paper, to maintain the Akshaya Centres in the same place for three years. The Centre owner has to secure the permission of Grama Panchayath and District Level Project Monitoring Agency and from the other entrepreneurs in the Panchayath for shifting the location.

Akshaya Centres Operate from 9 a.m. to 7 p.m. But during the e-literacy phase, there are centres which are working more than 18 hours a day. Akshaya Centres can open subcentres with the permission of local bodies to extend the services. During Malappuram e-literacy phase more than 2000 subcentres were opened by 630 Akshaya Centres.

Akshaya Centres required to use proper records. There are two registers, which is used for monitoring the e-literacy campaign by local bodies. Two types of registers will be made available, the main register has to be maintained in the main centre, and separate registers are to be maintained for sub centres.

Employment Creation

1. Staff in Akshaya Centres	1890 (3 persons appointed per Akshaya Centre).
2. Social Animators	105 Social Animators trained and positioned for campaign period
3. Project Staff	20 persons permanently employed for the project
4. Supporting Industries	
Hardware	261 employment in Hardware industry (Total 87 Local Hardware assemblers supplied the hardware for Akshaya Centres- 3 person per Hardware Unit)
Connectivity	25 persons working permanently with the Internet Service Provider.
Trainers Grid	100 persons in Computer Course programmes positioned in Trainers Grid (On going)
Companies	Nearly 40 companies/corporates operate through Akshaya network. Positioned nearly 50 persons on an average for coordination.Nearly 3000 direct employment opportunities created.

Employment Creation- Direct

1. Staff in Akshaya Centres	6665 (5 persons appointed per Akshaya Centre X 1333 Centres).
2. Social Animators	650 Social Animators trained and positioned for campaign period
3. Project Staff	80 persons permanently employed for the project (10 persons X 8 districts)
Total	**7385**

Delivering IT Power to Villages: Case of Rajiv Internet Village

Rajiv Internet Village (RAJiv) is an e-governance initiative, conceptualized and started by eGov-Services and powered by Sun Microsystems' open source infrastructure. Supported by government of Andhra Pradesh and leading banks such as State Bank of India, the initiative involves identifying local entrepreneurs and giving them complete support to set up an e-commerce kiosk. This kiosk offers a wide spectrum of services to the local residents, thus not only benefiting the entrepreneur to earn from transaction commissions, but also enables the entire village to benefit with services, which this kiosk brings to their doorsteps.

Around 73 kiosks have been set up under the project so far and already more than 2.8 lakh transactions have been implemented on the network. Right from project financing to the various services it offers, care has been taken to ensure that it is not dependant on any government subsidies. New and unique services have been introduced to ensure continuity and sustaina-bility of the business. Using revolutionary open source infrastructure from Sun Microsystems, we have ensured that the project is easily executable, replicable and easily scalable. It is being estimated that in next two years, all 2,000 villages will be brought under the ambit of this project and around two million villagers will benefit from the e-commerce success. The AP government has already studied the project in detail and plans to implement it throughout the state, thus benefiting around eight million citizens.

Project is offering:

- ☆ **G2C services**: Varied certificated from government departments like caste certificates, birth and death certificates, residence certificates and income certificates

- ☆ **B2C services**: Electricity bill payments of Andhra Pradesh Transmission Company Ltd; state bus tickets of Andhra Pradesh State Road Transport Corporation; telephone bill payments of BSNL, Reliance Info-comm and Tata Indicom; mobile bill payments of BSNL Cell One, Reliance Infocom, Tata Indicom and Airtel; railway reservations by IRCTC and courier service to transfer couriers from one mandal to another

- ☆ **Other Services**: Varied services like e-employment, information based services, Spandana for citizens petitions and grievances, e-education, computer aided learning, Azim Premji Foundation CDs, tele-health and entertainment will also be made available to the people

The Rural BPO Business Model

The TeNeT Group at Indian Institute of Technology, IIT-Madras has worked over the past 2 years on a rural BPO initiative that links urban clients with a rural workforce through the Internet kiosk network. The team identifies and trains workers in rural areas in various skills, relevant to the BPO industry. It liaises with urban clients and takes complete responsibility for the outsourcing and timely delivery of the projects undertaken, and ensures that quality standards are met. As a coordinating agency, the team protects the interests of both the clients and the kiosks. At the village-end it filters out unproductive kiosks from the system, and at the city-end it runs due diligence checks on the client to guard against fraudulent BPO activity. This unique initiative has a portfolio of services that includes typing in English and regional languages, data entry operations, web and multimedia development and regional language translation. More recently, engineering services such as 2D drafting and conversion of 2D to 3D for the manufacturing sector have also been introduced. While the model today runs on a smaller scale, a typical rural BPO centre, to be envisaged in the future, would consist of 10 PCs running in two shifts. Each centre would employ between 10-20 individuals, and the kiosk owner would be responsible for hiring and managing the staff, ensuring that timelines and quality standards are adhered to, and managing daily operations.

BPO activity in India is clustered around 5 main hubs today. These centres will continue to remain important in the future, but the industry is looking to expand to other locations for several reasons. Newer locations would imply access to a larger workforce, provide an opportunity to further reduce costs of operation, help acquire language-specific skills and mitigate overall business risk and ensure business continuity. Expansion to newer locations would also help to reduce the pressure on infrastructure, being faced in the current locations. Rural areas can be attractive outsourcing destinations for the BPO industry primarily because labour is less expensive than in the cities. Also minimal investment in infrastructure is required in the existing kiosks in order for them to serve as BPO centres. Rentals and overheads in these areas also tend to be low, further adding to the arbitrage. The most important leverage in this arrangement, however, is the existence of an entrepreneur running every kiosk, who is a trusted entry point into the village. Likewise, the Rural BPO model offers significant benefits to the rural population. With IT training, the youth in rural areas are exposed to skills that are highly valued in today's economy. As a result, their productivity and incomes increase, and so, also their personal confidence. The entire rural economy begins to thrive as more money flows into villages, allowing for more equitable economic growth at the national level.

An Internet Gateway to Promote Sustainable Livelihoods

The potential for ICT to bridge the gap between young people's self employment opportunities in local informal sector markets and the wider domestic and international economy is amply demonstrated by India's TARAhaat.com. TARAhaat or *Star Marketplace* is an Internet gateway that connects the village user to information about social services, health, entertainment, and to markets, through a network of franchised cyber centres, customised in the language of their choice. The website attracts between 5000 and 25,000 contacts per month. The project illustrates a number of best practice features, which won it the 2001 Stockholm International Challenge prize as best practice in the category of a Global Village. The first feature worth highlighting is that it is targeted at the poor by seeking to create sustainable livelihoods for people located in areas with limited economic opportunities and harsh living conditions. Second, it has been designed using extensive market research and socio-economic surveys, including a house-to-house survey of selected villages in the region. Third, its format aims to cater for the needs of people with wide variations in literacy, language, financial liquidity, and levels of understanding. Fourth, the project is supported by partnerships with enterprises in the public and private sector including the Indira Gandhi National Open University. Fifth, the project has support from youth organisations through the Association of National Youth Cooperatives. Sixth, the project is based on features that go beyond simply using the Internet to communicate with its target audience.

TARAhaat covers all three components for rural connectivity: content, access and fulfillment. Content in relation to law, governance, health and livelihoods is provided by the TARAhaat.com mother portal. Access is provided through a network of franchised local enterprises. Delivery of information, goods and services is provided by local courier services or franchised TARAvans. The revenue streams of TARAhaat provide for profit generation at each step of the supply chain, serving to further cement its networks. The project, although still in its pilot stage, is said to have increased the economic opportunities for the physically disabled and the franchisees, as well as to have improved access to education for rural girls. The main benefits include the generation of alternative sources of income for young people through desktop publishing.

ICTs in Developing and Sustaining Entrepreneurial Capabilities

Entrepreneurship is not an easy option and is best suited to those with the necessary skills and acumen. Some of these skills can be acquired, even via the Internet. However, some skills such as risk taking and self-confidence may be more deep seated. Young people starting their own businesses are likely to experience a range of problems.

One fundamental problem is the inability to secure start-up funds leading to under capitalization (starting a business without enough funds). Other problems commonly encountered are managing cash flow, especially dealing with bad debts and late payments; and coping with stress, especially without the support of friends who understand the demands of self employment. In such cases, ICTs may help young people in developing and sustaining their enterprise, using some concepts such as Business incubators. Nurturing and developing entrepreneurs may get some help from ICTs as discussed in these cases.

ICT Enabled Business Incubators

Throughout the developing world, innovative entrepreneurs are working to establish businesses that are 'ICT-enabled'–offering ICT services or, in some way utilising ICTs as a fundamental aspect of their business model. Technology entrepreneurship is key to innovation, employment, and national competitiveness. However, often the obstacles facing such start-ups seem insurmountable. Common barriers to converting an innovative idea to a viable business venture include limited and costly access to Information and Communication services, burdensome business regulations, the absence of basic business support services, the lack of advice, mentoring and best practice guidance, limited market knowledge, and lack of access to appropriate financial services. In 2002, in response to these challenges *info*Dev launched the Business Incubator Initiative, aimed at fostering ICT-enabled entrepreneurship and private sector development in developing countries. Some of them are discussed here.

IndiaCo.com Private Ltd

A business and technology incubator, and start-up resource centre with operations close to the innovation clusters in the Pune-Mumbai corridor. IndiaCo assists early stage companies by providing business infrastructure and office space, mentoring and coaching, and assistance in raising private equity capital. The goal is to increase the success rate of start-ups by operating a network that facilitates and motivates the use of local resources to commercialise available technologies.

Tiruchirapalli Regional Engineering College, Science and Technology Entrepreneurs Park

TREC-STEP is the first Science Park to be promoted in India since 1986, which aims to foster knowledge based ventures comprised of young science and technology entrepreneurs. TREC-STEP has been promoted by Department of Science and technology, Government of India, Government of Tamil Nadu, Industrial development Bank of India (IDBI), Industrial Credit and Investment Corporation of India (ICICI), Industrial Finance Corporation of India (IFCI) and state financial and Development organisations.

Society for Research and Initiative for Sustainable Technologies and Resources

It aims to build and enhance an ICT-enabled virtual incubator model for scaling up grassroots innovations.

Telecommunications and Network Group

TeNeT's expertise spans digital communications, wireless networks, computers protocols, optical communications, digital signal processing, speech, audio and video technologies, among others. The TeNet Group engages in teaching and training, product development, incubation of technology companies, telecommunications, IT policy studies and front-line research.

Vellore Institute of Technology-Technology Business Incubator

VIT-TBI was established at Vellore Institute of Technology with the sponsorship of the Government of India's Department of Science and Technology. VIT-TBI assists budding entrepreneurs with incubating new technology ventures.

Promotion of Youth Entrepreneurship through ICT in Schools

The Schoolnet Internet Learning Centres in Uganda have been set up by the country's Education Department to promote youth employment through giving young people entrepreneurship and leadership skills using ICT-based training and resources. The project was one of 100 finalists for the 2001 Stockholm Challenge. Some thirty ICT resource centres, each comprising ten networked computers and a server, with printers and modems, have been set up in Ugandan schools. The resource centres service between 200 and 1000 young people per month. The goals of the project are to: develop youth leadership, team building and business skills; promote youth employment through linkages with local industry/business; create new youth led business opportunities and encourage young people to exchange business ideas and information via e-mail. Youth who participate in the project are given an opportunity to develop business concepts and plans that draw upon the ICT resources available at the centres.

Assistance for Young People to Set Up Community-Based Businesses in Rural Areas

World Corps, an international non-profit organization based in Seattle that provides training to promising young business and community leaders worldwide. World Corps seeks to create jobs, sustainable social business ventures, and programs for social change that are easily replicable. Their Web site is http://www.worldcorps. org World Corps trains young men and women aged 21-28 to establish community-based businesses in rural areas of the developing world. These young people train together in multi-national teams, and return to their home communities (primarily in the developing world) to establish small businesses in areas such as Internet and renewable energy. World Corps is launching its first Pilot Program in India in the southern state of Andhra Pradesh. Starting in January, 25 young people (15 from India and 10 from five other countries) will train together for six months while establishing new community Internet centres in India. The new Internet centres will bring the resources of the Internet to poorer neighborhoods outside large cities. Training topics in other countries will focus on other sustainable, environmentally friendly business enterprises such as renewable energy.

However, other forms of mentoring could also be fostered, involving short-term visits (both ways) and ongoing contact through e-mail. There are a range of resources on the Internet in relation to online mentoring.

Way Ahead

The role of small and medium-sized enterprises in employment, growth and development is now recognized the world over. Entrepreneurship development has played a major role in the growth and development of economies. To achieve maximum gain from entrepreneurial abilities of rural Indians, there is a need to use cutting edge technologies such as ICTs. A more holistic ICT integration will help in networking and building of collaborative entrepreneurial clusters.

The focus of the paper is to show what could be possible with the ICTs blended entrepreneurship development. The cases discussed in the paper were proven to give some insights for future endeavours, there is a great need to work out ICT modules as how best ICTs can be utilized in the entrepreneurship development. The need for harnessing ICTs in entrepreneurship development is never acute than at present.

Success Stories of Entrepreneurship

Chapter 15
Prospects of Indian Food Industry for Entrepreneurship Development

Alok Jha[1] and Tanweer Alam[2]

[1] *Former Professor and Coordinator, Food Science and Technology*
[2] *Former Lecturer, Department of AH and Dairying,*
Institute of Agricultural Sciences, Banaras Hindu University, Varanasi, U.P.

Introduction

India's culinary tradition is constantly changing. With urbanization, rising incomes, more working women and a proliferation of fast food outlets, the acceptance of packaged and ready-to-eat food products is increasing, especially among the urban middle class. Demand for specialty and high-value processed or packaged and ready to serve/ready to cook or reconstitute food items has increased. India is the world's second largest producer of food next to China. India, with diverse agro-climatic conditions, has a production advantage in many agricultural goods, with the potential to cultivate a large range of agricultural raw materials required by the food-processing industry. India is a major producer of spices, spice oils, essential oils, condiments, and fruit pulps. Significant variations in food habits and culinary traditions across the country translate into a competitive advantage for small and medium local players, who are familiar with local food habits and markets. Some Indian food-processing companies have increased market share by decreasing product prices. High import duties on processed food and food ingredients make imports relatively costlier.

The total food production in India is likely to double in the next ten years and there is an opportunity for large investments in food processing, especially in areas of canning, dairy processing, specialty processing, packaging, frozen food/ refrigeration and thermal processing. The industry has witnessed fast growth in most of the segments. In India the food processing sector is one of the largest in terms of production, consumption, export and growth prospects. The important sub sectors in food processing industries are fruit and vegetable processing, fish-processing, milk processing, meat and poultry processing, packaged/convenience foods, alcoholic beverages and soft drinks and grain processing *etc*. The government has also accorded it a high priority to encourage commercialization and value addition to agricultural produce for minimizing post harvest wastage, generating employment and export growth.

Status of Food Industry

Processed Food Sector

The turnover of the total food market is approximately Rs.250,000 crores (US $ 69.4 billion) out of which value-added food products comprise Rs.80,000 crores (US $ 22.2 billion). Primary food processing is a major industry with lakhs of rice-mills/ hullers, flour mills, pulse mills and oil-seed mills. There are several thousands of bakeries, traditional food units *viz*. fruit, vegetable and spice processing units in unorganised sector. The size of the semi-processed and ready to eat packaged food industry is over Rs. 4000 crores (US $ 1 billion) and is growing at over 20 per cent.

In the organised sector, there are over 820 flour mills, 418 fish processing units, 5198 fruits and vegetable processing units, 171 meat processing units and 668 dairy processing units. India is the world's second largest producer of fruits and vegetables, but hardly 2 per cent of the produce is processed. India is the land of spices producing all varieties worth over Rs. 3500 crores (US $ 900million) amounting to 25-30 per cent of world production, which is processed for value-addition and export. It grows 22 million tonnes of oilseeds covering most of the varieties. Other important plantation products include tea, coffee, cocoa and cashew *etc*.

Dairy Processing

It is a matter of pride that India is the number one milk producing country in the world, maintaining the top position since 1988, thanks to successful implementation of the Operation Flood Programmes. World milk production is estimated at 613 million tons growing at a CAGR of 1.3 per cent. Indian production stands at 100 million tons growing at a CAGR of 4-5 per cent. Hence, India contributes 4 million tones to the world's incremental production of 7.5 million tonnes. Despite a higher growth rate, the per capita availability of milk in India (229 grams per day) is lower than the world average (285grams per day).Buffalo milk is now estimated to account for 57 per cent of the total milk production in India (Rajorhia, 2008; Bhasin, 2008).

India has a unique pattern of production, processing and marketing/ consumption of milk, which is unparallel. About 70 million rural households (primarily, small and marginal farmers and landless labourers) in the country are

Table 15.1: Consumption Pattern of Packaged and Processed Food products

Product (gms/person)	2003	2006	Change 2003 to 2006 (per cent/5 yrs)
Canned Meat and Meat Products	0.2	0.242	21.05
Canned Fish/Seafood	n/a	0.1	n/a
Canned Vegetables	0.3	0.317	5.56
Chips/Crisps	5.6	9.958	77.82
Extruded Snacks	5.6	7.568	25.15
Nuts	1.655	2.577	55.74
Butter	39.1	50.607	29.43
Cakes	41.5	57.179	37.78
Bread	1,355.60	1656.543	22.2
Breakfast Cereals	2.2	3.382	53.73
Baby Food	12.4	14.634	18.02
Biscuits	351.1	439.999	25.32
Chocolate Confectionery	19.3	25.939	34.4
Dried Food	283.5	407.871	43.87
Frozen Food	12.4	20.461	65.01
Ice Cream (ml/person)	29.8	47.546	59.55
Noodles	20	20.086	0.43
Oils	575.5	620.159	7.76
Other Sweet and Savoury Snacks	17.1	23.266	36.06
Other Fats	477.7	523.464	9.58
Pasta	0.4	0.705	76.3
Sugar Confectionery	58.3	78.145	34.04
Sauces, Dressings and Condiments	64.1	93.214	45.42
Soup	1	1.425	42.45
Spreads	14	15.672	11.94
Sweet and Savoury Snacks	29.9	43.277	44.74
Yoghurt	1.4	1.933	38.07

Source: Ministry of Agriculture, Govt. of India (2007).

engaged in milk production to change the dairy sector into viable self-sustaining organized sector. About 35 per cent of milk production, over 11 million farmers are organized into about 0.1 million village dairy cooperative societies (DCS). The cumulative milk handled by DCS across the country is about 18 million kg of milk per day. These cooperatives form part of a national milk grid which links the milk producers through out India with consumers in more than 700 towns and cities bridging the gaps on account of seasonal and regional variations in the availability of milk.

Table 15.2: Processing of Milk in India in Organized and Unorganized Sector

Degree of Processing	Type	Volume (MT)	Share (per cent)
Unprocessed	Retention by rural consumers/ sale to rural non-producers	39	45
Processed (organized)	Packed liquid milk	17	19
Value added milk products		5	5
Processed (unorganized)	Value added milk products	22	23
Total		91	100

Table 15.3: Share of Organized and Traditional (Unorganized) Market (Rs. in Billion) during 2005-06

Types of Milk	Organized	Unorganized
Packaged liquid milk	98	–
Ethnic sweets	62.5	455
Curd and curd products		160
Cheese	21 (Paneer)	
Ice Creams	8	–
Butter	60	
Ghee	210	
Milk Powder	38	–
Total	255	906

Source: Ministry of Food Processing Industry, 2006

In India current annual growth rate in milk production is pegged between 4 to 6 per cent. This is primarily due to the initiatives taken by the operation flood programmes in organizing milk producers into cooperatives to turn the dairy sector into viable self-sustaining organized sector. About 35 per cent of milk products in India is processed. The organized sector processes about 13 million tones annually, while the unorganized sector (halwaiis and vendors) processes about 22 million tones per annum. In the organized sector, there are 676 dairy plants in the cooperative, private and government sectors registered. There is huge potential for processing and value addition, particularly in traditional milk products, which are largely sold in unbranded form in the market. The key differences between the organized and the unorganized sector is with respect to investments in preserving the quality of milk, technology used for processing and compliance with food standards. The solution lies in promoting investment in quality control and developing scalable efficient technologies for the unorganized sector.

Health Food

Health food and health food supplements are another rapidly rising segment of food industry which is gaining vast popularity amongst the health conscious. India is one of the world's major food producers but accounts for less than 1.5 per cent of international food trade. This indicates vast scope for both investors and exporters. Food exports in 1998 stood at US $5.8 billion whereas the world total was US $438 billion. The Indian food industries sales turnover is Rs 140,000 crore (1 crore = 10 million) annually as at the start of year 2000. The industry has the highest number of plants approved by the US Food and Drug Administration (FDA) outside the USA.

Value Added Food Product

Although India is a world leader in the production of milk, rice, wheat, fruits and vegetables, livestock, fish and seafood, and eggs, India's current food processing capabilities are very small compared with its agricultural output. Only 8 per cent of total agricultural output is processed into value-added products. However, production of value-added products in India is forecast to grow by 27 per cent by 2025.

Fruits and Vegetable Processing

The installed capacity of fruits and vegetables processing industry has increased from 11.08 lakh tons (1993) to 24.74 lakh tons (2007). The utilization of fruits and vegetables processing is estimated to be around 2.20 per cent of the total production. Over the last few years, there has been a positive growth in ready to serve beverages, fruit juices and pulps, dehydrated and frozen fruits and vegetable products, tomato products, pickles, convenience veg-spice pastes, processed mushrooms and curried vegetables. The domestic consumption of value added fruits and vegetable products is also low compared to the primary processed food in general and fresh fruits and vegetables in particular. Hardly 2 percent of fruits and vegetables produced in India are processed.

Table 15.4: Major Products, Industry Size, and Major Players

Products	Industry Size (Million Rupees)		Key players in the organized segment
	Organized	Unorganized	
Jam	900	450	HLL, Marico, Mapro, Malas
Pickles	1,500	10,000	Priya Foods, Preveen, Desai Brothers, Cavin Kare, GD Foods
Sauce/ketchup	1,000	3,000	HLL, Nestle, Heinz, GD Foods, Bector Food Specialties
Pulp/concentrate	4,000	0	Foods and Inns, BEC, Claen Foods, Jain Irrigation, Usha International
Juices/fruit drinks	5,000	0	Pepsi, Dabur, Parle, Godrej, Mother Dairy
Squashes	1,000	2,000	HLL, Haldiram, Mapro
Ready-to-eat vegetables	1,000	0	ITC, MTR, Tasty Bite
Potato chips	2,500	3,000	Pepsi, Haldiram
Cooking paste	300	0	Dabur, HLL
Total	35,650		

Source: Ministry of Agriculture, Govt. of India (2007).

Meat and Meat Processing

India's livestock population is largest in the world with 50 per cent of world's buffaloes and 20 per cent of cattle, but only about 1 per cent of total meat production is converted to value added products. In meat and meat processing sector, poultry meat is the fastest growing animal protein in India. The estimated production is 1500 thousand tones growing at CAGR of 13 per cent through 1991-2005. Per capita consumption has grown from 870 grams in 2000 to about 1.68 kg in 2005. This is expected to grow to 2Kg in 2009. Buffalo meat production has been growing relatively less rapidly at a CAGR of 5 per cent in the last 6 years. The current production levels are estimated at 1.9 million MT. Of this about 21 per cent is exported. Mutton and lamb is relatively small segment where demand is outstripping supply, which explains the high prices in domestic market. The production levels have been almost constant at 950,000 MT with annual exports of less than 10,000 MT. This has restricted large processing companies from developing business interests in this sector. Indian consumer prefers to buy freshly cut meat from the wet market, rather than processed or frozen meat. A mere 6 per cent of production (about 100,000 MT) of poultry meat is sold in processed form. Of this, only about 1 per cent undergoes processing into value added products (Ready to- eat/Ready-to-cook). Processing of large animals is largely for the purpose of exports. The total processing capacity in India is over 1 million MT per annum, of which 40-50 per cent is utilized. India exports more than 500,000 MT of meat of which major share is buffalo meat. Indian buffalo meat is witnessing strong demand in international markets due to its lean character and near organic nature. India is the 5th largest exporter of bovine meat in the world. Indian buffalo meat exports have the potential to grow significantly. Due to emerging health threats of the diseases communicable to human through meat, the meat consumers are more vigilant towards the wholesomeness of the meat and demanding meat and poultry products processed in clean and sanitary environment. In metros and urban areas there are upcoming demands for "convenience items" such as semi cooked, ready-to-eat, ready-to-cook meat food products. Processing of meat products is licensed under Meat Food Products Order (MFPO) 1973. The main objectives of the MFPO, 1973 are to regulate production and sale of meat food products through licensing of manufacturers, enforce sanitary and hygienic conditions prescribed for production of wholesome meat food products, exercise strict quality control at all stages of production of meat food products, fish products including chilled poultry *etc.*

Table 15.5: Total Quantity of Meat Exported and Revenue Earned

Particulars	Total Value of Processed Meat in Organized Sector (Rs. In Crore)		
	2003-04	2004-05	2005-06
Meat Products	914	986	1025
Poultry meat	11645	12021	11118
Buff Meat	24038	25519	25290
Beef	2477	2830	2822
Mutton	8534	9226	9754

Based on figures available with NSSO, 2007.

Table 15.6: Status of Meat Processing Industry (Quantity in MTs and Rs. in Crore)

Meat Product	2002-03		2003-04		2004-05		2005-06	
	Qty.	Value	Qty.	Value	Qty.	Value	Qty.	Value
Buffalo meat	297897.26	1305.45	343817.08	1536.77	37777.65	1774.52	459937.63	2629.57
Sheep/Goat me	4973.55	39.95	16820.53	110.39	9024.49	81.27	7177.51	80.37
Poultry Products	26450.01	156.47	415228.17	202.40	277744.46	160.79	145889	167.58
Processed Meat	669.48	4.8	986.13	7.63	1359.7	9.45	256.04	2.43

Source: APEDA, 2007.

Fisheries Processing

India has large marine product and processing potential with varied fish resources along the 8041 km. long coastline, 28000 km. of rivers and millions of hectares of reservoirs and brackish water. At present, there are over 369 freezing units with a daily processing capacity of 10266 tons out of which 150 units are approved for export to EU. 499 units are engaged in production of frozen fish with a total storage capacity of 134767 tons. Apart from the above there are 12 surimi units, 5 canning units and 473 units are engaged in pre-processing and dry fish storage.

Table 15.7: Status of Fish Production in MT

Year	Marine	Inland	Total
2000-01	2.81	2.84	5.65
2001-02	2.83	3.12	5.95
2002-03	2.99	3.21	6.20
2003-04	2.94	3.45	6.39
2004-05	3.01	3.50	6.51

Source: Ministry of Agriculture Govt. of India 2007.

However, in 2004-05, export of marine products achieved a record 12 per cent increase in volume and 11 per cent increase in US$. Frozen shrimps constitute 63.5 per cent of total value of exports. USA is the largest importer of Indian marine products contributing 13.21 per cent in quantity and 29.8 per cent in value of total exports importing mainly frozen shrimps – HL Black tiger shrimps.

Table 15.8: Exported Revenue Earned from Fish and Fish Product

Year	Quantity (MT)	Value (Rs. Crore)
2000-01	440473	6444
2001-02	424470	5957
2002-03	467297	6881
2003-04	412017	6092
2004-05	461329	6647

Source: Ministry of Agriculture Govt. of India 2007.

Ethnic Food/Street Food

The concept of traditional food street has now been given a new meaning in developed countries, with food streets emerging as new tourist attractions. Major tourist destinations invariably have food streets offering exotic local foods. Street food in India has always been popular because of its affordability and convenient availability. This is an area with huge potential for increased income generation for vendors and for giving a boost to the tourism sector. This may also lead to popularity of local food traditions, offering fast food, which are very popular among Indians.

Strength Indian Food Industry

Although India's enormous population size makes it an attractive market for food manufacturers and suppliers, the majority of people do not have the necessary purchasing power to accommodate a large percentage of foreign foodstuffs. However, the strengthening Indian economy shows promise for potential exporters. According to Lynch (2007) reported that 22 per cent of Indian households (44 million households) are expected to be capable of purchasing packaged foods. In addition, there is demand for high-cost processed foods by India's ever-increasing consumer class.

Weakness Indian Food Industry

In an effort to comply with World Trade Organization (WTO) standards, India removed all import licensing restrictions effective April 1, 2001. The small-scale farming system in India, marketing problems, lack of grading and standards, poor distribution channels, and onerous government policies continue to pose problems for the processing industry to source the right type of raw materials and to discourage more investment in the sector

Opportunities of Indian Food Industry

Processed foods that have done well in market include cheese, butter, ice cream, bakery products, ready-to-make foods, biscuits, chocolate, tea and milk products. Other processed foods that have good market potential in India include condiments and sauces such as ketchup, jams, jellies and pickles, in addition to healthy foods and beverages, fruit juices, cereals, confectionary, snack foods and curry powders. Opportunities also exist for the export of processing and packaging technologies and storage and transportation alternatives as the Indian food processing industry strives to capitalize on current food surpluses and to increase its food processing capabilities to become a more prominent player in the global food market. However, the processed food market is expected to grow with continued strengthening of India's economy. India's significant economic growth of 8.2 per cent in 2003 and 6.0 per cent in 2004 is forecast to continue to rise another 6.5 per cent in 2007-8. Furthermore, increased urbanization, rising incomes and changing lifestyles in which time and convenience are valued will drive demand for packaged foods and easy-to-prepare meals. Expected with this increased demand is a corresponding demand for healthy ready-to-eat foods. The market for semi processed/cooked ready-to-eat meals in India grew 20 per cent during the 2003-2004 fiscal year. Between 2000 and 2003, demand for snack foods increased each year by 10 per cent. India's market for processed foods is expected to more than double its value by 2015, amounting to almost $340 billion. Several multinational companies, including US-based companies like Pepsi, Coca Cola, ConAgra, Cargill, Heinz, Kellogg's, IFF, and Mars (pet food only) have entered the Indian food-processing industry with significant investments. Indian food and beverage companies are expanding their operations to neighboring countries like Bangladesh, Nepal, Sri Lanka, Commonwealth of Independent States countries, and the Middle East. Takeovers and mergers are beginning to occur in the Indian food-processing sector, leading to consolidation.

Threat of Indian Food Industry

Even with import restrictions removed, imports face regulation, taxation and bureaucratic barriers. Import tariffs on consumer food items range from 35 per cent to 56 per cent. Moreover, India's retail sector is primarily unorganized, and its infrastructure (especially roads and refrigerated cold chains) is underdeveloped. Although imported products are affordable to only approximately 25 per cent of the Indian population, the number of mid- to high-income earners is growing, and they are developing a taste for imported products. Since 2001, Agri-food imports have increased steadily, rising 14.2 per cent through to 2003, reaching a value of $5.4 billion. Imported processed food products that have had success in this market were typically altered to cater to local tastes–in some cases, regional spices were used to make products more attractive to the Indian market. Furthermore, smaller packages are offered to accommodate Indian consumers' limited purchasing power and storage abilities. Examples of processed food items currently imported into India include ketchup, fruit juices, chocolates, biscuits, chocolate syrup, cake mixes, canned soups, pop corn, potato chips, canned fish, ice cream and canned corn.

Conclusion

The food-processing industry in India has undergone sea changes over the last few years, in terms of types, variety, quality, and presentation of products, which is mainly a result of the liberalization that led to foreign direct investment (FDI) in the processed food sectors. The growth in the food-processing sector has generated increased interest in high quality food ingredients in order to produce high quality processed and packaged foods. The ready-to-eat food sector is growing at a high rate due to the changing lifestyles of the middle-class consumers (both partners working, *etc.*). Thus increasing number of fast food chains. The recent trend toward a healthier lifestyle has generated a niche market for diet, healthy, low-calorie, and non-fat food products.

Chapter 16

Entrepreneurship of Ages - Quality Seed

A.K. Joshi

*Professor, Department of Genetics and Plant Breeding,
Institute of Agricultural Sciences, Banaras Hindu University,
Varanasi – 221 005, U.P.*

The search of good quality seed is as old as agriculture. During nomadic days, women used to collect quality seeds while men hunted. Gradually, good seeds of crops provided man a settled life. In India, older civilizations like Mohanjodaro and Harrappa used to have big stores for seeds and grains. For a long time Indian villages used to live like a self sufficient unit where seed played an important role. With the loosening of self sufficiency of Indian villages under British rule, there emerged a formal sector of germplasm dissemination and adoption. This process began in 19th century and got a vigorous expression through Green Revolution of 1960s. Although a substantial network of organized (both public and private) sector do exist for germplasm dissemination and adoption in India, the seed replacement rate is still less than 15 per cent. This figure is more distressing in the eastern parts of Indo-Gangetic plain. It is accepted that for proper dissemination and adoption of germplasm, both public and private sectors need to be strengthened. This would be achieved faster if entrepreneurship is encouraged in the seed sector. Since, quality seed production requires some knowledge of the intricacies of improved seed production and its marketing, it is believed that greater scientist-farmer interaction and creating of strong linkages with other stake holders could play a crucial role to meet this objective. Through participatory research conducted by Banaras Hindu, promising results have been obtained in the dissemination of seed production technology among farmers of eastern Uttar Pradesh and the bordering districts of Bihar. Quality seed production has been initiated in around 100 villages under participatory net work of the university.

Introduction

The history of improved seed is considered to be as old as agriculture. The first entrepreneurs of quality seed were probably those who fashioned our crops. In this venture, women played the leading role since they used to gather good quality seeds and fruits for their families, while men hunted. The FAO estimates that around 1.5 billion farmers still save and breed their own seeds and seed selection is often women's job (FAO, 2004). The importance of seed was known to Indian farmers since long. The "Manu Smriti" an old scripture of India says, "subeejam sukshetre jayate sampadyate" meaning there by "good seed in good soil yields abundant". The Indus valley civilization gives strong evidence of the presence of big seed stores to save food during distress years and also the precious seed for the next crop season. The famous explorers like Columbus (1451-1506) and Vasco de Gama (1469-1524) could also be called seed entrepreneurs as they searched for new seeds of valuable crops such as spices to earn more gold for their own country. Likewise, the development of new varieties, hybrids and composites by a number of crop researchers in last few centuries was also nothing but the emergence of new seed entrepreneurships. In modern agriculture, seed is a vehicle to deliver almost all agriculture-based technological innovations to farmers so that they can exploit the genetic potential of new varieties. The availability, access and use of seed of adaptable modern varieties is, therefore, determinant to the efficiency and productivity of other packages to enhance food security and alleviating rural poverty in a developing country like India. Hence, importance of seed is growing and growing.

History of Germplasm Adoption and Dissemination System

Historically, India's germplasm dissemination system through good quality seed used to be a part of the system where each village was like a self sufficient unit. This self sufficiency along with rich ecological variability across the Gangetic plains played a crucial role in sustaining agriculture. Since the beginning of farming in India, farmers have saved part of their harvest as seed and exchanged seeds with neighbours. Farmer's indigenous varieties have been the basis of Indian ecological and food security. However, with time changed the system. The self sufficiency of Indian villages no more existed under foreign rule. The Famine Commission report of 1810 indirectly generated the awareness about the importance of seed in Indian agriculture. This report also led to the establishment of Department of Agriculture in the centre, of Imperial Agricultural Research Institutes at Pusa (Bihar) and of five major agricultural colleges at different parts of the country. The Royal Commission on Agriculture (1928) is considered the first major effort to analyze the needs and problems of seed production in India.

After independence, agricultural scene of India got transformed. This change was propelled by improved seed of different crops and led to Green Revolution in mid sixties. In 1960, the first Agricultural University was established at Pantnagar, U.P. (now in Uttaranchal) which later provided an ideal example of improved seed production in India. During the same time several All India Coordinated projects were initiated, the first on being that of Maize in 1957. To promote quality seed production, National Seeds Corporation (NSC) was established in 1963. All these led to the emergence of Indian Council of Agricultural Research (ICAR) in 1965 as a central agency to coordinate and promote agricultural research and education in the country. The status of seed industry was reviewed by the Seed Review Team in

1968 and in 1975 the project report on National Seeds Project (NSP) was submitted. Consequently, NSP was launched in several Indian Universities and centres in 1978.

The first major effort in the direction of seed certification and seed law was taken in the year 1966 when Indian Seeds Act was enacted by the parliament of India. This law came into force on 2nd October, 1969. The objective of this act was to regulate the quality of seeds of all the crops being marketed for agricultural purposes. The two important features of this act were, compulsory labeling and voluntary certification. This act was amended in 1972 and again reviewed in 2004. Presently, the revised seed laws are under review by the Standing Committee on Agriculture of the Indian Parliament. The important points added in 1972 were, provision of notification of varieties, establishment of state seed certification agencies for seed certification and constitution of Central Seed Committee to advise government on the matters arising out of administration of this act. The 2004 amendment includes Farmers Rights, the first of its kind in any country. This act includes Compulsory Registration of varieties provided they meet the criteria of novelty, distinctiveness, uniformity and stability. So far, farmer varieties and indigenous agro-biodiversity has been registered by Local Biodiversity Committee through Community Biodiversity Registers (CBRs). The objective the 2004 Act states that it is aimed to replace farmers saved seeds with seeds from organized sector. However, it gives farmers the freedom 'to save, use, sow, resow, exchange, share or sell their farm produce, including seed (except for branded seed) of a variety protected under this Act in the same manner as they were entitled before the coming into force of this Act.

Current Scene

There exists two broad systems of germplasm dissemination and adoption in India, the formal (organized) and informal (unorganized). Modern varieties of crops are the backbone of the formal seed industry which is almost equally shared by public and private sectors. Private sector takes more interest in cross-pollinated and in low bulk crops where hybrids are common. National Agricultural Research System (NARS) plays a major role in germplasm conservation, variety development and for generating appropriate technologies to utilize the yield potential of new varieties. The new varieties are passed through a series of evaluation and release tests before farmers can use them.

Although, it is realized that new improved varieties developed by NARSs should be multiplied and made available to farmers in the shortest possible time to realize their benefits, it does not always happen in practice especially in areas having weaker linkages. Therefore, more than 80 per cent per cent of all seed in India is still saved by farmers. After planting seeds, every peasant in India says: "Let the seed be exhaustless, let it never get exhausted, let it bring forth seed next year". Farmers still have such pride in saying "this is the tenth generation seeds that I'm planting", "this is the fifth generation seed that I'm planting" (Shiva, 1998). This tendency of saving seed is more in self-pollinating crops like wheat.

Limitations of Seed Sector in Indo Gangetic Plains

The eastern part of Indo-Gangetic plains is considered to be lagging behind the western part in seed replacement rate. In this part, the availability of experienced seed growers, supported by public or private sector is limited. Farmers are resource poor and possess low profitability. Therefore, farmers saved seed is being used to the extent of 90 per cent or even more. Private sector does not take interest in crops like wheat and rice that are high-volume, low-profit seed crops and in organized sector it is produced primarily by subsidized government seed programmes. Since wheat and rice are self-pollinating crops and the grain can be used as seed, farmers tend to replant their own seed. It is expected that in future also, a large majority of resource-poor, small-scale farmers will have to rely on seed saved from the previous harvest of their own. However, this may not apply for hybrid rice varieties that are expected to gain more popularity in water limited areas.

Options for Germplasm Dissemination and Adoption

It is believed that three options exist for the betterment of germplasm dissemination and adoption. The first two pertain to strengthening of public and private sectors through a vigorous policy planning and implementation. The third option relies on the principle, "self help is the best help". However, this approach also needs an organized effort for the promotion of a new model that can provide a better flow of information to farmers and help them in becoming self reliant. It has been suggested that participatory research is capable of solving the problems of the farmers that very often are not realized due to vast diversity in farmer's fields (Ortiz-Ferrara *et al.*, 2000; Witcombe *et al.*, 2001; Joshi *et al.*, 2007a, b). The participatory research is expected to complement the on-going research in strengthening farmer's hands (Witcombe and Virk, 2001; Witcombe *et al.*, 2003; Kosina *et al.*, 2007). It is a known fact that unlike present days, early crop improvement research got initiated when farmers started selecting varieties of different crops based on their experience. Farmers not only developed useful varieties but also conserved this material for future generations. In addition, they also developed other useful technologies related to crop production, plant protection, seed storage and utilization of crop produce.

Some Points to Remember in Quality Seed Entrepreneurship

Any entrepreneurship needs some basic knowledge before taking it as a business. In case of quality seed, an entrepreneur must be clear about the meaning of improved seed. An Improved Seed is the seed of a notified variety and carries genetic and physical purity and germination percentage as per the norms of minimum seed standards. However, any other variety (may be land race or local variety) which is popular among farmers can be taken under seed production net work. Among various classes of seed, large scale seed production is generally targeted for two classes of seed - Foundation and Certified. It can also be the Truthful Seed in which a person has to state through a tag attached to the bag (called truthful tag), that the seed is of standard quality. All seed production programmes require a good source of seed as base material followed by its production in the field and processing. The Foundation and Certified seed fields must be registered with State Seed Certification Agency for its proper monitoring at appropriate stages and release of Foundation or Certified seed tags by the agency after verification of the quality of the processed seed. The entrepreneurship on seed becomes easier if there is good linkage with agencies associated with seed marketing such as National Seeds Corporation, private sector, NGOs, farmers cooperatives *etc*. According to new Seed Law, any marketing of quality seed shall demand registration of varieties and license for the sale.

Experimenting Seed Entrepreneurship through Participatory Research at BHU

At Banaras Hindu University, the farmers-participatory research got initiated in the *Rabi* (winter) crop season 1997-98 with a multi-pronged approach in which quality seed of better varieties was the central (Joshi *et al.*, 2003). In the participatory approach, greater stress was given on exchange of information between farmers and scientists. For this, a large number of Mother-Baby trials were conducted in different villages for providing better options to farmers. The mother trial usually consisted of a few (around 10-12) varieties previously not grown in the target environment. This included the popular check variety of the location. Around the site of mother trial, a number of baby trials of varieties identified by farmers were grown. In addition, farmers were also trained for seed production activities.

Our experience of working in the participatory mode proved highly successful. Through this approach, farmers got ample opportunity to judge varietal options and jumping technologies like zero-till which are proving a boon for eastern India. The seed sector was also strengthened by creating a net-work of seed villages. Farmers also got benefited through better marketing due to improved linkages. Following one such approach, farmers of District Mirzapur, UP, were able to select varieties providing better profitability (Ferrara, 2001).

The participatory work which has now spread to around a dozen districts included quality seed production at all the sites. At some locations farmers have started their own seed business. This has been facilitated by creation of strong linkage by the university with agencies such as National Seeds Corporation. For instance, at village Bhurkura, district Mirzpuar, two entrepreneurs Mr. Bhola Nath Singh and Mr. Indrasen Singh, started seed business (Shyam Seeds and Pooja Seeds, respectively) in the year 2007 by installing two seed processing plants, each of 10,000 qts per annum capacity. These farmers joined the participatory net work of Banaras Hindu University in the year 2001 and by the year 2007 their village got transformed in to a seed village with almost 100 per cent wheat area under quality seed production. Likewise, Mr. Anil Singh, a medium farmer of village Karhat (Mirzapur) has opened a seed center in the name of "Sooraj Seeds". At many places, farmer cooperatives have emerged. The notable examples are Pidkhir (Mirzapur), Raghunathpur (Chandouli) and Bhabhua (Bihar). At Pidkhir, a farmer seed society, "Malviya Krishak Samiti" (Malviya Cooperative Society) was established in the year 2004, where as at Bhabhua (Bihar), the seed cooperative KAFICO is strengthening farmers in a big way. The farmers society named "Chandrapabha Krishak Society" at village Raghunathpur, district Chandouli is targeting quality seed production in a range of crops including cereals, pulses and vegetables. Our survey of seed production activity in the Varanasi division indicates that from around a dozen seed processing plants around ten years ago, the number of such plants has risen to around sixty in the year 2007. This growth of seed sector in and around Varanasi is indicative of the huge awareness generated by Banaras Hindu University about the quality seed.

At most of the working sites, participatory research helped farmers to attain around one t/ha advantage through an appropriate combination of zero-till and suitable varieties. The most significant impact was felt in the district Chandouli, where wheat is grown in around 95,000 ha. In this district the average wheat yields showed an increase of 0.6 tons per ha in last five years – 1.6 t/ha in 2002 to 2.2. t/ha in 2007. It is also interesting to note that due to participatory approach where a package of technology was given to farmers along with quality seed, the area under zero till and surface seeding also increased significantly. In Chandouli district alone, where zero tillage was introduced in the year 2001-02, around 60 per cent wheat area came under zero tillage in the current crop season 2007-08. This district, which has round 95000 ha under wheat, has shown fastest growth in zero tillage adoption for wheat in the whole country. Likewise, surface seeding that was mostly known in pulses has also grown in acreage due to low cost and higher profit. In Mirzapur district, around 10,000 ha wheat area is under surface seeding in locations adopted by Banaras Hindu University. All this has enabled farmers to reduce their cost (of around Rs. 2000/ha) and get an extra profit (of around Rs. 3000/ha) through quality seed. This change has brought greater confidence, a change in the mindset and a better linkage with farmers. It was also felt that the average time of ten years taken in spreading a good agri-technology in the eastern Gangetic Plains can be reduced to 3-4 years through an effective interactive approach.

In the crop season *Rabi* 2007-08, Banaras Hindu University initiated a "Buy-Back Scheme" of quality seed from farmers. This approach is broadly based on the approaches followed by professional private and public sector organizations having expertise in quality seed production. The interested farmers are encouraged to procure Breeder or Foundation seed of different crops and their seed plots are monitored by a team of university scientists. The seed lots that meet "Minimum Seed Standards" are procured at appropriate prices and sold with nominal profit after proper processing. This gives benefit to seed producers as well as seed purchasers helping the farming community as a whole. In the first crop season itself, around one thousand quintals of quality wheat seed was procured from farmers and disseminated to remote areas. This approach is expected to meet out huge demand of quality seed in the eastern Gangetic Plains and thereby shall improve seed replacement rate and the national food security.

Conclusion

The present trends indicate that quality seed is one of the most simple and promising sector for entrepreneurship. It is profitable as well as satisfying as it provides social service to our farm sector where seed is so crucial and linkages so poor.

Acknowledgements

Support provided by our dearest farmers; ICAR, New Delhi; CIMMYT (Mexico); DFID (UK); CAZS (UoB, UK) and other agencies in helping us to initiate and harvest the gains of participatory research is gratefully acknowledged.

References

FAO (2004) Statistical database. www.fao.org. Verified 20 Feb, 2008.

Hobbs, P. and M. Morris, 1996. Meeting South Asia's future food requirements from rice-wheat cropping systems: Priority issues facing researchers in the post-Green revolution era. NRG paper 96-01, pp 46. CIMMYT, Mexico, D.F.

Joshi, A.K., R. Chand and V.K. Chandola, 2003. Ist Annual Report of CIMMYT collaborated, DFID funded project, "Participatory Research to Increase the Productivity and Sustainability of Wheat Cropping Systems in the Eastern Subcontinent of South Asia", P 3067, Banaras Hindu University, Varanasi, India.

Joshi, A.K., B. Mishra, R. Chatrath, G. Ortiz Ferrara and Ravi P. Singh, 2007a. Wheat improvement in India: present status, emerging challenges and future prospects. *Euphytica* 157:431-446

Joshi, A.K., R. Chand, B. Arun, R.P. Singh and Rodomiro Ortiz, 2007b. Breeding crops for reduced-tillage management in the intensive, rice-wheat systems of South Asia. *Euphytica* 153: 135 151.

Kosina, P, M. Reynolds, J. Dixon and A.K. Joshi, 2007. Stakeholder perception of wheat production constraints, capacity building needs and research partnerships in major wheat producing developing countries. *Euphytica* 157:475-483

Nagarajan S., 2005. Can India produce enough wheat even by 2020. *Current Sci* 89: 1467-71.

Ortiz-Ferrara, G., M.R. Bhatta, T. Pokharel, A. Mudwari, D.B. Thapa, A.K. Joshi, R. Chand, D. Muhammad, E. Duveiller and S. Rajaram, 2001 Farmers participatory variety selection in South Asia. Research highlights of the CIMMYT wheat program, 1999-2000, Mexico, D.F. pp 33-37.

Ortiz Ferrara, G., A.K. Joshi, R. Chand, M.R. Bhatta, A. Mudwari, D.B. Thapa, M.A. Sufian, T.P. Saikia, R. Chatrath, J.R. Witcombe and D.S. Virk, 2007. Partnering With Farmers to Speed Up the Adoption of New Technologies in South Asia. *Euphytica* 157:399-407

Siva, V., 1998. An interview with Vandana Shiva, In Motion Magazine, www.inmotionmagazine.com/shiva.html. Verified 20 Feb, 2008.

Witcombe, J.R. and D.S. Virk, 2001. Number of crosses and population size for participatory and classical plant breeding. *Euphytica* 122: 451-462.

Witcombe, J.R., A. Joshi and S.N. Goyal, 2003. Participatory plant breeding in maize: A case study from Gujarat, India. *Euphytica* 130: 413-422.

Witcombe, J.R., K.D. Joshi, R.B. Rana and D.S. Virk, 2001. Increasing genetic diversity by participatory varietal selection in high potential production systems in Nepal and India. *Euphytica* 122: 575-588.

Chapter 17

Project Formulation for Establishing an Enterprise

V.K. Yadav[1] and Basavaprabhu Jirli[2]

[1]*Senior Scientist, National Institute of Biotechnology, Ranchi*
[2]*Associate Professor, Department of Extension Education,
Institute of Agricultural Sciences, BHU, Varanasi, U.P.*

Entrepreneur needs project report for establishing an enterprise. A project report provides all the necessary information of the enterprise proposed to be set up. It is also required by various developmental agencies that helps in setting up of enterprise particularly demanded by bank for providing financial assistance. Project report provides following other advantages:

- ☆ A plant of action
- ☆ A blue print for the future
- ☆ Reduces risk of failure
- ☆ Helps in identifying market and customers
- ☆ Shows requirement of land, labour, capital *etc.*
- ☆ Saves time and energy of entrepreneur, banker *etc.*
- ☆ Gives confidence that enterprise will succeed.

The size and type of enterprise will be the deciding factors for the amount and type of information to be included in the project report. Generally following information should be given in the project report.

1. **General information:** The following aspects should be covered in it:

   a. **Bio-data of entrepreneurs:** It should include name, address, qualification, experience and other capabilities of the entrepreneurs.

   b. **Enterprise profile**: Past performance, present status of enterprise, the way it is organised, the problems it faces *etc.* should be mentioned.

    c. **Constitution and organisation**: Whether enterprise or firm is registered with the Registrar of Firms. Whether a registration certificate from the Directorate of Industries/District Industries Centre has been obtained or will be applied later on.

    d. **Product details**: The utility and range of the products to be prepared should be mentioned.

2. **Project Description:**

    a. **Site:** Location of site, whether owned or leased land, whether it is suitable to the type of enterprise being planned *etc*. should be mentioned. If the location is in a residential area then the copy of No Objection Certificate from the Municipal Authorities should be attached.

    b. **Physical Infrastructure**: It consist of following items:

      i. **Raw Material**: Source of raw material, what is the probability of getting it on a continuous basis at reasonable price *etc* should be mentioned.

      ii. **Labour**: Whether labour is available in that area? What arrangements have been made to train the labour in various skills?

    c. **Utilities**: It consists of following items:

      i. **Power**: Project report should mention the information regarding the power requirements, the load sanctioned, stability of supply of power and price at different consumption levels.

      ii. **Water**: Whether it is required or not for particular enterprise. If required, the source and quality of water should be clearly stated.

    d. **Pollution Control**: Many industries produce waste material or emissions (eg. gases, liquid, solid discharge *etc*.) that may cause problems. Therefore, scope of dumping wastes, sewage system, sewage treatment plant *etc* should be clearly mentioned in the project report.

    e. **Communication System**: Availability of telephone fax, teleconferencing *etc* should be stated in the report.

    f. **Transport Facilities**: Means of transportation of raw material, finished products *etc* should be mentioned.

    g. **Manufacturing Process**: A process flow chart should be presented. The details of production and the process involved should be mentioned.

    h. **List of Machinery and Equipments**: A complete list of machinery and equipment, its source of availability, prices *etc* should be mentioned.

    i. **Capacity of the Plant**: The installed licenced capacity should be stated.

    j. **Technology Selected**: Whether it is up to date and appropriate. How is the required know how proposed to be arranged?

    k. **Quality Control/Testing and Inspection**: Whether some system has been designed to check the quality of products on a continuous basis? Obtaining quality marks like 'ISI', 'Q' or Agmark help in creating confidence among consumers.

    l. **Research and Development**: Whether any cell to study improvement of quality is proposed to be formed in the enterprise?

3. **Market Potential**: It should cover following aspects in the project report:

    a. **Demand and Supply Position**: Gap in demand and supply of products should be mentioned. How much of this gap will be fulfilled by establishment of enterprise.

    b. **Marketing Strategy**: Marketing of products is essential. What strategy would be adopted for selling products.

    c. **After-sales Service**: It is very important in some cases. Customers find difficulties due to occurrence of small problems in machine. It has been found that money spent on after sales service is repaid many times in the long run due to faith of customer established in the product/machine.

    d. **Seasonality Factor**: If products have seasonal fluctuations in sales, the arrangement should be made for storage of goods in off-season.

    e. **Transportation**: Products would be transported on own vehicle or public carrier should be mentioned.

4. **Capital Costs and Source of Finance**: Capital should cover following items:

    ☆ Land and building

    ☆ Plant and Machinery

    ☆ Installation costs

    ☆ other assests like furniture/fixtures, vehicles *etc.*

    ☆ Preoperative expenses

    ☆ Contingency cushion against price rise/unforeseen expenses

    ☆ Working capital

The source of funds for above mentioned items should be mentioned in the project report. The sources would include the owner's fund together with loans and funds expected from financial institution/banks. This estimation should be realistic and correct.

5. **Assessment of Working Capital Requirement**: Any enterprise or unit will be able to function only when adequate working capital funds are available. Entrepreneurs should present working capital requirement in prescribed format. It will save time and energy of banker and entrepreneur both.

6. **Other Financial Aspects**: Projected profit and loss account, projected balance sheet, cash flow statements *etc.* should be prepared. Break-even-analysis should be worked out. Break even point is the level of production/

sales where the industrial enterprise shall make no profit no loss. This facilitates knowing the gestation period and likely moratorium required for repayment of loans.

$$\text{Break even point} = \frac{\text{Fixed costs}}{\text{Sales projected} - \text{Variable costs}} \times 100$$

In addition to this, following ratio should also be mentioned.

a. $\text{Profitability Ratio} = \dfrac{\text{Net profit}}{\text{Sales}} \times 100$

b. $\text{Return on investment} = \dfrac{\text{Net profit}}{\text{Capital Employed}} \times 100$

c. $\text{Debt, Equity Ratio} = \dfrac{\text{Debt}}{\text{Equity}} \times 100$

7. **Economic and Social Variables**: Whether the enterprise will have some socio-economic benefits. Following are few examples:

   a. Employment generation
   b. Import substitution
   c. Exports
   d. Local resource utilization
   e. Development of the area

8. **Other details:**

   a. **Project implementation schedule:** Preferably PERT/CPM chart should be appended to the project report.

   b. **Plant layout:** If possible, a copy of the plant layout should also be mentioned in the project report. It will determine sufficiency of area for present and future expansion requirements.

   c. **Annexures:** Following list should be appended along with a project report:

      i. Equipment required
      ii. Details of preliminary expenses
      iii. Estimate cost of production and profitability
      iv. Labour and staff requirement
      v. Break-even point
      vi. Working capital requirement
      vii. Cash flow statement

A Sample Project Proposal on Cultivation and Value Addition of Baby Corn

Introduction

Baby corn is a young finger like unfertilized cobs of maize with one to three centimeter emerged silk preferably harvested within 1-4 days of silk emergence depending upon the growing season. The size of the baby corn depends upon the variety, population density and the crop season. The desirable size of baby corn is 6 to 10 cm length and 1.0 to 1.5 cm diameter with regular row arrangement. The most preferred colour by the consumer/exporter/processing unit is generally

light yellow. Three to four baby corn crops can be taken in a year. Additional income may be generated through intercropping in baby corn. It is highly nutritious and delicious and used for preparation of several recipies *e.g.* salad, soup, vegetables, pickles, candy, kheer, burfy, murabba *etc*.

Market

It has domestic as well as international market. Generally raw baby corn is sold in mandis of big cities. It has demand even in rural areas during marriages and other functions for preparation of salad, soup, vegetables, kheer *etc*. Canned baby corn, pickles, candy *etc*. may be exported. Major baby corn importing countries are USA, Hongkong, Singapore, Australia, Malysia, Canada, Saudi Arabia, New Zeeland, European countries *etc*.

Requirements

Land, labour, capital, production technologies and inputs(*e.g.* seed, fertilizer, weedicide, irrigation *etc*), processsing plant, polythene packets and canning bottles are required for establishing this enterprise.

Technical Aspects

A. Baby Corn Cultivation

The cultivation of baby corn is more or less like maize grain crop with the following variations

☆ Higher plant population

☆ Higher dose of nitrogen application because of higher plant population

☆ Preference for early maturing single cross hybrid

☆ Harvesting within 1-4 days of silk emergence depending upon season

Impact Points for Obtaining Higher Baby Corn Yield

☆ **Selection of suitable varieties:** Short duration, prolific single cross hybrid with medium height. HM-4, Prakash, Pusa Early Maize Hybrid-3,4,5 *etc* are suitable varieties.

☆ **HM-4** hybrid posses the traits suggested for an ideal baby corn hybrid. This single cross hybrid has an attractive creamish to light yellow color with standard size baby corn having regular row arrangement. It has the most desirable medium height plant type. Three to four pickings without affecting size and quality of baby corn can be taken up. The pickings are almost completed within 10 days.

☆ **Sowing time:** Round the year throughout country except December-January in northern India.

☆ **Sowing method:** Sowing should be done on southern side of the ridges with plant spacing of 60 cm x 15 cm or 60 cm x 20 cm depending upon plant type(erect/spreading).

☆ **Seed rate:** 20-22 kg/ha depending upon the test weight of the variety.

☆ **Fertilizer application:** Application of **120:60:60:25** kg/ha NPK and $ZnSO_4$ with **8-10** tons/ha organic manures. In kharif, Nitrogen should be applied in three split doses: 1/3 N and all P, K and $ZnSO_4$ at the time of sowing, I/3 N after 25 days and remaining 1/3 N after 40 days of sowing. However, in rabi, Nitrogen should be applied in four split doses: 1/4 N and all P, K and $ZnSO_4$ at the time of sowing, I/4 N after 30-35 days, 1/4 N after 60-90 days and remaining 1/4 N after 80-110 days of sowing. In spring, Nitrogen should be applied in four split doses: 1/4 N and all P, K and $ZnSO_4$ at the time of sowing, I/4 N after 25 days, ¼ N after 40-45 days and remaining 1/4 N after 60-65 days of sowing.

☆ **Weed management:** Pre-emergence spray of solution of Atrazine @ 1000-1500 gm/hectare in 500-600 litre of water followed by 1-2 hoeing with khurpa depending upon the weed.

☆ **Irrigation management:** Apply irrigation as and when required. Pre-flowering is the critical stage for irrigation.

☆ **Pest control:** For the control of stem borer, 1-2 sprays of 625ml Endosulphan 35 EC in 625 litre of water/hectare and 900 ml in 750 litre of water/hectare after 10 and 20 days of sowing respectively.

☆ **Detasseling:** To maintain the quality of baby corn, detasseling is an essential operation. It is done by removing the tassel of the plant as soon as it emerges from the flag leaf.

☆ **Harvesting:** Picking should be done daily in kharif and on alternate days in winter season within 24 hours of silk emergence from the leaf sheath depending upon the variety. Ears on top should be reaped first then those below. In single cross hybrid plant, 3-4 pickings may be required. Yield of baby corn depends on potential of genotypes/climate/season. In a good crop on an average 15-19 q/ha baby corn (without husk) can be harvested. Additional income may be realised from the sale of green fodder. Green fodder yield is about 250-400 q/ha. Harvesting should be done when baby corn silk comes out 2.0- 4.0 cm from the top end of ears, preferably in the morning, when the baby corn moisture is highest and ambient temperature is low or in the evening.

☆ **Post Harvest Management:** The effort should be made to peal the baby corn on the same day and stored in cool place. It should be carried out in shady places having good ventilation and air circulation. De-husked baby corn should be put in containers like plastic baskets, bags, sacks and ensure that they are not heaped. Baby corn should be transported to the processing unit at earliest.

B. Processing of Baby Corn

The baby corn cobs are usually canned at processing factories. Flow diagram is mentioned below:

Peeled Baby Corn ⟶ Cleaning ⟶ Boiling ⟶ Soaking ⟶ Grading ⟶ Containing ⟶ Brine Soluiton ⟶ Exhaust ⟶ Lid Covering ⟶ Cooling ⟶ Quality Inspection.

 ☆ **Preservation:** After filling baby corn in container, brine and water are added in cans in the proportion of 2: 98 (Brine-2 per cent and water 98 per cent), alternatively, a solution of 3 per cent brine, 2 per cent sugar, 0.3 per cent citric acid and balance water can also be added.

 ☆ **Packaging:** The packing varies from unit to unit. It can be in tin, glasses and polybags. For longer time of preservation, glass packing is the best. Glass packing has 52 per cent baby corn and 48 per cent brine solution.

C. How Baby Corn is Consumed

 ☆ Baby corn soups

 ☆ Vegetable prepared by putting the ingredients like onion, ginger, tomato and spices to be consumed with rice or chapattis

 ☆ Deep fried baby corn with meat and rice and with other vegetables

 ☆ Pickles and candy

 ☆ Salads and in salad dressing

 ☆ Various dishes and snacks *viz.* Baby corn pakoras, kofta, tikki, burfi, laddoo, halwa, kheer *etc.*

D. Recipies from Baby Corn

Baby corn can be utilized for the development of a variety of recipies. Preparation of few recipes are given below.

Baby Corn Vegetable Soup

Ingredients

Baby corn	:	150g
French beans	:	25g
Carrot	:	25g
Butter	:	½ tsp
Corn flour	:	1 tsp
Salt and black pepper	:	to taste

Method

☆ Wash and chop baby corn, french beans and carrot and boil till soft.

☆ Strain boiled vegetable and baby corn.

☆ Make a paste of corn flour with water and add to strain soup.

☆ Add salt and pepper in the soup and boil on slow flame for 5 minutes.

☆ While serving, add butter and ¼ cm long slices of blanch baby corn, french beans and carrot.

Baby Corn Mixed Vegetable

Ingredients

Baby Corn	:	100g
Capsicum	:	25g
Cabbage	:	25g
Carrot	:	25g
Peas	:	25g
Tomato	:	50 g
Onion	:	50g
Garlic	:	2 cloves
Ginger	:	small piece
Green coriander	:	few leaves
Red chilli powder	:	½ tsp
Garam masala	:	½ tsp
Salt	:	as per taste
Oil	:	20ml

Method

☆ Blanch baby corn for 5 minutes and cut into round pieces (½″ long).

☆ Chop all other vegetables, onion and tomato.

☆ Heat oil in a frying pan and add ginger and garlic paste.

☆ Add onion and fry till light brown in colour and add all other spices followed by chopped tomatoes.

☆ Finally add baby corn and all other vegetables and cook till completely done on a slow flame.

☆ Decorate with coriander leaves and serve hot.

Baby Corn Candy

Ingredients

Baby corn	:	500g
Sugar	:	1.25kg
Citric acid	:	2.5g
Water	:	750ml

Method

★ Select good quality baby corn and cut into 1" long pieces.

★ Prick baby corn gently with fork and blanch for three minutes.

★ Add 750g sugar to 750 ml water and make sugar syrup.

★ Add 2.5g citric acid to syrup and boil and remove dirt from syrup surface.

★ Cool the sugar syrup till TS percent is 55 using refractometer.

★ Add pricked and blanched baby corn in syrup solution and leave overnight.

★ Next day, remove baby corn from syrup solution.

★ Add 250g sugar and boil till TS percent is 57 percent.

★ Add baby corn to syrup and leave overnight.

★ Next day again, remove baby corn and note the TS percent which will be nearly 55 percent.

★ Add 250g sugar and boil for five minutes till TS percent is 75 percent.

★ Add baby corn to syrup and leave overnight.

★ Next day, note the TS percent as 70 percent, boil and cool.

★ Remove baby corn from syrup and dry in oven at 50°C overnight.

★ Pack candy in airtight polyethylene packets.

Baby Corn Pickle

Ingredients

Baby Corn	:	1kg
Aniseeds	:	25g
Red chilli powder	:	as per taste
Mustard seeds	:	50g
Saunf	:	50 g
Salt	:	180g
Mustard oil	:	150ml
Vinegar	:	10ml

Method

☆ Wash fresh small sized baby corn in hot water and cut into long slices (1½") and dry to remove excess water.

☆ Grind mustard seeds, aniseeds and saunf coarsely.

☆ Heat half of oil in a pan and add aniseeds, mustard seeds and saunf and red chilli powder and fry.

☆ Remove from fire and add baby corn followed by vinegar and salt and oil.

☆ Mix well, cook and fill in clean airtight jar and keep in sun for 5-6 days.

D. Production Capacity (per annum)

Quantity of Baby Corn: 18 quintals x 100 hectares x 4 crops in a year = 720 tonnes

(a) Value of Baby Corn: 720000 kg x Rs. 50/- kg = Rs 3.6 crores

(b) Value of intercrops (radish cauliflower/broccoli/gladiolus/marigold) = Rs. 10,000/- x100 ha = 10 lakhs

Quantity of green fodder: 300 quintals x100 hectares x 4 crops in a year

(c) Value of green fodder: Rs. 10,000/- x 100 hectares x 4 crops in a year = Rs. 40 lakhs

Total income(a+b+c) = Rs 4.1 crores

Fixed Capital

Land: 100 hectares on lease/contract

Machinery and Equipment:

a) Canning Machine = Rs. 49 lakhs

b) Polythene packing machine=Rs 1 lakh

Total Fixed Capital = Rs. 50 lakhs

Working Capital (per year)

a) Leased value of land = Rs. 20 lakhs

b) Cost of labour:Rs. 100/-x 100 mandays x250 acres x 4 crops in a year = Rs. 1 crore

c) Cost of seed: Rs. 100/- x 10 kg x 250 acres x 4 crops in a year = Rs. 10 lakhs

c) Cost of Fertilizer: Rs. 2000/- x 250 acres x 4 crops in a year = Rs. 20 lakhs

d) Cost of Irrigation: Rs. 4000/- x 250 acres x 4 crops in a year = Rs. 40 lakhs

e) Cost of Pesticides: Rs. 100/- x250 acres x 4 crops in a year = Rs. 1 lakh

g) Cost of Polythene and bottles = Rs. 20 lakhs

h) Miscellaneous = Rs. 19 lakhs

Total Working Capital = Rs. 2 crores 30 lakhs

Financial Analysis (per annum)

1. Total Investment =Total Fixed Cost + Total Working Capital = Rs. 2.80 crores
2. Turn over = Rs. 4.10 crores
3. Net Profit = Rs. 1.30 crores
4. Rate of return = Net Profit x100/Total Investment = 46.43 per cent
5. Break even point = Total fixed cost x100/Total fixed cost + Net Profit =27.78 per cent

Entrepreneur may earn more through supplying canned baby corn in big cities. Regular supply of baby corn may be made to hotels and restaurants for making products like salad, soup, vegetables, kheer *etc*. Pickles, jam, murabba, candy *etc* may be exported to USA, European countries *etc*.

References

Gupta Vinod (2000). Formulation of a Project, paper published in reading material of Short Course on Entrepreneurship Development in Agriculture, organised by Division of Agril. Extension, IARI, Pusa Campus, New Delhi from November 14-21, 2000, pp.140-149.

Sain Dass *et al.* (2007). Baby Corn: Cultivation and Value Addition, published by DMR, Pusa Campus, New Delhi, pp.10-12.

Success Stories of Agri-business

Chapter 18

Agri-Clinics: A Tool of Agri Entrepreneurship

G.M. Wani

Former Director Extension Education and Director,
State Agriculture Management and Extension Training Institute (SAMETI)
Sher-e-Kashmir University of Agricultural Sciences and Technology of Kashmir,
Shalimar, Srinagar – 191 121, J&K
e-mail: wanimohyuddin@yahoo.com

Introduction

Agricultural development over years has been the result of continuous agri skill generation and its popularization. The earliest agriculture was animal domestication over thousands of years ahead, man domesticated wild fowl, dog, goat and smaller animals, whom he could overpower easily and subjugate to his sub-ordination. Agriculture thus since beginning has been the results of trails, experiments and experiences over years, learned first though behavioral changes, psychic reoccurrences, memories passed through parents to children and later on through doing and learning and now through sharing experiences and writing them or dotting them as an Entrepreneurship concern. (Dipak, 2004)

Entrepreneurship Concept

The Entrepreneurship adds economic profits and cost-benefit ratios to Agricultural Output. Entrepreneurship is dominated by four factors like:

a. Social systemic changes

b. Support system availability and use

c. Resource base and its utilization

d. Self confidence, exploration work capacity and intellectual potency.

An entrepreneur has to have a thinking of his own, a capacity building interest in acquiring needed technique. An explorative and analytic faculties to judge the way of procuring cheap raw material. He must be equipped with "knowledge" and mindset to use and benefit out of it (Wani, 2004).

Farm Business

A potential entrepreneurship must strive from getting maximum output. Decades back agricultural development and industrial setups was a public sponsored and heavily subsidized but over time "knowledge" explosion in India Agriculture, have brought us on threshold of a system, where wide distances exist between industry and farm business. Where huge subsidies are benefiting Agro-Industrialists. The Farmers who use fertilizers or agro-chemical are crushed under economic pressures. The gaps between technology generated and technology use at farmers door is increasing day after day. The farm technology adoption rates are not more than 20-30 per cent by any higher prospectives. The use of information and communication technology (I CT) for reducing the gaps and increasing productivity is the need of the hour (Wani, 2005). The modern technology and knowledge flow is fast expanding and bringing change. It demands more educated and trained farmers. Our education system has produced more literates but not educationally trained youth to earn their own bread. They after attaining graduation in agriculture and allied sectors, beg for job. The system has to be corrected to make these graduate as employers and not employees. I wrote a treatise as back as 1992, emphasizing a system. Germans are smart to have Farmers school, Farmer business training institutes, practical agri-farmers training centres and like, where every farmer or animal husbandry man is essentially a trained fellow. The banking system is so organized that they are on the door of convocation hall to sell their agri-business and agri-clinics to graduates, without any personal investments. Banks are so smart, that they have surveyed the villages who need vets or agri-graduates or have attained land and all facilitation, so that agricultural or veterinary or even other medico-biological graduates are used as bank investment. This is what is envisaged in India under agriclinic, Agri-business venture. We have trainings not in the hands of banks but universities (Wani, 2004).

Success of Agri-business

A systematic liaison and support system between Govt. banking and University culture has made this otherwise an remunerative and lucerative programmes into a failure inspite of its personal monitoring of PMO. The success of Agri-business and Agri-clinics success rates are shown in Table 18.1. The universities involved and their success stories are shown in Table 18.2. Both these details are distressing inspite of huge moral, financial support from Govt. This is inspite of subsidiary support Table 18.3.

Various ventures are listed in Table 18.4–18.7. Agricultural professionals are getting converted into Agri-business and agri-clinical experts. More than 14,000 applicants and 615 agriclinics came to existence in Indian 12 states. The agri clinic trained persons in J&K many number in hundreds. Among them 34 have registered agri clinics earning a handsome profit annually.

Table 18.1: MANAGE Progress of Agriclinics and Agri-business Centres Scheme Period from 01-04-2002 to 30-10-2015

S.No.	Name of the State	No. of Applications Received	No. of Candidates Trained	No. of Agri-ventures established
1	Andhra Pradesh	883	825	306
2	Arunachal Pradesh	35	32	3
3	Assam	601	597	200
4	Bihar	3387	3282	1208
5	Chandigarh	3	3	1
6	Chattisgarh	638	527	250
7	Delhi	25	17	3
8	Goa	10	9	4
9	Gujarat	1326	1274	503
10	Haryana	601	552	202
11	Himachal Pradesh	423	418	108
12	Jammu and Kashmir	1332	1276	175
13	Jharkand	619	618	155
14	Karnataka	3105	3037	1244
15	Kerala	203	183	51
16	Madhya Pradesh	1354	1294	521
17	Maharashtra	10272	10189	4723
18	Manipur	416	413	126
19	Meghalaya	11	11	3
20	Mizoram	34	34	0
21	Nagaland	177	174	21
22	Orissa	549	506	106
23	Pondicherry	112	111	68
24	Punjab	534	531	197
25	Rajasthan	2769	2623	971
26	Sikkim	10	9	0
27	Telangana	1037	982	361
28	Tamil Nadu	5191	5116	2690
29	Tripura	4	2	1
30	Uttar Pradesh	9437	9171	4574
31	Uttarakhand	418	392	114
32	West Bengal	777	725	216
		46293	44933	19105

Table 18.2: Agri-clinics and Agri-business Centres Cells National Institute of Agricultural Extension Management (MANAGE), Hyderabad, Activity-wise Categorisation of Agri-Ventures, Period from 01-04-2002 to 17-06-2016

Name of the Agri-venture	AP	ARN	ASS	BHR	CHA	CHD	DEL	GOA	GUJ	HP	HRN	JAM	JHA	KAR	KER	MAN	MEG	MP	MS	NAG	ORS	PON	PUN	RAJ	SKM	TG	TN	TRI	UP	UTC	WB	Total
01. Agri-Clinics	127		21	155	20	1			39	12	19	11	72	191	16	6		61	881	1	27	16	29	146		86	595		610	18	25	3,185
02. Agri-Clinics and Agribusiness Centres	67	2	30	494	63		1	1	253	25	97	49	26	426	7	25	2	169	1,422	3	20	17	32	163	1	87	606		2,611	26	47	6,772
03. Agro-Eco Tourism								1	1					1					7								1					11
04. Animal Feed Unit	1		2	1					1			2		6				1	16							1	5		9	1		47
05. Bio-fertilizer production and Marketing	3		1	3					4		1			15	1				30		2	1		9		9	9		12	2		102
06. Contract Farming	5													13				1	1							6	21		11	1		66
07. Cultivation of Medicinal Plants	1		1	16	2					1	3			14				3	6			1	1	31		5	10		12			112
08. Direct Mkt	1			6					6	3				16				4	66				4	5		2	8		38	2	1	168
09. Farm Machinery Unit	8		3	11	8				32	16	9	8	1	28	1	2		46	207		3	2	3	14		8	89		196	7	11	713
10. Fisheries Development	4		4	104	34				1	1	3	2	7	12	14	10		3	40		5	7		2		4	43	1	29	1	18	349
11. Floriculture	3				1				3	5	7	7							33			2	4	3		11	14		8	4	3	108
12. Horticulture Clinic	3			6	3					7	2	3		23				5	62			2	1	1		11	26		11			170
13. Landscaping + Nursery	2		1	1				2	1	1	1	4		15	3				46			4				3	24			1		113
14. Nursery	11		5	21	3				13	8	7	11	2	45	3	8		8	164		6	1	11	15		10	91		53	2	8	506
15. Organic Production/Food Chain				3	3				4	1				20				1	15		1	2	5	8			13		10	1		88
16. Pesticides Production and Marketing				1					4		1	1	1	10				2	14							2	1		2			40
17. Value Addition	4		2	38	6				5		11			10				8	120	3		3	3	10		1	22		25	3	2	276
18. Fishery clinic	1												1		1				3	3							6					15
19. Seed Processing and Marketing	13		1	28	10		1		19	3	12			56	1	1		13	23		3	2	26	14		31	31		39		7	335
20. Soil Testing Laboratory	2		1	23	1		1		6		1		1	11				2	16		1	1	1			1	25		5		3	102
21. Tissue Culture Unit	2													11				1	8								2					28
22. Vegetable Production and Marketing	1			1	21				3	9	7	2	8	7		2		7	33	6	2		13	6		7	21		56	3	35	245
23. Vermicomposting / Organic manure	7		5	125	3				6	1	3	2	6	39	2	2		6	82		1	2	17	71		12	43		53	6	1	495
24. Veterinary Clinics	2		91	66					17	2	6	23	16	11		28	1	1	323	7	2	3		83		6	152		22	3	10	875
25. Crop Production	3			3	11				3			5		27				11	41		11	3		49		2	13		11	2	2	197
26. Dairy/Poultry/Piggery/Goatery	50	1	35	107	56				130	8	18	24	13	253	3	18		219	1,593	6	4	12	55	385		65	1,041		1,216	54	77	5,443
27. Rural Godown				5										2				2	8					1			3		28			49
28. Production & Marketing of Bio-Control Agents														4				1	4								1		8			18
29. Agriculture Journalism	1								2					3				1	3							2			2	2		16
30. Sericulture	1			1	1								1	7		19						5					7				1	49
31. Mushroom Cultivation				16						2	14	6	2						3			5		1			39		5	2	5	99
32. Apiary	1			8						2	14	3							3					2			2		65	1		101
Total	321	3	206	1,246	253	1	3	4	557	108	205	176	163	1,302	51	128	3	576	5,283	21	106	77	203	1,024	1	363	2,964	1	5,148	140	256	20,893

Table 18.3: Under Agri-clinics/Agri-business Norms Thereof

Sl.No.	Means of Finance	General Category	Weaker Category
1.	Promoter's contribution	10 per cent	5 per cent
2.	Soft loan from NABARD	0 per cent	5 per cent
3.	Total promotion contribution	10 per cent	10 per cent
4.	Subsidy	25 per cent	33.33 per cent
5.	Eq.	355	43.33 per cent
6.	Term loans	65 per cent	56.67 per cent
	Total	100 per cent	100 per cent

Table 18.4: List of Ventures

1.	Soil and water quality cum inputs testing labotatories (with Atomic Absorption Spectrophotometers)
2.	Pest surveillance, diagnostic and control services
3.	Maintenance, repairs and custom hiring of Agricultural implements and machinery including micro irrigation systems (sprinkler and drip);
4.	Agri Service Centres including the three activities mentioned above (Group activity);
5.	Seed Processing units
6.	Micro-Propagation through Plant tissue Cultural Labs and hardening Units

Table 18.5: List of Ventures

1.	Setting up of Vermiculture units, Production of bio fertilizers,bio-pesticides, bio-control agets;
2.	Setting up of apiaries (bee-Keeping) and honey and bee products' processing units;
3.	Provision of Extension Consultancy Services;
4.	Facilitation and agency of agricultural insurance services;
5.	Hatcheries and production of fish finger-lings for aquaculture;

Table 18.6: List of Ventures

1.	Provision of livestock health cover, setting up of veterinary dispensaries and services including frozen semen banks and liquid nitrogen supply;
2.	Setting up of Information Technology Kiosks in rural areas for access to various agriculture related portals
3.	Feed Processing and testing.
4.	Value Addition Centres
5.	Setting up of Cool Chain from the farm Level onwards (Group Activity)

Table 18.7:List of Ventures

1.	Post Harvest Management Centres for sorting, grading, standardization, storage and packaging;
2.	Setting up of Metallic/Non-Metallic Storage structures (Group activity)
3.	Retail marketing outlets for processed agri-products.
4.	Rural marketing dealerships of farm inputs and outputs.
5.	Projects in any other service oriented activities in agriculture and other allied areas can also be considered.

We visited Bandipora district and unregistered Agriclinics were earning a handsome salary, more than the Rahbar-e-zerat or Agriculture Asstt. A visit documentary is enclosed and shall be shown. It consisted of Agri-business *viz.* sale of pesticides, cattle feed, poultry feed and agri-extension services. At a small village in Papchan, one agri graduate Mr. Iqbal Shah earns Rs. 10,000/- per month by selling the services and input. At a distance of few kms. In same district one Mr. Khyatlani owns a big poultry farm and earns around Rs. 20,000/- per month. Both these entrepreneurs employ 2-3 persons at present. Similarly, the success shown by one Mr. Shah at Malangam in Agri products and pesticide sale and one Mr. Bhat in Dairy production and milk product sale earn a handsome income besides generating employment for poor.

Farmer as Entrepreneur

Indian Farming and farmer has to change if proper WTO recommendation and GATT agreements are to be followed. The present day poultry scenario has emerging high profile agri-business prospects in India.

The conversion of poultry farmer's into poultry entrepreneurs shall make the present day 6 per cent contribution of poultry products to 25 per cent share of Global market from India and China. This when translated into action shall increase employment generation by manifolds. The introduction of rural based Vanraja, Gramapriya, Giriraja, Cari Gold and vast other locally grown varieties of poultry have adopted well to our agri-rural base. The market acceptability is higher than exotic poultry concerns. Research to farmers doors in generating free-rang-poultry like BT cotton hybrid spreading through villages of India and assuring high returns and exports (Wani, 2007).

1. Poultry as Agri-business

Dr. Gordon Butland, president of Global poultry strategies presents "Backyard poultry production" as a tool of alleviating poverty and malnutrition. We have tried to distribute "birds" under free-rang system in all our KVK's our results were excellent and income generation was totally in favour of the Agri-business and agri-clinics as will be shown in case histories and success stories.

A grand show of using poultry, rabbit meat processing introduction at SKUAST-K have innovated white meat usage. This all will need the involvement of Agri-Veterinary and food processing technoractes to develop rural-based establishments so as to foster export and fast returns.

2. Holistic Vision for Livestock Enterprise

Improving income, employment and self-reliance among educated graduates and un-employed youth especially women, needs fostering community development, women empowerment, environmental protection. Rural-based backyard poultry subscribes to all these norms and could be a rich resource for developing agri-entrepreneurship. Govt. of India is liberally financing such agri-business ventures and a proposed infrastructure cost set-up can be seen in Table 18.8. A vast and finance assured schemes are available for agri-graduates for establishing poultry ventures (Table 18.9). An initial allocation of 107 crores for initiating nucleus breeding farms. Further more provision of hatcheries to provide chicks to more than 2 lac farmers and farm women will need many agri-business centres for providing basic germplasm, medicine and above all training.

Some of the success stories in animal husbandry section can be reproduced as follows:

a. Backyard Poultry and Incubation

Though the Vanraja are the most suitable for back yard poultry, they do not have habit of broodiness. There is a problem among the farmer to get a broody hen in all season. KVK solve this problem of hatching by installing small unit of hatchery. Every month 15-20 farmers are benefited by purchasing chicks for backyard poultry. There are 200 back yard poultry units of Vanraja. Each farmer is rearing 10 to 25 in the backyard. There is a good demand and response for the chicks and eggs of Vanraja. KVKs are now planning to expand this unit.

b. Semi-Stall-Fed Goat Rearing

KVK's made an intervention to improve this enterprise by conducting short durational training programmes for rural youth. Similarly exposure visit were organized on goat feed, breed and health management. More emphasis was given on Osmanabadi goat and up-gradation in selected non-descript goat breed by Osmanabadi pure buck and given the knowledge about semi stallfed goat rearing concept.

c. Broiler Production

KVK has conducted many durational training programmes for 165 trainees. Due to training and demonstrations awareness was increased about contract farming in broiler production with private sector which provide chicks, feed and medicine and after 40 days purchases Rs. 3 to 3.50 per kg on live weight and FCR basis and changed their attitude. They acquired skills through learning by doing at KVK demonstration unit.

The technology has been adopted by 10 percent of youths now in the radius of 20 km there are 27 poultry units having capacity of 5000-10000 poultry birds on contract farming basis. These self employed rural youth earning Rs. 10000-15000 per lot (Bishnupada, 2007).

Recently a seminar-cum-farmer's meet was arranged at SKUAST-K on 26-27th of Oct.2007. The knowledge –sharing and use for making agricultural graduate and

Table 18.8: Poultry Agri-business

Sl.No.	Component/Cost	I	II	III	IV	V	Total	Remarks
1.	Satellite Hatcheries	100	50	–	–	50	200	Assuming cost of each Satellite unit around Rs.10.00 lakhs and Proposing 10 and 5 units during Ist and 2nd year respectively and another 5 units during 5th year.
2.	Mother Units	420	60	300	–	–	780	Assuming Single Mother Unit cost to be around Rs 1.2 lkhs during Ist and 2nd year and Rs. 1.5 lahks during 3rd year and Proposing for Establishment of 350,50 200 unts during 1st, 2nd and 3rd year respectively.
3.	Poultry traning cum – extension Centres for traning of trainers, poultry link workers *etc.*	50	30	30	–	–	110	Assuming cost of each Centre to be around Rs.2.00 lakhs and proposing 25.15,15 Centres during Ist, 2nd and 3rd year respectively.
4.	Poultry Processing Units	–	–	500	500	500	1500	Assuming cost of each Mini Plants with dressing capacity of 500 birds per hour and proposing 5 units to be set up each year during the 3rd,4th,and 5th year.
5.	Skill up gradation and training of beneficiaries	420	400	450	500	500	2270	Assuming 70,000 beneficiaries are trained during the 1st year and 50,000 each during the subsequent years,. Further only honorarium is proposed @ Rs. 600, 800, 900, 1000, 1000 during 1st, 2nd,3rd,4th and 5th year respectively.

Table 18.9: Poultry Business Targets

Sl.No	Component/Costs	Year I	Year II	Year III	Year IV	Year V
1.	Cumulative no. of nucleus breeding farms	35	40	40	40	40
2.	Cumulative no. of mother units	350	400	600	600	600
3.	No. of beneficiary families ('000)	70	120	180	180	180
4.	Financial assistance towards fixed cost of bene ficiary families (at the rate of Rs.1300) Rs in lacs	910	650	780		
5.	Subsidy towards input costs of beneficiary families (assuming average batch size of 30 birds)- Rs. in lacs	1,607	1,683	1,760	459	
6.	Subsidy towards fixed cost of mother units (at the rate of Rs. 18,000 i.e 20 per cent of Rs. 90,000) Rs. in lacs	63	9	36		
7.	Interest free loan (net of recovery) to mother units (at the rate of Rs. 27000)- Rs. In lacs	95	14-19	54-23 = 31	Rec. of 32	Rec. of 32
8.	Poultry training cum extension centres for training of trainers, poultry link workers etc. (25, 15 and 15 centres to be set up in the 1st, 2nd and 3rd years at the cost of Rs. 2 lacs each) Rs. in lacs.	50	30	30		
9.	Skill up gradation and training of beneficiaries (70,000 to be trained in lst year and 50,000 in each subsequent year) Rs. in lacs	420	300	350	350	400
10.	Administrative expenses, monitoring and evaluation – Rs. in lacs	120	130	140	160	200
	Total cost	3265	2797	3127	937	568

scientists was emphasized by our worthy Chancellor. A vision of poverty alleviation through backyard poultry intervention was the theme of the seminar. Many belts in Gurez, Tangdar, Telail and Zanskar are rearing native livestock species. Who are better suited and need improvement and identification. The cooking methods will need more expansion and scientific intervention for export. More emphasis has to be made on:

☆ Safe feed and food.

☆ Organic fodder and food.

☆ Operational excellence and modern mechanization to improve quality of indigenous enterprises.

☆ Local family management to farm business management and seller-buyer medal adoption.

High Value Agri-business

Rapid growth rate in high value commodities in Indian agriculture promises 40 per cent total output. The sectors assuming importance for export earnings are Fruits, milk vegetable and poultry. Thus Agriclinic training centres should focus on these commodity oriented training. The sector may need more than 1 lac young entrepreneurs to achieve national goals and not the mere 14000 applicants (Sidiqui, 2006).

What is needed:

☆ Openness and transparency.

☆ Simple banking.

☆ Credit facility.

☆ Mission and Training.

References

De Dipak, 2004. Workshop Background and recommendations In communication support for sustaining extension services edited Dipak De, Jirli, B and Ghadi,K. Deptt. of Extension Education, Institute of Agric. Sci, BHU, Varanasi.

Wani, 2004. Role of communication in enhancing agricultural production – a Vision. In Communication support for sustaining extension services edited. Dipak De, Jirli, B and Ghadi,K. Deptt. of Extension Education, BHU, Varanasi, pp. 1-26.

Wani,G.M.2007. Message, "Backyard poultry Production". Seminar on Backyard poultry Farming for women empowerment and nutritional security Oct. 26-27, 2007, SKUAST K, J&K.

Sidiqui,M.A.A, 2004. Plant Resource Management for Entrepreneurship Development. Agri Clinics and Agri-Business, Deptt. of Extension Education, SKUAST-K, Shalimar, Srinagar- J&K.

Wani, G.M. 2004. Agriculture Production and Entrepreneurship – Preface to the proceedings on Plant Resource Management for entrepreneurship development, Directorate of Extension Education, SKUAST-K Shalimar, pp. 1-4.

Bishnapada, Sehthi, 2007. Backyard Poultry in Orissa. Orissa Review, January, 2007. pp. 48-52.

Chapter 19
Entrepreneurship Development in Agriculture in India

T. Prabhakara Sastry

Former Professor and Head, Department of Extension Education,
Acharya N.G. Ranga Agricultural University,
S.V. Agricultural College, Tirupati

Agriculture is lifeline of India's economy. This sector contributes about 20 percent of India's G D P and provides livelihood for over 60 per cent percent of its population. In the past Indian agriculture was considered as a way of life but now it is transforming from traditional to modern commercial enterprise. A commercial enterprise involves rational decisions on investment after assessing risk, other alternatives and possibility of profit or loss. An entrepreneur is an economic man who tries to maximize his profits by identification and adoption of innovations. However, entrepreneurs are not only innovators but they are persons with a will to act to assume risk and bring about a change through organization of human efforts. Entrepreneurship is a firm behaviour covin and slevin (1991). It is a phenomenon of emergence, it evolves over time Gartner *et al.* (1992). Entrepreneurship is a way of life, a thought process to bring any sustainable change, the effort has to be more broad based. It is risk taking ability of the individual broadly coupled with rational decision making to increase production in agriculture, agricultural business and industry *etc.*, Entrepreneurial behaviour of farmers has gained importance since farmers have to change their knowledge, skills and attitudes towards their selected enterprise. The new way of doing things that is already done in a new way is part of entrepreneurial behaviour. Entrepreneurial behaviour consist of different components like 1) decision making, 2) Innovativeness, 3) risk taking ability, 4) achievement motivation, 5) information seeking, 6) Knowledge of farming, 7) Assistance of management services, 8) Coordination of farm activities, 9) cosmopoliteness and, 10) Leadership ability according to Nandapurkar (1982). Effective linkage of production systems, marketing, Agro processing and other value added activities would play an increasingly important role in the diversification of Agriculture.

Enhanced Roles of Extension Personnel for Improving Entrepreneurship among Farmers

Swot Analysis of Market has to be done by Extension Personnel for Improving Entrepreneurship:

1. **Strength:** Suitable soils for taking diversified farming, Good human resource, 9000Kms. Longest coastal line congenial atmosphere.

2. **Weakness:** Lack of knowledge, skills, and illiteracy lack of scientific temper.

3. **Opportunities:** Export of processed foods marine foods, flowers, fruits, vegetables *etc.*, 52 Agril., Commodities are now being exported to global markets.

4. **Threats:** Lack of technical know how, High cost of inputs, lack of storage felicities, Lack of market facilities, Dominance of M.N.C's.

Organization of farmer interest groups (FIGs) on commodity basis and building their capabilities in organic products with regard to management of their farm enterprise. Establishing agro processing units by linking Farmers interest groups with private processors to improve entrepreneurship among farmers. Advice on product planning selection of commercial crop production suiting to their soils and marketability of products will be the starting point of Agricultural enterprise. Educating farmers to treat agriculture as an entrepreneurial activity and accordingly plan various phases of crop production, Aquaculture, sericulture Horticulture *etc.* Increasing Entrepreneurship of farmers in terms of establishment of Agro prosing units and storage facilities and linking with export markets. Production of Video films of success stories of Agricultural Entrepreneurs. Creation of web sites of successful entrepreneurs in the field of Agriculture and allied enterprises with all the information to help other Farmers to achieve success.

Required Information to Extension System and Farmers

Sustainability of landholding to various commercial crop enterprises. Market prices of Agricultural Products after processing should be known. Credit facilities for the establishment of small scale enterprises in the village itself. Supply of Desired quality products to consumer after value addition. Network of storage facilities for storing perishable products. Post harvest management by farmers *Viz.* processing, grading, standardization of produce, value addition, packing certification *etc* with reference to fruits, Vegetables, prawns, fish *etc.*

Flow Chart of Agriculture as an Enterprise

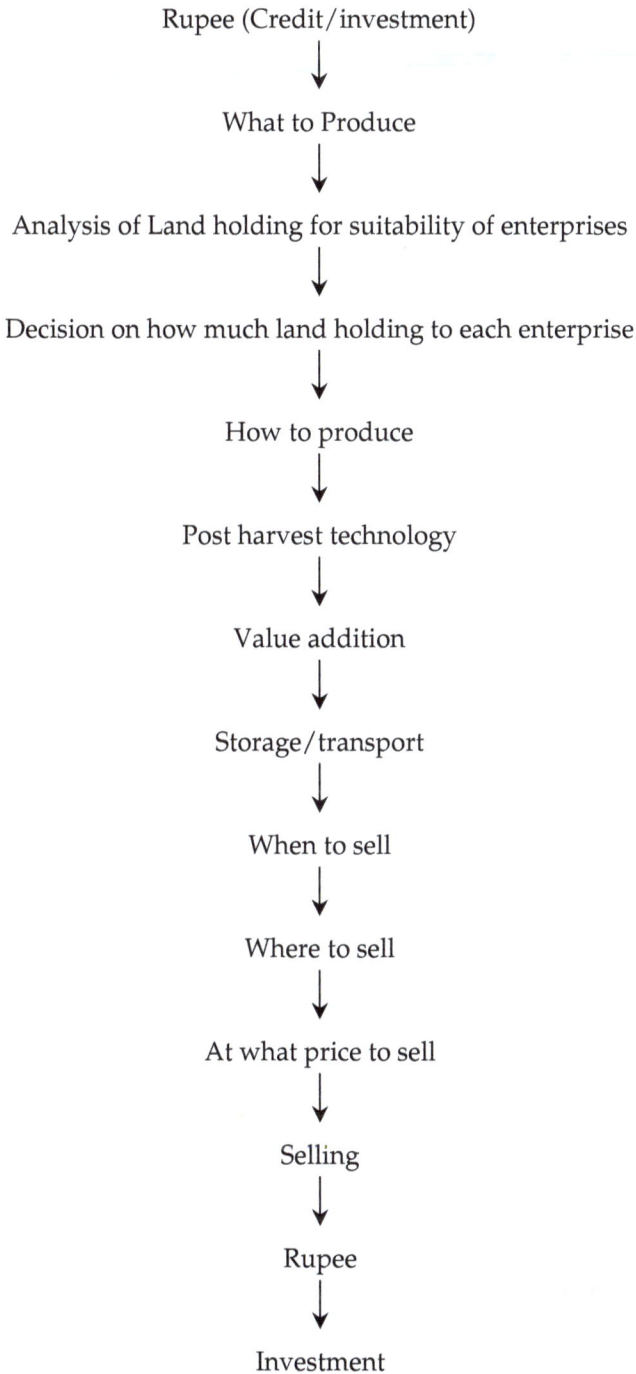

Rupee (Credit/investment)

↓

What to Produce

↓

Analysis of Land holding for suitability of enterprises

↓

Decision on how much land holding to each enterprise

↓

How to produce

↓

Post harvest technology

↓

Value addition

↓

Storage/transport

↓

When to sell

↓

Where to sell

↓

At what price to sell

↓

Selling

↓

Rupee

↓

Investment

Paradigm shift of Farmers from production to Entrepreneurship

Sl.No.	Aspects	Production Aspects	Aspects of Entrepreneurship
1.	Objective	Transfer of production technologies	Enable farmers to follow demand driven approach for getting high returns from the enterprise
2.	Farmers seen as	Progressive farmers high producers	Farmer as an "entrepreneur or Agripreneur"
3.	Focus	Seed to seed	Money to money
4.	Extensionist role	Dealing with Farmers problems feed back to research system	Extensionist role as Transfer of Technology functionary besides establishing Agro processing linkages between farmers, processors and linking to global markets.
5.	Maintenance of Records	Not having much importance as the focus was more on production	Very Important as agriculture is viewed as an enterprise based on cost benefit ratio.

Promotion of Entrepreneurship in Farmers

Besides Agriculture, Blue revolution was lunched in the recent past with a view to increase production and productivity of aquatic resources. Aquaculture has emerged as a major source of earning and employment to millions of people. Prawn culture as an enterprise have received maximum importance because they earn valuable foreign exchange to developing counties like India. Aquaculture has become an inseparable key farming activity of coastal Andhra Pradesh Providing employment to nearly a million people of the state out of the total marine exports valued at 6500 crores. The results of Vidyadhari (2007) on entrepreneurial behavior of prawn growers in Nellore district of Andhra Pradesh indicated that majority of prawn growers had medium level of entrepreneurial behaviour. Education, Social participation, extension contact, income, scientific orientation, management orientation and value orientation had positive and significant relationship with entrepreneurial behaviour of respondents. Age, marketing facilities and credit orientation had non-significant relationship with entrepreneurial behaviour of prawn growers.

Problems and Suggestions for Better Entrepreneurship According to Vidyadhari (2007) in her Study

See Figure 19.1.

Important Success Stories of Farm Entrepreneurs in India

A) Success Stories of Farmers of Punjab in North India

Dr. Kaur Singh belongs to village Duggan started cultivation of Tomato, Chilli and Potato in 2.5 acres during 1995 – 96 on Contract basis. He decreased 50 per cent area under rice and earned 2-3 times more income from this farming system. He got training in hybrid seed production in chilli and started production in ½ acre

Problems	Suggestions
1. Lack of quality seed and feed	1. Provision of quality seed and feed
2. Lack of market support price	2. Provision of minimum support price
3. High cost of inputs	3. Provision of inputs at reasonable price
4. Lack of power supply	4. Supply of power for about 12 to 15 hours per day
5. Lack of cold storage facilities	5. Establishment of cold storage units
6. Inadequate knowledge of diseases	6. Strengthening of training and research facilities to update knowledge
7. Inadequate credit facilities	7. Credit with low interest rate
8. Lack of subsidies	8. Provision of subsidies on inputs

Better Entrepreneurship

Figure 19.1: Paradigm of Problems and Suggestions for Better Entrepreneurship.

and later increased to 3 acres with CH.1 and CH-3. Later he showed interest in stud farming and he him self got expertise in it and now he has 2 horses and 15 mares at his farm and he says that a mare can give return Rs. 50,000/- per annum. The local farmers and adjoining state farmers visited his modernized stud farm house and got Inspired from his ideas.

B. Horticulture in Sangrur District

Sh. Gopal Dev Singh, village nomol, near sheron, a well known premier ber grower in the district, planted 2 acres of ber variety Umnan. During 1995 he earned only a net income of Rs. 16,000/-. With the advice of Agriculture Advisory Agencies he improved the management of fruits by giving training on pest management and earned net income of 50,000 to 80,000 during 1990 to 1993 respectively. He started inter cropping of Vegetables like cucurbits, Cowpea and okra and earned Rs. 50,000 annually. His annual income was doubled by adopting diversified agriculture farming system.

C. Hybrid Seed Production in Vegetables

S. Balurinder Singh Sohi is a prominent hybrid seed producer among the growers who belongs to village Bhari Mansa near Dhuri. He has been cultivaling diversified crops at his 20 acres land. After getting training he started hybrid seed production of Chilli, Tomato, Brinjal and Muskmelon on small scale and with increasing demand of hybrid seeds. He increased acreage of these crops. In addition he produce certified seeds of sunflower, moong, wheat, Rice and sarson. His net income from the sale of hybrid seeds and seedling per acre is Rs. 4 lakhs.

D. Success Stories of Farm Enterprises of Chittoor District of A.P. in South India

Horticulture (Pudina Cultivation)

Smt. Padmavathamma, farm women of bangarapalem mandal of Sheshapuram Village has cultivated Pudina in her 3.00 acre from 1997. She has invested Rs. 48,000 and got net profit of Rs. 24,000. By seeing her success three neighboring farmers have adopted the above innovation.

Chicory Cultivation

A farmer Sri Siva Sankar Raju started chicory cultivation in 1999 in 1 acre of land with an investment of Rs. 12,000 and got net profit of Rs. 36,250. His neighboring Village farmers have adopted this innovation.

High Density Plantation of Mango

Sri G. Padmanabha Naidu farmer of G. Gollapall Village of Thavanampalli mandal started mango plantation in 1991 in 5 acres of land. He has invested Rs. 30,000 and got net profit of Rs. 1,20,000. By seeing his success about 10 to 15 neighboring farmers have adopted the same technology and are getting high profits.

References

Capacity building of extension scientists to meet the WTO challenges for sustainable agriculture - Short course organized by ICAR and Directorate of Extension, ANGRAU, Hyderabad.

Covin J. and Slevin D.P. (1991). A conceptual model of entrepreneurial theory and practice Vol.(16) No. (1): 7-25.

De Dipak and Rao Mandava Srinivasa (2001). *Entrepreneurial behaviour of Farmers: An axiomatic theory.* Ganga Kaveri Publishing House, Jangamawadi Math,Varanasi.

Gartner W.B., Bird B.J. and Starrj.A (1992). Acting as if differentiating entrepreneurial from organizational behaviour. *Entrepreneurship theory and practice* Vol (16) (3): 13-31.

Strategic Research and Extension Plan of Sangrur district of Punjab, Manage SREP series 14.

Strategic Research and Extension Plan of Chittoor district of A.P., Manage SREP series 19.

Vidyadhari and Sastry (2007). Entrepreneurial behaviour of prawn growers in Nellore district of Andhra Pradesh M.Sc., thesis 2007.

Chapter 20

Entrepreneurship in Agriculture: An Experience of Working with Farmers

Ramesh Chand

Department of Mycology and Plant Pathology,
Institute of Agricultural Sciences, Banaras Hindu University,
Varanasi – 221 005, U.P.
e-mail: rc_vns@yahoo.co.in

Introduction

Agriculture is primary industry and directly related to the survival of man kind. The first entrepreneurship started when agriculture produce was exchanged with the goods manufactured by the village artisans. This system continued from generation to generation and knowledge of crop raising the basic inputs like seed and planting materials started moving to different places. Seed and planting materials are such inputs that can be multiplied within short period thus monopoly in agriculture was not possible like other industries. Agriculture being one the oldest industries most of the production technology simplified to such a level that any one can start with little information and experience.

Being primary industry profit in agriculture is not like secondary and tertiary industries. In the recent years many new innovations helped to developed many new products that made agriculture profitable. In spite of this, development of entrepreneurship is poor in rural area. However, entrepreneurship is possible in rural area by imparting training to the providing inputs, infrastructure and services to the farmers.

Scope in Agricultural Entrepreneurship

1. Technologies those reduce the cost of production and increased benefit of the farmers will open new opportunities for the entrepreneurship.

2. New technologies that are simple, time saving and keep away farmers from the drudgery of labour.

3. Technologies that provides social and psychological benefits to the farmers.

Seed

Improved and hybrid seeds are in great demand. Total seed trade estimated worth of several hundred corers and it is likely to increase with the seed replacement rate. Seed is only a business that is produced and used by the farmers. All the national and multinational seed companies produced their seed on the farmer's fields. Companies only follow seed regulation act, process it scientifically, maintain the quality control and market it. Most of the companies produce their seeds at few locations and transport them to different parts of the country. This adds to the cost of production. If the same seed with the similar quality will produce locally will be more competitive in the market. Several experiments conducted by the by individual and groups of farmers in this direction and doing well in the seed business.

Hybrid seed of vegetables, maize, rice, sorghum, bajara, cotton and many other crops are in great demand. The technique of hybrid seed production is technical and any farmers can produced the hybrid seed of rice and other crop after taking essential training and materials.

Research seed a new term introduced by the seed companies. Research seeds are not released but many seed companies start selling it at very high cost in the market. However, most of these seeds are not superior to released varieties in most of cases.

KAIFICO: Kaimur Farmers Cooperative a group of farmers started production of certified and truthful seeds of rice,wheat, gram and pea in Kaimur district of Bihar Farmers of this society are collecting breeder seeds from the different Universities and organizations and raising the crop as per guide lines of the certification agency. This cooperative is producing around 100 tons of seed of different crops. Members of the society are regularly getting a dividend of average 9 per cent on their investment. Society is arranging the breeder and foundation seeds from the authentic sources and give it to local farmers those are members of the society. Quality control and other standards are met by consulting experts time to time. Seed produced by the KIFICO are very competitive and become popular among the growers.

National Seed Corporation (NSC), Terai Development Corporation (TDC) and many private agencies are out sourcing the seed of cereals, pulses and millet to farmers. After completing necessary formalities seed growers enter in the agreement for the production of breeder and foundation seeds to these agencies. Under this scheme benefit is not much; however, farmers can learn quality seed production and seed business.

A Journey of Grain Producer to the Owner of Seed Companies

Bhola Singh, a small farmers of Village Bhurkura, Chuanar, Marzipan (U.P.) produced rice and wheat grain from the generations. A team of Banaras Hindu University motivated the farmers of this village to adopt the zero till technology for wheat cultivation instead of traditional ploughing that take 7 to 10 days in the land preparation. By advancing the sowing date of wheat though the zero till technology, farmers harvested 15 to 20 per cent more than the traditional method at reduced cost of production. Farmers of this village adopted new cultivars of wheat and rice and got their yield increased by 10 to 15 per cent. In order to increase the income few farmers of this village grown wheat and rice for seed and enrolled themselves as seed grower with NSC and other companies. They learned many skills of the quality seed production and processing. Their close contact with many government officials during this period helped to good rapport. These officials encouraged them to take up the seed business. In the year 2007, Bhola Singh, established a new seed company named Pooja. In the first season itself he obtained the good order for wheat seed from Banaras Hindu University. Seed was also sold from his processing plant. A similar successful story in the same village was also made by Mr Chandra Sen Singh. These two farmers now received the orders from many public and private companies for the seed of rice, wheat and others.

Several NGOs and farmers cooperative are also engaged in the seed business under the participatory seed production and seed villages programme. State Agriculture Department and Seed certification agency are also helping them in the seed production and seed trading.

Hybrid Seed

Hybrid of maize and rice are becoming popular among the farmers. Hybrid seeds of rice and maize are produced at few locations in India mainly in Andhra Pradesh. Sorghum and bajara hybrid seeds are produced mainly in the Karnataka and Maharastra. There is good scope of entrepreneurship if hybrid seeds of these crops could be produced by the local farmers. Efforts are made by Dr Arun Joshi and his team for the hybrid rice seed production at different places of Varanasi, Mirzapur, Chandauli and Ghazipur districts of U.P with the farmers. They standardized the production technology under local conditions and proved that hybrid seed production of rice can be taken up by the farmers after acquiring necessary skills. This again created an opportunity of entrepreneurship among the growers. Similarly there is great demand of hybrid seed of maize that is monopolized by the private companies. Hybrid seed production of maize can be taken up by the farmers after acquiring required skill and material.

Hybrids of vegetable are very much popular among the growers. Hybrid seed of vegetables are produced manually and need large numbers of trained man power from crop caring, emasculation, pollination and processing. Hybrids seed production of tomato, okara and cucurbits can be taken up by the farmers after obtaining appropriate parents and skills. Hybrid seed of these vegetables produced locally by employing local labours will be more competitive and can be sold at lower price. Thus, it has great potential for new entrepreneurship.

Planting Materials

In the recent year many new horticultural crops are introduced and they are in great demand. Many farmers selected their own materials and commercialized it by multiplication.

A Success Story in Maharashtara

Tas-e-Ganesh, a bud mutants of Thompson seedless selected by Arve Brothers at Tasgaon, Sangali of Maharastra. Tas – e – Ganesh is dual purpose variety good for table as well as raisin making. This material stands better under saline conditions of the Mahrastra. It responds well to gibberelic acid and berry got elongated. Just after the identification of Tas- e- Ganesh huge demand was came for the planting materials. This material was multiplied at large scale and all the nursery growers earned the money out of this material. Success story of Tas- e- Ganesh, motivated a large number of grape growers for the selection of new planting material that could meet the desired quality and other requirement. SONAKA a table grape clone selected by Mr. Nana Kade from Nanaj Village of Solapur become popular among the grower for its elongated berries and good market price as table grape. However, it is not good for raisin making. Nursery growers multiplied this material and benefited. Similar type of story was replicated by many and selected specific materials that addressed the specific problem.

New Root Stocks for Grape

Grape industries suffered a lot due to salty water that is commonly available in the grape growing areas of Maharastra and Karnataka. Salty water damaged the vineyard and many vineyard of Sangali district uprooted. This matter was discussed in the grape growers association and association approached to Indian Institute of Horticultural Research, Bangalore, for the salt tolerant root stocks. Two root stocks Saltcrick and Dogridge imported from Australia on the demand of growers. These root stocks tested at different concentration of salty water. Mr. Basant Rao Arve, took 4- 5 cuttings after completion of one year data and start multiplying the cutting with the hope that it may click in the Maharastra. Mr. Basant Rao Arve multiplied several thousands of planting material by the time result of third year testing came. Mr. Arve sold the cutting and proved that business by new planting materials with the fellow nursery men and farmers. Similarly there is great demand of many new fruit crops in the semiarid and arid areas for quality planting materials. There are many commercial fruit crops where lot of variability exists for different characters (shape,size, colour, keeping quality, processing value *etc.*) that can be selected and commercialized at regional level.

Production of Bio-agents and Bio-fertilizers

Alternative plant protection mainly depend on the use bio agents and cultural practices. Most of the bio agents registered for the marketing are produced at few places *i.e.* Mumbai, Delhi and Hyderabad and send to different places by train and trucks. Bio agents are living organism and need a special care in handling and transportation. Improper handling and transportation bring the population count low and finally the desired quality impaired and expected result from bio agent could not obtain. These limitations give advantages to the local bio agents producers. Reduced production cost and better quality help them to compete in the market in comparison to existing product. Most the formulations available in the market is talcum powder based and containing 0.1 to.2 per cent of total weight. A new granular and reusable formulation developed for the production of an established bio agent *Trichoderma* that provide protection to the plants against a number of diseases and promote growth of the plants. A pilot programme was initiated at Rahia village (Chunar), District of Mirzapur, here the women farmers are producing the granular *Trichoderma* that are used for the seed treatment of their own crops. Surplus produced is distributed among the members of group. This group is supplying to others after receiving the order. This story was validated by many individual farmers of Jaunpur, Mirzapur, Varanasi after getting necessary training and culture of *Trichoderma* during year 2006 and 2007.These farmers adopted the technique, produced and marketed *Trichoderma* successfully. State Agriculture Department purchased the bioagent worth of Rs. several million for the distribution of among the growers. Part of this order can be obtained by the farmers co operative for the supply of bioagent.

Similarly several bio fertilizers can be produced by the farmers in their village. Required amount of BGA, Rhizobium, Phosphorus solubilizing bacteria (PSB), PGPR and other biofertilizers are in great demand. However, the quality product and timely supply to the farmers is a major problem. A new innovative formulation with the good quality and timely supply may create good business. Thus there is good scope for entrepreneurship in the microbial business if some farmers adopt it.

Custom Hiring/Services

Services are in great demand by the farmers for the farm operations, transportation, harvesting and processing. These required substantial investment. Many farmers purchased tractor with cultivators and providing ploughing and seeding services to the farmers. There are many new tools introduced recently that are in great demand *e.g.* zerotill seed drill that is in great demand by the farmers of eastern and others regions in order to advance their wheat sowing. Many tractor owners purchased the zero till drill and providing services to the farmers. Within one wheat season every drill owner earned Rs. 20 to 25 thousand.

Rotavator is another new tool that is used for the bed preparation. Farmers calculated that field preparation by rotavator is profitable. There is great demand of service of rotavator for wheat and rice both. Many service providers those who purchased the rotavator making good profit on their investment.

Bed Maker

raised bed cultivation are becoming popular among the growers due to, increased production, high quality produce, less infestation of diseases and pests and reduced water requirement. Bed maker is a simple tractor drawn implements cost Rs. 12 to 15 thousand. There is good demand of bed maker for planting rainy season vegetables *i.e.* cabbage, cauliflower, tomato and brinjal. Raised bed saves plants from water lodging. In the main season it save water due to furrow irrigation that save lot of water and give good quality of fruits due to dry bed that create unfavorable conditions for the growth of many pre and post harvest disease and pest. Thus it increased keeping quality of produce when transported for long distance.

Ridge sown pegionpea helped in maintaining the adequate plant stand during heavy rains. It conserves water during winter rain thus helps to retain sufficient moisture in the field during flowering and pod filling. By ridge sowing pegionpea farmers can harvest 15 per cent 20 per cent extra. Services for ridge making are needed by most of the farmers during sowing operation. A tractor owner can add substantial income by providing this kind service.

Transport

Villages those are situated for of places from the market need transport facility daily to send their vegetables and fruits to the market. Most of the tractor owners are proving transport services to the villagers for short distance. National Horticulture Board is also providing subsidy to purchase the vehicles for the quick transportation of horticultural produce to long and short distance market.

Small Scale Milling

In the eastern India lot of rice are produce and milling is required in every village. Small scale rice mill can be operated by an investment of Rs. 1.5 to 2 lakhs. Milling starts during lean season and tractor and labour can be used effectively. It provide job to 2 to 3 person for few months. Rice bran and rice husk come as an additional benefit.

High Tech Nursery

Raising nursery of early cabbage and cauliflower and difficult and need special skill and experience. There are few farmers who are specialized and developed infrastructure for raising good quality of seedling during the rainy season. They procure good quality of seed of cauliflower and raise the nursery for the small and medium farmers who can not afford high tech nursery facilities. These nursery growers book the order of the farmers and supply them good quality of seedling of desired variety in time. Nursery growers earn good amount from a small piece of land. Farmers those who once took the planting materials and fully satisfied with the quality are permanent customers of these nursery growers. In and around Bangalore it has developed as small scale industry with high tech infrastructure.

Cut Flowers

Demand of cut flowers is increasing every year for the various occasions. A special skill is required about the regulation of flowering. All, the grower regulate the flowers in such a way so that it could be ready to supply to specific occasion. Small and marginal farmers of periurban are earning their livelihood by adopting this.

Duck Rearing

It is becoming popular among the landless farmers belonging to schedule tribe specialized in profession of duck rearing. They moved with their flock of duck in the canal for perching. They bring them to their nest in the evening and give some ration. During the time of Christmas all the ducks are sold to the traders coming to them form North Eastern region. This has brought the confidence among the landless farmers and they are earning their lively hood. These ducks are improving the quality of water by cleaning and adding the oxygen in the water. It has emerged a good business among Beaar Tribe of Mirzapur and adjoining districts where the canal net work is well developed.

Beekeeping

it is also a profitable business to the farmers the demand of honey is more than the supply and this can be adopted by the farming after learning the skill of bee keeping. Honey bee also increased the yield of most of the cross pollinated vegetables and fruits. In every village at least two to three farmers can earn Rs 12000/pcr year by selling honey and additional income can also be generated by supplying the colonies of bees to other growers. This area is still untapped.

References

Chand R. and. Kishun. R (1990).Out break of grapevine bacterial canker disease from India. *Vitis.* 29, 183 –188.

Chand, R.Patil B.P and Kishun. R.(1991). Management of grapevine bacterial canker disease by pruning. *Indian J. Agric. Sci.* 61, 220 – 222.

Chand, R.(2007). A process for producing reusable granular formulation of Trichoderma. *The Patent official J.* 12/1/2007 pp. 717.

Joshi, A. K, Chand R and. Chandola V. K (2007). Reaping the Benefits: Assessing the Impact and Facilitating the Uptake of Resource Conserving Technologies in the Rice-Wheat Systems of the Indo-Gangetic Plain Impact Studies On Zero Tillage Technology and Replacement of Varieties in Eastern Uttar Pradesh. Study by A Institute of Agricultural Sciences, Banaras Hindu University, Varanasi, India CABI Ref: U3013 (WWW.cabi.org) pp. 69.

Ortis, Ferrara G. Joshi AK, R.Chand,. Bhatta, M.R. Mudwari, A. Thapa D.B Sufian, M.A. Sakia, T.P. Chatrath, R. Witcombe J.R, Wirk and. Sharma R.C (2008) Partnering with farmers to accelerate adoption of new technology in South Asia to improve wheat productivity. *Euphytica* (in Press).

Chapter 21

Entrepreneurship Development in Agriculture: Case Studies

Gautam Goswami

Director (Scientist E),
Technology Information, Forecasting and Assessment Analysis (TIFAC),
Vishwakarma Bhavan, Saheed Jeet Singh Marg, New Delhi – 110 016
e-mail: goswamig@hotmail.com

Innovation: Key Component of Entrepreneurship Development

Innovation – Why?

☆ Because the world is changing (fast) and so must you

☆ Because you can't be too happy with yourself

☆ Because the old tricks don't work any more

☆ Because you are on the right place to do it

☆ Because innovation is a business there are some selling arguments....

Case Studies

☆ Agriculture

☆ Fisheries

☆ Carbon trading

Agriculture

Case – 1 (Through seed production)

Location: Block Garer, Dist: Deoria

Background

TIFAC supported project on "Demonstration of Systems Approach in Low Productivity Areas of Deoria District"

Project impact (3 yrs)

☆ Productivity of Rice and Wheat increased.

☆ Revolving Fund of Rs.1,00,000 generated

☆ Farmers got trained on seed production

After completion of the project

☆ The project manager took up seed production activity as business

☆ Revolving Fund of Rs.1.00 Lakh – Capital

☆ Own Investment of Rs.0.70 Lakh – Capital

☆ Total Capital – Rs.1.70 Lakh

Business Details (in Rs.)

Crop	Seed Procured (Ton)	Amount (Rate in Rs./t)	Processing Bagging, Tagging etc. (Rs.4,000/t)	Selling Price (Rate in Rs./t)	Total Profit	Total Capital
Rice 1	15	1,12,700 (7500)	60,000	2,16,000 (16,000)	43,300	2,13,300
Wheat 1	15	1,50,000 (10,000)	60,000	2,70,000 (20,000)	60,000	2,73,300
Rice 2	20	1,60,000 (8,000)	80,000	2,88,000 (16,000)	48,000	3,21,300
Wheat 2	20	2,50,000 (10,000)	80,000	4,00,000 (20,000)	70,000	3,91,300
Rice 3	25	2,00,000 (8000)	1,00,000	3,60,000 (16,000)	60,000	4,51,300
Wheat 3	25	2,50,000 (10,000)	1,00,000	4,50,000 (20,000)	1,00,000	5,51,300

Case – 2 (through Ago Services)

Location: Nayagarh District, Orissa

Background

☆ High rainfall (1400 – 1600 mm)

☆ Mostly mono crop (rice in *kharif* season, practically no crop in *Rabi* season)

☆ Rice productivity is very low (< 1t/ha)

☆ No concept of water conservation, no mechanization

☆ Very poor economy

TIFAC supported project on micro water resources development linked with agriculture development

Project components

☆ Micro water resources development

☆ Crop demonstration (with improved technology)

☆ Emphasis on *Rabi* crop cultivation

☆ Mechanization

☆ Agro Service Center

Project Impacts

☆ Productivity of rice increased (>2 t/ha)

☆ Cropping intensity increased

☆ Mechanization started

☆ Three agro Service Center created

☆ Efficient distribution of water demonstrated

After completion of the project

☆ Water resources were handed over to Gram Panchayat.

☆ Agro machines (Multi purpose power tiller, Thresher–cum-Grain Cleaner, Pump *etc.*) were sold to Agro Service Centers.

☆ Out of three, only one Agro Service Center took up the work as entrepreneurship.

☆ Agro Service Centers earned the confidence of local bank (became credit worthy).

Business Details

Year	Power Tiller	Thresher-cum-Grain Cleaner	Pump	Advisory Services	Total income
1	1000–Kharif 1000–Rabi	8000–*Kharif* 8000–*Rabi*	20,000	Complementary	38,000
2	1250–Kharif 1250 – Rabi	10000–*Kharif* 10000–*Rabi*	25,000	– do –	47,500
3	1500–Kharif 1500 – Rabi	12,000–*Kharif* 12000–*Rabi*	30,000	– do –	57,000

Case – 3 (through Agro Processing)

Location: Almora district of Uttarakhand.

Background

☆ The food grain production in hills is not sufficient for year round consumption of the households, thereby compelling the farmers purchase food grains from nearby markets.

☆ Coupled with this the losses due to traditional processing further aggravates the problem.

☆ Wheat is major crop in Uttarakhand (48.6 per cent area) followed by rice (35 per cent area), finger millet(17.2 per cent) and barnyard millet (8.5 per cent).

☆ In spices, the percentage area under chilies is maximum (45.90 per cent) followed by ginger (18.12 per cent), garlic (10.48 per cent), turmeric (10.13 per cent), coriander (6.4 per cent), fenugreek (5.36 per cent)and large cardamom (1.27 per cent).

☆ The number of processing units for food grains is inadequate.

☆ The establishment of APC is seriously felt need in the region.

Action

☆ An agro-processing center (APC) was established with the institutional support.

☆ The establishment cost of INR 2,19,000 was borne by the institute.

☆ The entrepreneur invested INR 1,66,255 for purchasing raw food grains.

☆ Total annual cost (Operational cost + Rental value of APC and storage building) was INR 27956.

☆ Annual income of INR 95,145/- was generated from sale of the processed products in brand name of "HIMRAJ" apart from generating income of INR 27,867 by providing processing service to the nearby farmers.

☆ This way he earned the INR 1,23,012 as gross income.

Annual Income Generation through Sale of Processed PDTS

Commodity	Quantity Purchased, kg (Unprocessed)	Net Input Rs. (Rate: Rs./kg)	Quantity Sold	Net Output Rs. (Rate: Rs./kg)	Total Income
Wheat	14310 (0)	78,705 (5.5)	14310	143100 (10)	64395
Paddy	9000 (0)	76500 (8.5)	6000	90000 (15)	13500
Finger Millet	3969 (3469)	2500 (5)	500	6000 (12)	3500
Barnyard Millet	1821 (821)	5000 (5)	600	15000 (25)	10000
Turmeric	200 (160)	1600 (40)	40	2800 (70)	1200
Dhaniya	70 (40)	1050 (35)	30	2400 (80)	1350
Chilli	50 (20)	900 (30)	30	2100 (70)	1200
Total		1,66,255		2,61,400	95,145

Annual Income Generation from Processing Services

Processing	Quantity (kg)	Rate (Rs./kg)	Income (Rs.)
Wheat grinding	7300	2.0	14,600
Rice milling	3402	2.0	6804
Oil extraction	1241	5.0	6205
Spices grinding	86	3.0	258
Total	12029		27,867

Fisheries

Case – 4

Location: Village Garer, Dist: Deoria.

Background

☆ Eleven unemployed youth got frustrated as there was no chance of getting employment.

☆ They formed a Self Help Group.

☆ They took lease of a pond of 66.5 acres from UP Govt. for a period of 10 yrs.

☆ Lease amount: Rs.8.00 Lakh.

Business Details

Capital

☆ Contribution from each member @ Rs.40,000/- = Rs.4,40,000/-

☆ Contribution from SHG   = Rs.50,000/-

☆ Loan taken from Grameen Bank   = Rs.1,00,000/-

   Total Capital   = Rs.5,90,000/-

Business Details (Contd.)

Year	Investment (Rs.)	Amt Recd. (Rs.)	Net Income (Rs.)
1	2,00,000 – to Govt. 1,50,000 – Seed	2,10,000	–1,40,000
2	80,000 – to Govt. 2,00,000 – Seed 35,000 – Cleaning 24,000 – Maintenance	2,50,000	–89,000
3	80,000 – to Govt. 2,50,000 – Seed 40,000 – Poisoning 24,000 – Maintenance 20,000 – Return to Bank	3,10,000	–1,04,000
4	80,000 – to Govt. 24,000 – Maintenance	7,00,000	+5,96,000
	Net Income		+2,63,000

Carbon Trading Kyoto Protocol

The Kyoto Protocol is an agreement made (11th Dec 97) under the United Nations Framework Convention on Climate Change (UNFCCC). Countries that ratify this protocol commit to reduce their emissions of carbon dioxide and five other greenhouse gases. and entered into force on 16 February 2005

The objective of the Kyoto Protocol is to achieve "stabilization of greenhouse gas concentrations in the atmosphere at a level that would prevent dangerous anthropogenic interference with the climate system.

Governments are separated into two general categories: developed countries, referred to as Annex I countries (who have accepted greenhouse gas emission reduction obligations and must submit an annual greenhouse gas inventory); and developing countries, referred to as Non-Annex I countries (who have no greenhouse gas emission reduction obligations but may participate in the Clean Development Mechanism);

Kyoto includes "flexible mechanisms" which allow Annex I economies to meet their greenhouse gas emission limitation by purchasing GHG emission reductions from elsewhere. These can be bought either from financial exchanges, from projects which reduce emissions in non-Annex I economies under the Clean Development Mechanism.

Comparison of Emissions

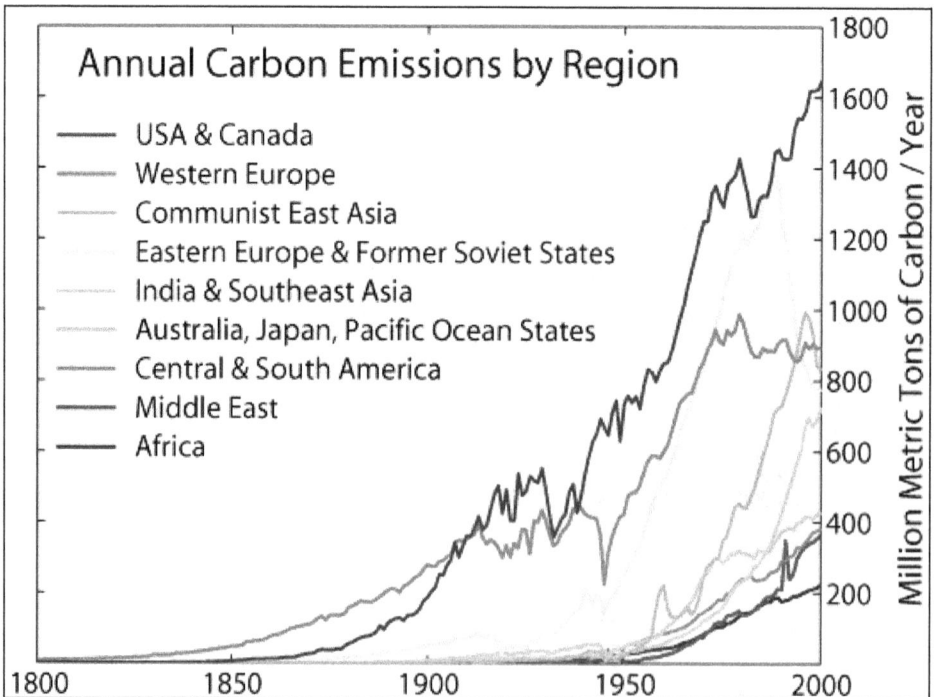

Annual Carbon Emissions by Region

— USA & Canada
— Western Europe
— Communist East Asia
  Eastern Europe & Former Soviet States
  India & Southeast Asia
  Australia, Japan, Pacific Ocean States
— Central & South America
— Middle East
— Africa

Chicago Climate Exchange (CCX) Worlds first and North America's only active voluntary, legally binding integrated trading systems to reduce GHGs Eligible projects include:

- ☆ Agricultural methane
- ☆ Landfill methane
- ☆ Coal mine methane
- ☆ Soil carbon
- ☆ Forestry
- ☆ Renewable energy
- ☆ Resource Conservation Technology (RCT)

CDM Project Cycle

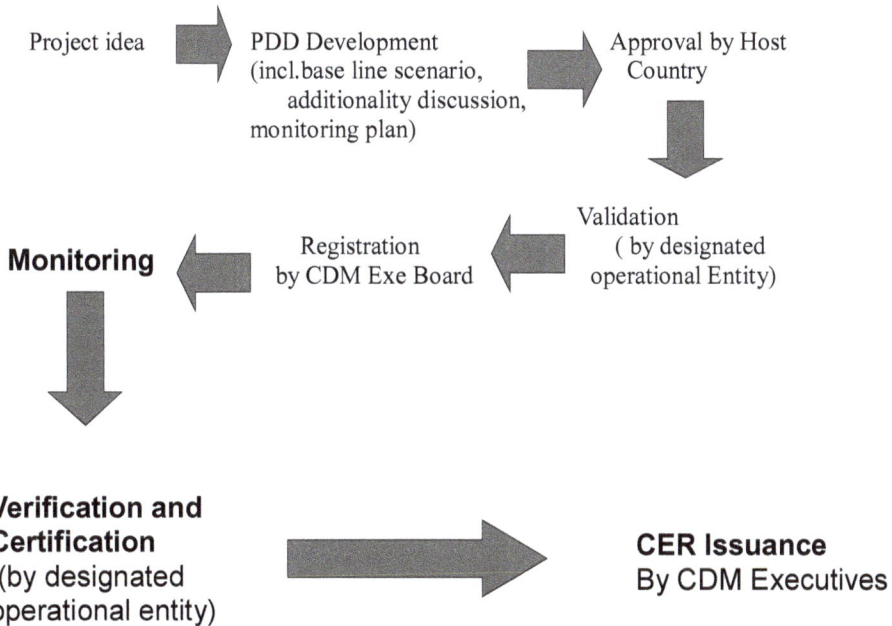

Project idea → PDD Development (incl. base line scenario, additionality discussion, monitoring plan) → Approval by Host Country

↓

Validation (by designated operational Entity)

←

Registration by CDM Exe Board

←

Monitoring

↓

Verification and Certification (by designated operational entity) → **CER Issuance** By CDM Executives

Case Studies

ITC Ltd, in a bid to achieve 'greening of wastelands' through its ongoing social forestry programme executed by the Bhadrachalam Paper Division, has, as on September 15, promoted 74,427 hectares of plantations (both Eucalyptus and Subabul trees) by distributing 313 million saplings in Andhra Pradesh.

In the process, ITC has succeeded in registering as many as seven CDM (clean development mechanism) projects (three large scale and four small), accounting for nearly one million CERs (Certified Emission Returns).

ITC, working with PwC (on the CDM projects), is able to convince UNFCCC with its satellite picture data of the increase in green cover and reforestation of degraded lands, and thereby affirm carbon sequestration, it would be a breakthrough in securing carbon credits under the Kyoto Protocol.

Chapter 22

Mushroom: A New Enterprise for Women and Rural Youth

A.K. Chaubey

Professor,
Department of Agricultural Extension,
CSK Himachal Pradesh Krishi Vishwavidyalaya, Palampur, H.P.

As per the census 2001 report, out of the country's total 402.51 million workers 235.08 million are directly dependent on agriculture. A majority of them consist of small and marginal farmers and agricultural workers, and a large number of them are either unemployed or under employed. Although the percentage of work force dependent on agriculture has declined from 64.8 per cent in 1991 to 58.40 per cent in 2001, the absolute number has significantly increased. If we look at the contribution of this sector to the total GDP (about 22 per cent in 2003-04), the percent dependence on agriculture seemed to be very high and indicates to the intensity of disguised unemployment absorption capacity of the sector may further decline due to the application of labour saving technology, declined public investment and lifting of quantitative restriction on imports of agricultural products. Similarly urban organized manufacturing sector does not appear to be promising in absorbing the growing work force. This is evident from the decline in the absolute number of jobs in the public sector enterprises due to retrenchment and disinvestment policy and a slow growth in employment in Private sector due to application of more capital intensive technology. Although urban informal sector is growing very fast and a large number of workers are being absorbed by it but the exodus of rural workers in urban areas has been creating serious problems to the carrying capacity of the urban sector. Under these constraints, the viable option for providing gainful employment to the growing rural workforce and to lessen the burden of manpower on agriculture seems to be in generating more employment in the rural non-farm sector which constitutes mainly three components service such as hotels and dhabas, transport and communication, repair shop, tailoring and hair cutting, cobblers, *etc.*, trade and commerce such as wood, iron, cloth, groceries shop, medical stores and other business and manufacturing such as shoes making, farm implements, bakeries, pottery, basket making, soap making, rice-mills, sugar and khandsari, fruit and vegetable processing units, dairy, mushroom, honey bees, meat processing unit and other small and cottage industries. This paper is restricted to the mushroom cultivation for rural youth and women.

Demand for mushroom is increasing at 7 per cent per annum resulting in further widening gap between demand and supply and as such immense opportunities for rural women and youth to capitalize on this gap. The demand is not only for the button mushroom but for other specialty mushrooms also, *viz.*, Oyster (*Pleurotus* sp.), Block ear mushroom (*Auricularia polytricha*), shiitake (*Lentinula edodes*), *etc.* In recent years, there has been an increasing demand for cultivation of medicinal mushroom *viz.*, *Ganaderma lucidum*). In India, commercial mushroom cultivation started in 1960's when button mushroom production started in Himachal Pradesh. However, it was in 1990's when the spread of seasonal production of button mushroom began in northern states of Haryana, Punjab, Uttar Pradesh and Uttarakhand and the total production increased from 4000 tonnes (1985) to 30,000 tonnes (1995) and at present it is estimated to be 50,000 tonnes/annum. However, 85 per cent of it is still dominated by the button mushroom. In India, all types of mushrooms *viz.*, temperate, sub tropical and tropical mushrooms can be produced.

Essentials for Entrepreneurship Development

Skill Formation

For the cultivation of mushroom, improved technical and other skill among the women associated with the rural enterprises are of prime importance for enhancing the productivity and marketability of their products in the competitive market environment. We have a infrastructure for vocational trainings in the state that need to be upgraded and made demand driven so that trainees may have access to those skill which are required in the changing labour market.

Table 22.1 Vocational Trainings on Mushroom Cultivation
(Year 2003-2007) at Krishi Vigyan Kendra, Una (H.P.)

Year	No. of Trainings	No. of Participants		
		Male	Female	Total
2003	1	11	7	18
2004	2	21	69	90
2005	4	87	57	144
2006	3	43	33	76
2007	2	22	30	52

It has been observed that the training provided in the most of the KVK's and Institutes for the skill that has demand. The curriculum is not attached to current market requirement due to lack of revision. The training is oriented towards the setting up of self oriented enterprises. Trainees have multi skills and are generally trained in skill to having access to information, credit and market knowledge. The training is a continuous process and the post training services provided to the trainees as and when required.

Technical Guidance to Rural Women on Mushroom Cultivation

Mushroom cultivation is an indoor activity that provides much scope for women employment. The enterprise requires little land and thereby less initial investment. But technical knowledge and access to spawn (seed) are important to start the enterprise. Mushrooms are excellent source of vitamins and minerals, popular for their delicacy and flavour and thus have more demand in homes and restaurants. In urban vegetable markets, mushrooms are sold at fairly high prices and have a good market demand.

The most popularly cultivated mushroom in India are the white button mushrooms and Oyster mushrooms. The cultivation of these mushrooms depends on the suitability of the region and season as they need different temperature ranges for growing. While for white button mushroom, optimum temperature is 16-20°C, the temperature range for Oyster Mushroom is 20-33°C.

Steps in the Cultivation of White Button Mushroom

The compost on which the mushrooms grow is prepared by microbial degradation of organic wastes. During the training courses, the detail process about mushroom production is being communicated to the participants.

☆ Beds of the compost are prepared.

☆ Spawn (mushroom mycelium) is spread uniformly on the compost bed.

☆ The compost is then covered by casing soil to meet nutrient requirement and moisture conservation.

☆ The cropping cycle begins when the mycelium starts to come out. Cropping includes pinning (appearance of mushrooms as pinhead) and cropping *i.e.* maturing of mushrooms which are 3-5 days old after pinning.

Steps in the Cultivation of Dhingri Mushroom Production

Paddy straw which is easily available throughout the year is the substrate for this variety. This variety of Mushroom is often cultivated in warmer climate and the investment is almost negligible. The steps involve in the cultivation of this mushroom are:

☆ Fresh well dried paddy straw is cut into 3-4 cm pieces and soaked overnight in clean water.

☆ The straw is then filled in big vessels and boiled in water for 10-15 minutes.

☆ The water is then drained and the straw is spread out for drying.

☆ The straw is then filled in polythene bags and sterilized at 10 lbs per sq. inch for 99 minutes.

☆ When the straw is neither too wet nor too dry but moist to hold it is filled in polythene covers.

☆ Spawning is then carried out either on the surface or in layers.

☆ The spawned bags are kept in a dark and cool room.

☆ Once bags are covered with mycelium, polythene cover are removed and the bags are kept in rakes and spread with water twice a day.

☆ After 2-3 days pin heads appear.

☆ When mushrooms are 6-8 cm in size they are plucked.

Mushroom Production – Economics

Space required – 10x12 Sq. foot

Time/Duration – 28 days

Expenditure involved

☆ 10-12 Bottle seed (Rs. 15/Bottl) = Rs. 150.00

☆ Compost Cost = Rs. 800.00

Income

☆ Total production 80 kg – 1qt. = Rs. 3200.00 - 5000.00
(Rate Rs. 40-50 per kg)

☆ Net Profit = Rs. 2000.00 – 4000.00

Future Prospects of Mushroom

In view oaf the post WTO India has tremendous potential for mushroom production and all commercial edible and medicinal mushroom can be cultivated there is a an increasingly demand for quality products at competitive rate both in domestic and export market. However, the economic reform liberalization privatization and emergence of WTO have changed the agricultural trade.

Developing Effective Market Mechanism

Experience shows that many of micro and small enterprises in the country could not survive due to lack of market support. Their growth decline largely depends on markets links with large firms and access to technology, credit skill and marketing. Therefore there is a need to develop a suitable institutional market mechanism for the products of these enterprises. In case of mushroom and other perishable products, the existing marketing infrastructure of co-operative, KVK, and other government outlets could be used. Village market could also be developed at village level. What is more required is to build brand equity for these products.

Conclusion

Agriculture can not provide gainful employment to the entire rural workforce. The mushroom cultivation is now a new enterprises in rural areas adopted by the women and youth. The sharp increase in the income of middle and higher income groups in rural and urban area simultaneously, the poverty level reduced from 38.6 per cent in 1987-88 to 26.7 percent in 1999-2000.

The social trend also changed with increasing nuclearisation of families where both members earning and also there is an increase in health consciousness and quality awareness among the middle class, due to which mushroom cultivation boomed up.

Chapter 23

Rural Entrepreneurship: An Example of Income Generating Activities of Women Self Help Groups in Karnataka

N.R. Gangadharappa, M.S. Jyothi and T.N. Anand

Department of Agricultural Extension,
College of Agriculture, UAS, GKVK, Bangalore – 560 065, Karnataka

ABSTRACT

The study was conducted during 2010-11 in Tumkur, Chitradurga, Kolar and Bangalore Rural districts of Karnataka. Total sample size of the study was 120. The results indicated that the non-farm activities like business and rural artisans (39.32), dairy (25.7), sheep rearing (14.32) etc., has generated additional employment (mandays) and the additional average income generated/annum from non-farm activities is higher than farm and off-farm activities. The higher income, has been generated from business (Rs.4168) followed by animal husbandry (Rs. 4003), sheep rearing (Rs. 3065) and Rs. 287 from collection of NTFP's. It has resulted in changes in cosmopoliteness (61 per cent) followed by extension orientation/contact (54 per cent), extension participation (24 per cent), mass media participation (23 per cent), family education (22 per cent) and organizational participation (16 per cent), also the percentage change in dressing pattern has increased by 41 per cent followed by celebration of rituals/ceremonies and festivals (14 per cent) and food habits increased by 8.2 per cent after taking up Income Generating Activities (IGA) under the Karnataka Sustainable Forest Management and Biodiversity Conservation Project (KSFMBC).

Women have been deprived of socio-economic and cultural independence. The empowerment of women and improvement of their status and economic role needs to be integrated into economic development programmes, as the development of any country is inseparably linked with the status and development of women. One of the powerful approaches to women empowerment and rural entrepreneurship is the formation of self help groups especially among rural women. In this regard, there are many programmes/ schemes/projects use women self help groups as a vehicles to move forward women socially and economically. The Karnataka Forest Department, in order to improve the tree cover and livelihood of forest dependents of the forest fringe village, implemented Karnataka Sustainable Forest Management and Biodiversity Conservation (KSFMBC) Project. This project gives importance to income generating activities to improve the livelihood of resource poor Village Forest Committee (VFC) members. Hence, it is important to know the impact of Income Generating Activities (IGA) on the socio-economic and cultural conditions of rural women beneficiaries of Karnataka Sustainable Forest Management and Biodiversity Conservation project. In this context the study was undertaken analyse socio-economic and cultural impact due to Income Generating Activities (IGA) on the beneficiaries.

Methodology

The study was conducted in the southern Karnataka, *viz.*, Tumkur, Chitradurga, Kolar and Bangalore Rural districts during 2010-11. The unit analysis of the study was exclusively on Self Help Groups (SHGs) women who were also the members of village forest committees. The overall development of rural women from income generating activities regarding their socio-economic and cultural aspects was the key factor in the operational area. In each selected districts, depending on the number of village forest committee villages, and activeness of the SHGs in income generating activities, one village forest committee was selected purposively. Under each village forest committee, 2-3 SHGs were selected randomly to constitute 120 respondents.

Results and Discussion

Occupations of Rural Women Beneficiaries Before and After IGA

A cursory look at Table 23.1 shows that, there was greater increase in the percentage of people shifting towards agriculture + livestock (19 per cent) followed by labour + business (16 per cent), labour + livestock (15 per cent) and a combination of all the three occupations *i.e.* agriculture + livestock + labour (12 per cent). In contrast, there was a profound decrease in the percentage of farmers depending on agriculture alone (4 per cent), agriculture + labour wages (8 per cent) and there was 100 per cent discontinuance of people depending upon labour wages alone. As the very motto of the IGA was to extend financial assistance to the beneficiaries engaged in agriculture or labour wages, the beneficiaries have been benefited much by KSFMBC revolving fund and thus, able to take up livestock rearing or to setup their own independent business. Among the various enterprises, much emphasis has been given livestock particularly for labourers and hence, their enterprise combination *viz.*, agriculture + labour wages+ livestock has increased considerably in the present investigation. The reasons given above holds good for the finding relating to the decrease in the percentage of farmers depending on agriculture or labour wages alone as their occupation. By utilizing the financial assistance of the project, the percentage of people depending on either labour wages or agriculture has decreased as abserved under the results. The findings were similar to the findings of Umashankar (2004).

Table 23.1: Occupations of Rural Women Beneficiaries Before and After IGA

(n=120)

Categories	Before		After	
	No.	Per cent	No.	Per cent
Agriculture	21	17	5	4
Agriculture+Labour	14	12	10	8
Agriculture+Livestock	15	12	23	19
Agriculture+Business	5	4	7	6
NTFP+Agriculture	2	2	2	2
Agriculture+Livestock+Labour	12	10	15	12
Labour	20	16	0	0
Labour+Livestock	11	9	19	16
Labour wage+Business	5	4	20	16
NTFP+Livestock	3	2	7	6
NTFP+Labour	4	3	4	3
Other combinations	6	5	8	7
Total	**120**	**100**	**120**	**100**

Distribution of Beneficiaries Based on Employment Generation from different Sources Before and After IGA

It could be seen from Table 23.2 that, the total human-days of labour generated before IGA were 233.53 days as against 317.55 days after IGA. Thus, an additional 83.82 human-days were generated due to IGA as such. The results in the table also depict that the highest average human-days was generated from business and rural artisans (39.32), followed by dairy (25.7), sheep rearing (14.32), collection of NTFP (9.77) and only 6 human days from agriculture. Whereas, labour wages has been reduced to 11.29 human days after taking up IGA. The percentage change in farm activities due to IGAs has increased only by 6 per cent and it was observed to be nonsignificant as revealed by 't' test.

Table 23.2 also depicts that, the average humandays generated from non-farm activities *viz.*, dairy (30.80 days), sheep rearing (12.21 days), business and rural artisans (23.68 days) and collection of NTFP (12.30 days) before IGA were less than that of after IGA. The highest average human-days was generated from business was (63.00 days), followed by dairy (56.50) sheep rearing (26.53) and from collection

Entrepreneurship in Agricultural Development

Table 23.2: Distribution of Beneficiaries Based on Employment Generation from different Sources Before and After IGA

(n=120)

Particulars	Average Human Days		Additional Average Human Days/ Annum	Percentage Change Due to IGA Increase/ Decrease	Paired 't' Test
	Before IGA	After IGA			
Farm					
Agriculture	91.53	97.53	6.00	6	1.20[NS]
Non-farm					
Dairy	30.80	56.50	25.71	83	4.26**
Sheep rearing	12.21	26.53	14.32	117	2.97**
Business and rural artisans	23.68	63.00	39.32	166	4.50*
Collection of NTFPs and others	12.30	22.07	9.77	79	2.49*
Off-farm					
Labour	63.01	51.72	-11.29	17	3.75**
Total	**233.53**	**317.35**	**83.82**	**35**	

One human-day is equivalent to 8 hours of working.

NS: Non-significant; *: Significant at 5 per cent level; **: Significant at 1 per cent level.

of NTFPs (22.07) after taking up IGA. This accounts to 166 per cent increase in business and rural artisans followed by sheep rearing (117 per cent), dairy (83 per cent), collection of NTFPs (79 per cent) and agriculture (6 per cent). Whereas, labour wages has decreased by 17 per cent. The calculated 't' values for non-farm activities like diary, sheep rearing were found significant at one per cent level. Whereas, off-farm activities were also found to be significant. Average human days generated from off-farm activities before IGA is more (63.01 days) than that of after IGA (51.72 days). After taking up IGAs, nonfarm activites were reduced by 11.29 average humandays per annum. The percentage change observed after taking up IGAs has decreased by 17 per cent and the 't' value was found to be significant. The very nature of these enterprises *viz.*, dairy, sheep rearing, petty shops, tailoring, hotel business etc. requires individual attention and requires more time on the enterprise. The beneficiaries of the dairy will have to spend more hours on feeding, washing the animal, sale of milk and milk products. While, the Sheppard has to take the animals for grazing for a considerable length of time and the person doing petty business will have to sit in his shop for the entire day and hence, these activities have generated higher employment. In contrast to this, human days devoted to actual labour work on others fields for earning wages has decreased. The findings were similar to the findings of Puhazhendi (2000), Shenkut (2002), Sudheendra (2003) and Nirmala *et al.* (2004).

Annual Income of Beneficiaries Before and After the Intervention of KSFMBC Project

It could be seen from Table 23.3 that, 47 per cent of the beneficiaries belongs to the annual income range of Rs. 20,001 to 40,000 before and 40.00 per cent after IGA followed by 26.66 per cent of the beneficiaries before IGA belongs to the income range 1000 to 20,000. Whereas, significant increase (35.00 per cent) is observed in the percentage of beneficiaries in the income range 40,001 to 60,000 after IGA. Around 23 per cent ofbeneficiaries before IGA were in the annual income range 40,001 to 60,000 followed by four per cent in the 60,001 to 80,000 annual income. Decrease in the number of beneficiaries (14.16) is observed in the income range of 1000 to 20,000 after IGA and only 7 per cent of the beneficiaries had annual income of Rs. 60,001 to 80,000. There were no beneficiaries in the income range of above 80,000 before IGA, whereas, 4 per cent of the beneficiaries had income above 80,000 after IGA. The possible reasons might be that all the beneficiaries have takeup income generating activities *viz.*, dairy, sheep rearing, business etc, in addition to their earlier occupations like agriculture, labour, livestock etc. these activities has generated additional income.

Table 23.3: Annual Income of Beneficiaries Before and After the Intervention of KSFMBC Project

(n=120)

Annual Income (Rs.)	Before		After	
	No.	Per cent	No.	Per cent
1000-20,000	32	24.67	17	14.17
20,001-40,000	56	46.67	48	40.00
40,001-60,000	27	22.50	42	35.00
60,001-80,000	5	4.16	8	6.67
Above 80,000	0	0.00	5	4.16
Total	**120**	**100.00**	**120**	**100.00**

Distribution of Beneficiaries Based on Income Generation from different Sources Before and After IGA

Table 23.4 depicts that, the total amount of income generated before IGA was Rs. 38675 where it raise to Rs. 48578 after IGA. The average differential amount generated due to IGA was Rs.9903, with 25 per cent change due to IGA. Out of which, the average income generated from farm activities was Rs. 27338 among the beneficiaries before taking up IGAs and it increased by 1.16 per cent after taking up IGA (Rs.27660). It is also seen from the table that additional income from farm activities was Rs. 322 after taking up IGAs. The calculated 't' value found non-significant.

Average annual income generated from off-farm activities mainly through labour wages before IGA is more (Rs. 6121) than that after IGA (Rs. 4179). It has reduced by Rs.1942. The percentage change after taking up IGAs has decreased

Table 23.4: Distribution of Beneficiaries Based on Income Generation from different Sources Before and After IGA

(n=120)

Particulars	Average Amount in Rs.		Additional Differential Annual Income	Percentage Change Due to IGA Increase/ Decrease	Paired 't' Test
	Before IGA	After IGA			
Farm					
Agriculture	27338	27660	322	1	0.36[NS]
Non-farm					
Dairy	2297	6300	4003	174	4.91**
Sheep rearing	1661	4726	3065	184	4.95**
Business and rural artisans	1058	5226	4168	393	5.84**
Collection of NTFPs and others	200	487	287	143	2.30**
Off-farm					
Labour	6121	4179	-1942	31	2.26**
Total	**38675**	**48578**	**9903**	**25**	

NS: Non-significant; *: Significant at 5 per cent level; **: Significant at 1 per cent level.

by 31 per cent and the calculated 't' value found negatively significant. Table 23.4 depicts that the average annual income generated from non-farm activities *viz.*, diary (Rs. 2297), sheep rearing (Rs.1661) and business(Rs.1058), collection of NTFP (Rs. 200) before IGA were less than that of after IGA. In contrast, the annual income generated from animal husbandry (Rs. 6300) followed by business (Rs. 5226), sheep rearing (Rs. 4726), other activities like business, rural artisans etc. (Rs. 4726) and Rs. 487 from collection of NTFPs. The additional average income generated/annum from non-farm activities is higher than farm and off-farm activities. After taking up IGA the highest income has been generated from business (Rs. 4168) followed by animal husbandry (Rs. 4003), sheep rearing (Rs.3065) and Rs.287 from collection of NTFPs, which accounts to 393 per cent increase in business followed by sheep rearing (184 per cent), dairy (174 per cent) and 143 per cent in collection of NTFPs. The calculated t values of nonfarm activities like diary, collection of NTFP and sheep rearing were found significant at 1 per cent level.

Distribution of Beneficiaries Based on their Social Change Before and After IGA

It is evident from the Table 23.5 that, there was an increase in percentage change due to IGA in the determinants like cosmopoliteness increased at 61 per cent followed by extension orientation/contact (54 per cent), extension participation (24 per cent), mass media participation (23 per cent), family education (22 per cent) and institutional/organisational participation (16 per cent). The calculated 't' value of mass media use was found to be significant at 5 per cent level while those of cosmopoliteness, education, extension participation, extension orientation/

contact and organizational participation found significant at 1 per cent level. Availing the IGA, naturally requires contacting the officials of the forest and other development department officials, meeting the local forest officials and NGO representatives, meeting the bank officials etc. hence, their cosmopoliteness has increased subsequently. The lesser change noticed in organization participation may be probably due to the lower Level of literacy. The findings were similar to the findings of Gaonkar (2001); Punithavathy Pandian and Eswaran (2002) ; Vanitha Chethan (2002) and Joshi (2004).

Table 23.5: Social Change Due to IGA among the Beneficiaries and their Families

(n=120)

Determinants	Mean Score		Difference	Per cent Change	Paired 't' Test
	Before IGA	After IGA			
Education	4.50	5.50	1.00	22	3.12**
Cosmpoliteness	4.74	7.64	2.90	61	17.57**
Extension orientation/contact	4.69	7.25	2.56	54	17.96**
Organization participation and membership	11.85	13.85	2.00	16	7.80**
Extension participation	6.07	7.55	1.48	24	8.124**
Mass media use	6.06	7.50	1.44	23	2.44*

*: Significant at 5 per cent level; **: Significant at 1 per cent level.

Distribution of Beneficiaries Based on their Cultural Change Before and After IGA

The findings of Table 23.6 revealed that, percentage change in dressing pattern has improved by 41 per cent, followed by celebration of rituals/ceremonies and festivals (14 per cent) and food habits increased by 8.2 per cent after IGA. The calculated t values of food habits was found to be significant at 5 per cent level and the 't' values of cultural traits like dressing pattern and celebration of rituals found significant at 1 per cent level. The beneficiaries might be probably having a satisfactory level of food habits and hence there was lesser percentage of change in this respect. The highest change noticed in the dressing pattern is due to the additional income generated from IGA and the greater attitude/attachments towards dressing. The findings were similar to the findings of Puhazhendi (2000) There was an increase of 83.82 human days of labour as a result of introduction of IGAs. The employment generation in terms of human-days before IGA were 233.53 days, as against 317.55 days after IGA. The highest average human-days were generated from business and rural artisans, followed by dairy, sheep rearing, collection of NTFP and from agriculture. The annual income of the respondents before IGA was Rs. 38675 which raised to Rs. 48578 after IGA, which works out to 25 per cent change due to IGA. The highest income was generated from business and rural artisans, followed by dairy, sheep rearing, agriculture and from collection of NTFP. The IGAs has brought above changes in family education, cosmopoliteness, extension orientation, organizational participation and mass media use, food habits, dressing pattern and celebration of ritual/ceremonies.

Table 23.6: Cultural Change Due to IGA among the Beneficiaries and their Families (n=120)

Determinants	Mean Score		Difference	Per cent Change	Paired 't' Test
	Before IGA	After IGA			
Food habits	7.55	8.22	0.67	8	3.36*
Dressing pattern	4.11	5.82	1.71	41	17.98**
Rituals/ceremonies	9.09	10.44	1.32	14	10.09**

*: Significant at 5 per cent level; **: Significant at 1 per cent level.

References

Puhazehendi, V., 2000. *Evaluation Study of SHGs in Tamil Nadu*. National Bank of Agriculture and Rural Development, Mumbai : 20-24.

Shenkut Ayele Tegenaw, 2002. Study on Success of Developmental Policy in Dharwad District of Karnataka State, India: A Case of Employment Assurance Scheme, *M.Sc. (Agri.) Thesis (Unpub.)* University of Agricultural Sciences. Dharwad.

Sudheendra, M., 2003. A Critical Analysis of Joint Forest Mangagement Programme on Knowledge and Perception among Beneficiaries In Northern Karnataka, *M.Sc.(Agri.) Thesis*, University of Agricultural Sciences. Dharwad.

Nirmala, V., Shambha, K. and Bhuvaneshwari, P., 2004. SHG for poverty alleviation in Pondicherry. *J. Rural Devel*. **23** (2): 203-215.

IPR Issues in Enterprise Development

Chapter 24

Intellectual Property Right and Entrepreneurship

Yashawant Dev Panwar

Senior Scientific Officer,
Patent Facilitating Centre (PFC),
Technology Information Forecasting
and Assessment Council (TIFAC),
Department of Science and Technology, New Delhi

Nature of Intellectual Property Rights

☆ IPR are largely territorial rights except copyright which is global under Berne convention.

☆ IPR can be held only by legal entities *i.e.*, who have the right to sell and purchase property.

☆ IPR can be assigned, gifted, sold and licensed like any other property.

☆ IPR are monopoly rights.

☆ IPR have to be renewed from time to time for keeping them enforced except in case of copyright and trade secrets. Renewal essential to maintain up to the term.

☆ IPR go hand in hand with trade and commerce.

☆ IPR have fixed term except trademark and geographical indications which can have indefinite life provided these are renewed after a stipulated time under the law by paying official fees.

☆ This is perhaps the only right which can be simultaneously enjoyed in more than one country.

The New Environment

1. Globalization essential for creating large markets leading to collapse of geographical barriers.
2. New market needs emerging in short time leading to shorter product cycles.
3. Tough competition.
4. Need for rapid changes in technology.
5. High innovative risks.
6. High investment in r and d, production and marketing.
7. Need for highly skilled human resources.

Many Technologies Needed in a Product

Case of battery storing electrical energy mechanically:

1. Innovation by American Flywheel System Company (ex scientists from EPA).
2. Three different technologies required.
3. Light weight but strong material (kevlar, Technora, fused silica).
4. Computer power for simulation of many prototypes.
5. Magnetic bearings for supporting electromagnetic fields.

Uruguay Round the Final Text

☆ Agriculture
☆ Application of sanitary and phytosanitary measures
☆ Textile and clothing
☆ Establishing WTO
☆ Technical barriers to trade
☆ Trade related investment measures
☆ Preshipment inspection
☆ Rules of origin
☆ Import licensing
☆ Subsidies and countervailing measures
☆ Safeguards
☆ Trade in services
☆ Trips
☆ Trade in civil aircraft
☆ Government procurement
☆ Dairy agreement
☆ Bovine meat

World Trade Organisation

Basic Features

☆ Multilateral trade agreement

☆ National treatment

☆ Most favoured nation

☆ No action should adversely affect domestic/international market of other member

☆ Harmonisation of measures/standards

☆ Transparency

☆ Dispute settlement system

Intellectual Property Rights (IPR)

☆ Patents

☆ Copyright

☆ Trademarks

☆ Industrial design

☆ Geographical indications

☆ Lay out design of integrated circuits

☆ Protection of undisclosed information

Present Legal System

☆ Patents: The Patents Act. 1970. Amended in March, 1999, early 2002 and 2004 by 2005 by ACT.

☆ Design: A Design Act. 2000

☆ Trade Marks: A new Trademarks Act, 1999

☆ Copyright: The Copyright Act, 1957 as amended in 1983, 1984, 1992, 1994 and 1999 and the Copyright Rules, 1958.

☆ Layout Design of integrated Circuits: The Semiconductor Integrated Circuit Layout Design Act 2000.

☆ Protection of undisclosed information: No exclusive legislation exists but the matter would be generally covered under the Contract Law (Contract Act 1872).

☆ Geographical indications: The Geographical Indication of Goods (Registration and Protection) Act 1999.

☆ Plant Variety Protection and Farmers Rights Act 2001.

Copyright

☆ Literary, dramatic and musical work including software/programmes

☆ Artistic work

☆ Cinematographic films including sound track and vidiofilm

☆ Record

☆ Software, engineering drawings, plant layout design *etc.*

☆ Provides protection for expression and not for the idea/concept Copyright

☆ 1709 First Copyright Act (Uk)

☆ 1957 Indian Copyright Act

☆ Amended 1983, 1984, 1992, 1994.

International Convention

☆ Bern Convention 1886, 1908, 1928,1948, 1989, 1990

☆ Universal Copyright Convention 1952

Term Life+60 Yrs

Design

☆ Original and novel in India

☆ Relates to feature of shape, configuration, pattern, ornament or composition of lines or colours applied to any article in 2-d or 3-d or both forms

☆ Should apply to any article by industrial process (for replication) and be an integral part of article.

☐ Stamps, labels not covered

☐ Paintings, sculptures and like not covered

☆ Not Registered As Design

☐ Design Not New Or Novel

☐ Disclosed To Public In India Or Abroad In Tangible Form Or By Use In Any Other Any

☐ Not Significantly Distinguishable From

☆ Known Designs

☐ Comprises Or Contains Scandalous Or

☐ Obscene Matter

Duration 15 Years

The Semiconductor Integrated Circuits Layout - Design Act, 2000

☆ Provides Protection of Semiconductor IC lay out design.

☆ Layout - Design - A layout of transistors and other circuitry elements and includes lead wires connecting such elements and expressed in any manner in a semi conductor IC.

☆ Semiconductor IC -A product having transistors and other circuitry elements which are inseparably formed on a semiconductor material or an insulating material or inside the semiconductor material and designed to perform an electronic circuitry function.

IC Layout-Design

Not Registrable as Layout-Design

Lay out Design:

☆ Not Original

☆ Commercially exploited anywhere in India or in a convention country

☆ Inherently not distinctive

☆ Inherently not capable of being distinguishable from any other registered

☆ Lay out - design

Note: Design not exploited commercially for more than 2 years from date of registration of application shall be treated as commercially not exploited for the purpose of this Act.

The Geographical Indications of Goods (Registration and Protection) Act, 1999

Definition: Geographical indications, in relation to goods, means: an indication which identifies such goods as:

☆ Agricultural goods,

☆ Natural goods or

☆ Manufactured goods

as originating, or manufactured in the territory of a country, or a region or locality in that territory, where a given quality, reputation or other characteristics of such goods is essentially attributable to its geographical origin,

The GI Act, 1999

and in case where such goods are manufactured goods:

one of the activities of either the production or of processing or preparation of the goods concerned takes place in such territory, region or locality, as the case may be.

China Banarasi Sari India

Punishment for falsifying GI:

Imprisonment between 6 months to 3 years, and

Fine between Rs. 50,000/- and Rs. 2 lakh

Registration

Controller General of Patents, Designs and Trade Mark shall be the Registrar of GI.

Who can Apply

Any association of persons or any organization or authority under law representing the interest of procedures of concerned goods.

Duration

☆ 10 years (Renewed from time to time after payment of prescribed fee)

☆ Can be kept alive for an indefinite period

Indian Legislation on Protection of Plant Varieties and Farmers' Rights Act, 2001

National Plant Variety and Farmers' Rights Protection Authority (PVFRPA)

Essential Requirements Qualifying

To Secure Plant Breeders' Rights

Novelty

Distinctness,

Uniformity, and

Stability Acronym NDUS.

Main Mandates of the PVFRPA are:

1. Registration of plant varieties,
2. Developing characterization and documentation of registered varieties,
3. Documentation, indexing and cataloguing of farmer's varieties,
4. Providing compulsory cataloguing facility for all plant varieties,
5. Ensuring that seeds of all registered varieties are made available to farmers,
6. Collection of comprehensive statistics on plant varieties,
7. Maintenance of National Register of Plant variety.

Main Features of PPVFR

1. Registration to be allowed for:
   Plant Breeder's Varieties,(PBV)
   Extant varieties (EV) and
   Farmer's varieties, (FV)
2. PBV to satisfy the NDUS conditions,
3. EV and FV to satisfy DUS conditions,
4. Allows elaborate farmers' rights

What is Patent?

☆ Right awarded by a country for exclusive use of invention.

□ For a limited period

❐ Right applicable within the country (There is nothing like global patent).

❐ In return inventor has to disclose the invention to public

THREE CRITERIA. Novelty

Inventive Step

Industrial Application

Novelty

☆ Not part of state of the art.

☆ State of the art comprise product, process, information in public domain.

Inventive Step

☆ Not obvious to person (s) skilled in the art.

☆ Not governed by smallness/quality of step.

Capable of Industrial Application

Patent Granting File Application with Complete Specification

18 months Published in Gazette

Opposition

Examination by Patent Office

Granted/Rejected

Microorganism

☆ Micro-organism per se

☆ Micro-organism isolated from nature for the first time.

☆ Gmos

☆ Products made by using micro-organism for the first time.

☆ New uses of micro-organism.

☆ Virus, bacteria, fungus, mushroom, protozoa, unicellular algae, cell lines, hybridoma (fused cells), variants, dna (eukaryotic and prokaryotic origin)

Budapest Treaty

☆ International convention regarding deposit of strains, microorganism.

☆ Set up in 1973

☆ Officially approved culture collection centres known as designated international depository authority (IDA)

☆ Ida stores deposited microorganism for at-least five years after the most recent request for a sample and for atleast 30 years from original date

Simple Things

- ☆ Coke Cane Opener
- ☆ Doggy Bell
- ☆ Needle
    - ❏ Simple needle
    - ❏ Machine needle

Storage Stable Azadirachtin US 5124349

Prior Art

- ☆ Extraction of azadirachtin from neem seeds using methanol, ethanol, water, chloroform hexane, ether, acetone, diethyl carbonate, *etc.*
- ☆ By increasing polarity of solvents, *i.e.*, from hexane to ethanol, ethanol to methanol, methanol to water, *etc.* extraction efficiency can be increased.
- ☆ No attention paid to shelf life known to degrade in heat and UV radiation. Concentration of Azad. in solution and pH of solution affect shelf life.

Storage Stable Azadirachtin US 5124349

- ☆ Storage stable azad. formulation which retains 80 per cent efficiency after one year at room temperature (25°C)
- ☆ pH between 3.8 to 4.2
- ☆ Use of aprotic solvents that do not contain acidic or basic functional groups

Storage Stable Azadirachtin US 5124349

Claims

- ☆ Storage stable azadirachtin composition having at least 50 per cent by volume aprotic solvent, 1.0 to 10 per cent surfactant, less than 15 per cent by volume water
- ☆ Many different formulations are part of claims
- ☆ *e.g.*: 0 to 40 per cent neem oil, 0 to 1 per cent para amino benzoic acid or its esters with pH between 3.8 to 4.2, 50 per cent by volume alcohol and less than 2 per cent water, solvents selected from nitriles, substituted aromatics, chlorinated aliphatics, aromatic aldehydes, sulfones, ethers, esters, amides, sulfoxides, alkyl carbonates, ketones and their mixtures

Process for Producing 5-Methyl Uridine (5 MU) US 5547857 (1996)

Use

5-Methyl Uridine is a useful intermediate for anti-AIDS drugs.

Prior Art

5-MU is produced by reacting a nucleoside or ribose-1-phosphoric acid with 5-methyl uracil in presence of a micro-organism *Micrococus luteus* Ferm P-7399.

Draw Backs

☆ Purification step not described

☆ Size of 5-MU crystals obtained very small (20 - 30 micrometer)

☆ Separation rate decreased

☆ Increased size of separator

Process for Producing 5-Methyl Uridine (5 MU) US 5547857 (1996)

Present Invention

☆ Uses same chemicals and microorganism

☆ Change in process

☆ Crystal size of 50-550 micrometer (high purity 5 MU)

☆ Increased separator efficiency

Process for Producing 5-Methyl Uridine (5 MU) US 5547857 (1996)

Claims

1. A process for producing 5-MU, comprising the steps of:
   i) Culturing a microorganism in a culture medium
   ii) Removing 50-90 per cent of said culture medium from microorganism
   iii) Adding buffer to microorganism
   iv) Reacting a nucleoside or ribose-1-phosphoric acid with 5-methyl uracil in remaining culture medium and buffer containing micro-organism
   v) Crystallizing 5-MU formed, by forming 5-MU crystals of average particle diameter - 50-550 m and impurity crystals of particle size of 5 to 50 micrometer
   vi) Separating 5-MU crystals based on difference in sedimentation velocity

Vitamin Based Fortification of Milk US 5480661

Prior Art

1. Add 400 International Units (IU) of vitamin D3 to whole milk prior to packaging. No need to add Vitamin A
2. Add 2000 IU of Vitamin A and Vitamin 400 IU of D3 per quart of low fat and skim milk.

Methods Known

1. Inject water soluble emulsion of the vitamin preparation into milk
2. Inject oil soluble vitamin preparation into milk.

Vitamin based Fortification of Milk

Disadvantages

1. Oil Based Method

Can not be added prior to centrifugal separation as specific gravity of the preparation (0.948) is almost same as that of the cream (< 1.0). Vitamin D_3 separates out with the cream.

To be added after separation through processing line on the suction side. Has to face high back pressure 0 to 80 PSI.

Metering of quantity is difficult.

Risk of contamination

1 ml will fortify 100 quarts (100 litters)

Vitamin based Fortification of Milk

Disadvantages

2. Water Soluble Emulsion

May oxidize over a time causing Vitamin A

Palmitate to emit strong and offensive smell

- ☆ Short shelf-life
- ☆ Needs refrigeration for storing
- ☆ 1 ml will fortify only 40 quarts (40 litres)

Presention Invention

- ☆ Vitamin A and D additive having specific gravity greater than 1.0
- ☆ Add oil based additive prior to centrifugal separation
- ☆ Additive which can be added anywhere during processing before packaging
- ☆ Easy to mix and economical
- ☆ No refrigeration required
- ☆ Additive contains the following: Corn oil, polysorbate, triglycerol mono oleate, ester gum, Brominated vegetable oil, Vitamin A Palmitate, Vitamin D_3 resin.

Claims

- ☆ Oil based additive for whole milk and low fat milk having specific gravity greater than 1.0
- ☆ 1 per cent to 3 per cent by weight corn oil and 1 per cent to 3 per cent by weight Easter gum
- ☆ 25 per cent to 30 per cent corn oil and 8 per cent to 15 per cent Brominated vegetable oil

 ☆ 25 per cent to 35 per cent corn oil and 10 per cent to 15 per cent Easter gum and 4 per cent to 10 per cent Brominated vegetable oil

 ☆ Different applications: Fortified milk, beverages and milk products

PATENT

TRADE MARKS

The RingG-PEN™

Flair, Raynolds, Cello, luxure *etc*

Copyright (Slide 50)

Slogans

....idea needs an expression.

.......likhte likhte love ho jaye.

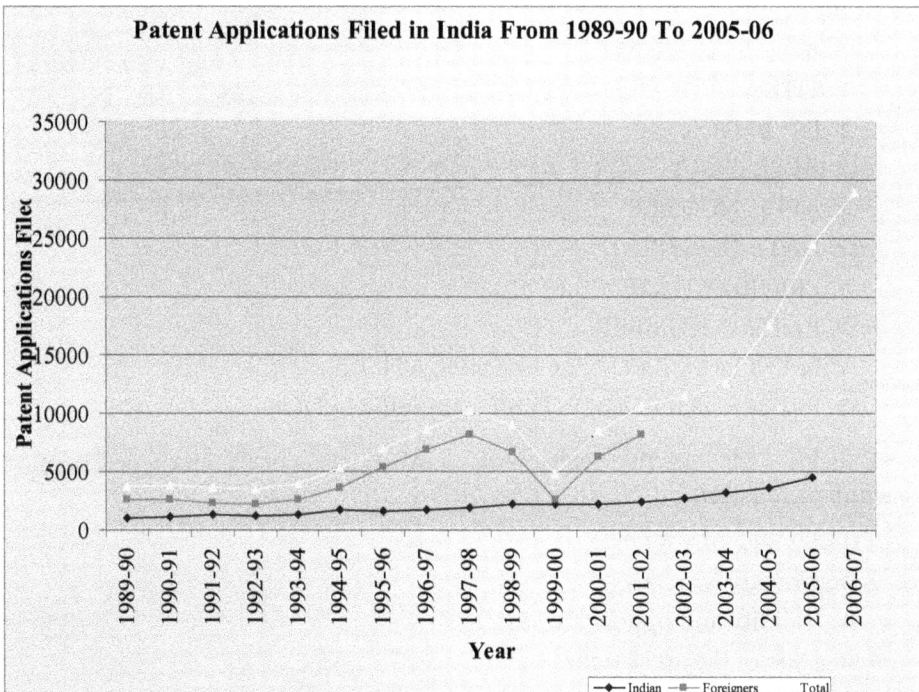

Patent Applications Filed in India From 1989-90 To 2005-06

Industrial Design

Costs Involved in Protection (Patents)

 ☆ Patent search

 ☆ Filing

 ☆ Translation

 ☆ Objections and Opposition

- ☆ Maintenance/Renewal
- ☆ Intelligence to locate infringement
- ☆ Legal

Institutional Support

- ☆ Technical Patent Search, Patent Documents, Analysis
- ☆ Legal Engaging Attorney, Follow-up Actions, Infringements, Revocation
- ☆ Financial Fees (Patent, Attorney), Costs of Searches, Patent Documents, Litigation Charges
- ☆ Procedural Clearances from University, correspondence, Licensing, Actual Buying of Info., Record Maintenance
- ☆ Administrative Ownership, Sharing of Benefits

Component of patent document

Bibliographic Databases

- ☆ TITLE
- ☆ INVENTOR
- ☆ APPLICANT
- ☆ APPL. NUMBER
- ☆ DATE OF APPL.
- ☆ PRIORITY DATE
- ☆ PATENT NUMBER
- ☆ NATIONAL PATENT CLASSIFICATION
- ☆ INTERNATIONAL PATENT CLASSIFICATION

Databases for Full Text Documents or Abstracts + or Claims + or Chemical structure or Images

Types of patent searches

- ☆ Novelty searches
- ☆ Validity and opposition searches
- ☆ State of the art searches
- ☆ Infringement searches
- ☆ Alerting searches
- ☆ Family and equivalent searches
- ☆ Citation searches
- ☆ Competitor monitoring
- ☆ Patent no. Search
- ☆ Quick search
- ☆ Boolean search

☆ Advanced search

☆ IPC classification search

Bibliographic

Abstract

Applicant/inventors

Full text

International patent classification

5 level system

Section 8

Class

Subclass

Group

Subgroup 69,000

International patent classification

Class	each section is divided into classes
Symbol	two digit number *e.g.* H 01
	a01 agriculture
	b24 grinding, polishing
	c12 biochemistry
	f42 ammunition
	h01 basic electric, elements-relays, switches, semiconductor devices
Subclass	each class is divided into subclass
Symbol	alphabet a, b, *etc.*
	a01b soil working
	h01q aerials
	h01s devices using stimulated emission
Group	each subclass is divided into groups
Symbol	1/00, 3/00 *etc*
	a01b 1/00 hand tools
	h01s 3/00 lasers
	h01q 9/00 short aerials
Subgroup	each subgroup is divided into subgroups

Symbol   1/02, 3/04, *etc*

h01s3/085   construction details of semiconductor lasers

h01q9/04   resonant aerials

Websites Providing Free Access To Patent Databases

1. http://www.indianpatents.org.in Ekaswa-A and Ekaswa-B
2. http://pk2id.delhi.nic.in Access to EPIDOS-INPADOC database Bibliographic and Equivalent searches Coverage more than 65 countries
3. http://www.uspto.gov Allow searches in bibliographic field, abstract, claims and full text of US patents since 1976. Full text images from 1970
4. http://ep.espacenet.com/ European, PCT, Japanese abstract and EPO's worldwide collection database. In the worldwide data collection data for most of the countries are from 1970 and for few from 1920 also.

Infringement Costs

☆ Honeywell vs Minolta $ 166 m

☆ Litton vs Honeywell $ 1.2 b

☆ Polaroid vs Kodak $ 873 m

☆ Hughes tool vs smith Intel $ 200 m

Dow corning vs sumitomo sumitomo out of fibreoptics business.

What could be implemented for next level of TT

Methods of Marketing Technologies

☆ Customized tools for TT in India

☆ Method of marketing through advertisement paper and online

☆ Membership to TT forum and web resources

☆ Advanced planning for system for maintenance of IPR and TT

☆ Creation of HR pool in the area

☆ Policies for different issues

Practice Patent

☆ Register as patent agent

❏ Degree in science or technology

❏ Exam conducted by patent office

1. Patent Department in Industry and Govt.
2. KPO
3. Own enterprise

❏ Attorneys

❏ Infringement analysis

❏ Advisory services
❏ IP intelligence
❏ Strategy formulation

TOYOTA Story

☆ 1896 Sakichi Toyota obtained patent for power loom

☆ 13 year later Sakichi succeeded in patenting automatic loom

☆ In 1924 Toyota type G automatic loom reach the market

☆ Kiichiro agreement with Platt Brother and Co paid £100,000 (equivalent to US$25 million today)

☆ Now you know Toyota co. selling cars

Knowledge

It is what we think we know already that often prevents us from learning.

Appendices

Appendix - I
Intellectual Property Rights (IPR): Questions and Answers

1. What are Intellectual Property Rights (IPR)?

IPR is a general term covering patents, copyright, trademark, industrial designs, geographical indications, layout design of integrated circuits, undisclosed information (trade secrets) and new plant varieties.

2. What are the legislations covering IPRs in India?

Patents: The Patents Act, 1970 as amended in 1999, 2002 and 2005

Design: The Designs Act, 2000

Trade Mark: The Trade Marks Act, 1999

Copyright: The Copyright Act, 1957 as amended in 1983, 1984 and 1992, 1994, 1999 Layout **Design of Integrated Circuits:** The Semiconductor Integrated Circuits Layout Design Act, 2000.

Protection of Undisclosed Information: No exclusive legislation exists but the matter would be generally covered under the Contract Act, 1872.

Geographical Indications: The Geographical Indications of Goods (Registration and Protection) Act, 1999.

Plant Varieties: The Protection of Plant Variety and Farmers' Rights Act, 2001.

3. Who are responsible for administration of IPRs in the country?

Patents, designs, trademarks and geographical indications are administered by the Controller General of Patents, Designs and Trademarks which is. under the control of the Department of Industrial Policy and Promotion, Ministry of Commerce and Industry. Copyright is under the charge of the Ministry of Human Resource Development. The Act on Layout-Design of Integrated Circuits is administere9 by the Ministry of Telecommunication and Information Technology. Protection of Plant Varieties and Farmers' Rights Authority, Ministry of Agriculture administers the Act on Plant Variety.

4. What is a patent?

A patent is an exclusive right granted by a country ~o the owner of an invention to make, use, manufacture and market the invention, provided the invention satisfies certain conditions stipulated in the law. A Letters Patent (a kind of certificate) is issued to the owner of the invention by the patent office of the country conferring this right. Exclusivity of right implies that no one else can make, use, manufacture or market the invention without the consent of the patent holder. This right is available to the owner of the invention only for a limited period of time. However, the use or exploitation of a patent may be affected by other laws of the country which has awarded the patent. These laws may relate to health, safety, food, security *etc.*

A patent in the law is. a property right and hence, it can be gifted, inherited, assigned, sold or licensed. As the rights are conferred by the State, it can be revoked by the State under very special circumstances for the benefit of public even if the patent has been sold or licensed or manufactured or marketed in the meantime. The patent right is territorial in nature meaning thereby, that a patent granted in India can only be enforced in India. In case the owner of the invention wishes to obtain patents in other countries, the owner will have to file separate patent applications in countries of his interest, along with necessary fees.

5. What is expected from patentee as an obligation to the State?

A patentee must disclose the invention in a patent document for people to practice it after the expiry of the term of the patent or after the patent has lapsed due to nonpayment of maintenance fee or 'practice it with the consent of the patent holder during the life of. the patent. Disclosure of an invention is a legal requirement for obtaining a patent.

6. How is invention defined in the Indian Patents Act, which can qualify for grant of a patent?

Invention means a new product or process involving an inventive step and capable of industrial application. Capable of industrial application means that the invention is capable of being made or used in an industry.

7. How do you explain the term 'a new product or a process'?

An invention is considered new or novel if it is not known to the public through publication or prior use anywhere in the world. The invention should not be a part of the existing global state of the art. Information appearing in magazines, technical journals, books, newspapers, published patents *etc.* constitute the state of the art. Oral description of the invention in a seminar/conference can also spoil novelty. Novelty is always assessed in the global context. An invention will cease to be novel if it has been disclosed in public through any type of publications anywhere in the world before filing a patent application in respect of the invention. Prior use of the invention before the filing date can also destroy the novelty. Novelty is determined through extensive literature and patent searches. It should be realized that patent search is essential and critical for ascertaining novelty as most of the information reported in patent documents does not usually get published anywhere else or it is published in technical literature after a fairly long gap.

8. How is inventiveness defined in the Indian Patents Act? Explain the meaning of the term.

Inventive step is defined as a feature of an invention that involves technical advance as compared to the existing knowledge or having economic significance or both and that makes the invention not obvious to a person skilled in the art

A person skilled in the art means a person having knowledge and skill in the subject matter of the patent application. This term has not been defined in terms of educational qualifications and experience. Inventiveness is different from.novelty in the sense that an invention must lead to a technical advance. Replacement of a wooden knob in an almirah by a metallic knob is not considered inventive. A substantial degree of interpretation is called for in determining the inventiveness. Inventiveness cannot be decided on the basis of the material contained in unpublished patents. This would mean that while deciding r inventiveness of an invention one has to examine it in the light of published information on the date of filing of the patent application. The complexity or the simplicity of an inventive step does not have any bearing on the grant of a patent. In other words a very simple invention can qualify for a patent. A mere 'scintilla' of invention is sufficient to found a valid patent.

9. What are the types of inventions which are not patentable in India?

An invention may satisfy the condition of novelty, inventiveness and usefulness but it may not qualify for a patent under the following situations:

(i) An invention which is frivolous or which claims anything obviously contrary towel established natural laws;

(ii) An invention the primary or intended use or commercial exploitation of which could be contrary to public order or morality or which causes serious prejudice to human, animal or plant life or health or to the environment;

(iii) The mere discovery of scientific principle or the formulation of an abstract theory or discovery of any living thing or non-living substance occurring in nature;

(iv) The mere discovery of a new form of a known substance which does not result in enhancement of the known efficacy of that substance or the mere discovery of any new property or new use for a known substance or of the mere use of a known process, machine or apparatus unless such known process results in a new product or employs at least one new reactant;

Explanation: For the purposes of this clause, salts, esters, ethers, polymorphs, metabolites, pure form, particle size, isomers, mixtures of isomers, complexes, combinations and other derivatives of known substance shall be considered to be the same substance, unless they differ significantly in properties with regards to efficacy.

(v) A substance obtained by mere admixture resulting only in the aggregation of the properties of the components thereof or a process for producing such substance; hylactic (diagnostic therapeutic) or other treatment of human beings or any process for a similar treatment of animals to render them free of disease or to increase their economic value or that of their products;

(xi) Plants and animals in whole or any part thereof other than microorganisms but including seeds, varieties and species and essentially biological processes for production or propagation of plants and animals;

(xii) A mathematical or business method or a computer program per se or algorithms;

(xiii) A literary, dramatic, musical or artistic work or any other aesthetic creation whatsoever including cinematographic works and television productions;

(xiv) A mere scheme or rule or method of performing mental act or method of playing game;

(xv) A presentation of information;

(xvi) Topography of integrated circuits;

(xvii) An invention which, in effect, is traditional knowledge or which is an aggregation or duplication of known properties of traditionally known component or components;

(xviii) Inventions relating to atomic energy.

10. When should an application for a patent be filed?

Filing of an application for a patent should be completed at the earliest possible date and should not be delayed. An application filed with provisional specification, disclosing the essence of the nature of the invention helps to register the priority by the applicant. Delay in filing an application may entail some risks such as (i) some other inventor might file a patent application on the said invention and (ii) there may be either an inadvertent publication of the invention by the inventor himself/ herself or by others independently of him/her.

11. Can a published or disclosed invention be patented?

No. Publication of an invention in any form by the inventor before filing of a patent application would disqualify the invention to be patentable. However, a grace period of 12 months is available in India for filing a patent application after disclosure of the invention through publication or display. This grace period is available only when such publications or displays are done in government notified exhibitions or paper read before learned society. A patent application must be filed before the grace period ends. A practical advice to inventors is that they should try to utilize the grace period only under special circumstances. As a matter of sound practice they should not disclose their invention before filing the patent application. The invention can be considered for publication after a patent application has been filed. Thus, it can be seen that there is no contradiction between publishing an inventive work and filing a patent application in respect of the invention.

12. What is considered as the date of patent?

The date of patent is the date of filing the application for patent (whether provisional or; complete). The term of the patent is counted from this date.

13. What is the term of a patent in the Indian system?

Term of the patent is 20 years from the date of filing for all types of inventions.

14. How does one keep a patent in force for the full patent term?

A patent has to be maintained by paying the maintenance fees every year. If the maintenance fees are not paid, the patent will cease to remain in force and the invention becomes open to public. Anyone can then utilize the patent without the danger of infringing the patent.

15. What are the essential documents to be generated and submitted by a potential patentee?

There are two types of patent documents usually known as patent specification, namely (i) Provisional Specification and (ii) Complete Specification.

Provisional Specification

A provisional specification is usually filed to establish priority of the invention in case the disclosed invention is only at an early stage and a delay is expected in giving final shape to- the invention. Although, a patent application accompanied with provisional specification does not confer any legal patent rights to the applicant, it is, however, a very important document to establish the earliest ownership of an invention. The provisional specification is a permanent and independent scientific cum legal document and no amendment is allowed in this. No patent is granted on the basis of a provisional I specification. It has to be followed by a complete specification for obtaining a patent for the said invention. Complete specification must be submitted within 12 months of filing the provisional specification. This period is non-extendible. It is not necessary to first file an application with provisional specification and then the complete specification. An application with complete specification can be filed right at the first instance.

Complete Specification

Submission of complete specification is necessary to obtain a patent. The contents of a complete specification would include the following:

1. Title of the invention.
2. Field to which the invention belongs.
3. Background of the invention including prior art giving drawbacks of the known inventions and practices.

4. Complete description of the invention along with experimental results.

5. Drawings *etc.* essential for understanding the invention.

6. Claims, which are statements related to the invention on which legal proprietorship is being sought. Therefore, the claims have lobe drafted very carefully.

7. Abstract of the invention.

8. If a biological material is mentioned in a specification; then the source and geographical origin of the same is to be disclosed in the specification,. For new biological material, registration number of the same given by an International Depositary Authority (IDA) is to be included in the specification along with its address. [Institute of Microbial Technology (IMTech), Chandigarh is a recognized IDA in India]

16. What is request for examination and when it is to be filed?

An applicant has to make a request to the Patent Office for examining the patent application by paying the requisite fees. The request can be made at the time of submitting complete specification or within 48 months of filing the application. The request can also be made by any other interested person.

17. When will a patent application be published?; How will it be accessible to public?

A patent application will be published in the Official Journal of the Patent Office on expiry of eighteen months from the date of filing or date of priority of the application, whichever is earlier. It can also be published earlier, if such a request is made by the applicant.

An application will not be published in cases where directions have been given to secrecy, until the term of those directions expires. If an applicant wishes to withdraw the application on his own without any secrecy consideration, he can do so by withdrawing his application at-least three months before the date, of the publication. In that case the application will not be published. The publication of every application includes the particulars of the date of application. Application number, name and address of the applicant, an abstract, and is open for public inspection. However, the whole patent document can be obtained from the Patent Office upon payment of the requisite fees.

18. What is opposition under the Indian Patents Act 1970?

The Act now provides 'for pre-grant and post-grant opposition. Pre-grant opposition can be filed after the publication of patent application and before the grant of patent. Post-grant opposition can be filed within one year of the grant of the patent. An opposition board will be constituted for each of the opposition notifications accepted by the Controller for the post grant opposition proceedings. Opposition in both cases will be allowed on all grounds specified in the Act.

19. What are the grounds for opposition?

Pre-grant and post-grant opposition can be filed only on the following grounds:-

i. Claimed invention or its part wrongfully obtained;

ii. Claimed invention is published in a patent or any other document before the priority date;

iii. Claimed invention is published in a patent after the priority date having earlier priority date;

iv. Claimed invention was publicly known or publicly used before the priority date;

v. Claimed invention is obvious and does not involve clearly any inventive step, as regards to the matter published or used (in India) before the priority date;

vi. Claimed invention is not an invention within the meaning of the Patents Act or is not patentable under the Patents Act;

vii. The complete specification does not sufficiently and clearly describe the invention or the method by which it is to be performed;

viii. The applicant has failed to disclose to the Controller the information regarding foreign applications filed by him for the same invention or has furnished the information which in any material particular was false to his knowledge;

ix. In case of convention application if the application is not filed before the expiry of 12 months from the date of first application in convention country;

x. The complete specification does not disclose or wrongly mentions the source and geographical origin of biological material used in the invention;

xi. Claimed invention was anticipated having regard to the knowledge, oral or otherwise available within any local or indigenous community in India or elsewhere.

20. Is a patent granted in one country automatically enforceable in other countries?

No. There is nothing like a global patent or a world patent. Patent rights are essentially territorial in nature and are protected only in a country (or countries), which has (have) granted these rights. Therefore, the patent could only be enforced in such countries. In other words, for obtaining patent rights in different countries one has to submit patent applications in all the countries of interest for grant of patents. This would entail payment of official fees and associated expenses, like the attorney fees, essential for obtaining patent rights in each country. However, there are some regional systems where by filing one application, one could simultaneously obtain patents in the member countries of a regional system; European Patent Office is an example of a similar system.

21. What is the cost of filing a patent application in India?

Some important fees are given below:

Sl.No.	Action	Limits, Conditions and Timeliness	Official Fees	
			Individual	Legal Entity
1.	Filing of patent application along with complete/provisional specification	For maximum numberof pages 30 and maximum number of claims 10	Rs.1000/-	Rs.4000/-
		For each additional sheet	Rs.100/-	Rs.400/-
		For each additional claim	Rs.200/-	Rs.800/-
2.	Request for examination of patent		Rs.2,500/-	Rs.10,000/-
3	Request for early publication		Rs.2,500/-	Rs.10,000/-
4.	Renewal fee (every year)	2nd year to 6th year	Rs.500/-	Rs.2,OOO/-
		7th year to 10th year	Rs.1500/-	Rs.6000/-
		11th year to 15th year	Rs.3000/-	Rs12,0001-
		16th year to 20th year	Rs.5000/-	Rs.20000/-

22. Does grant of a patent in one country affect its grant or refusal in another country?

Each country is free to grant or refuse a patent on the basis of scrutiny by its patent office. This means that grant of a patent in one country does not guarantee that some other country will also grant patent for the same invention. Similarly, the refusal of the patent in one country does not mean that it will be refused in other countries.

23. What is 'mail box' provision?

TRIPS requires that countries, not providing product patents in respect of pharmaceuticals and agro-chemicals have to put in a mechanism for accepting product patent applications for pharmaceuticals and agro-chemicals w.e.f. 1st January 1995. These applications would be taken up for examination after the countries concerned put in force a product patent regime. The time limit for putting product patent regime in force was January 1, 2005 for developing countries and January 1, 2016 for least developed countries (LDCs). This mechanism of accepting product patent applications is called the' Mail Box mechanism. Such applications will now be examined in India for grant of patents, after the amendments in the national patent law made by the Patent Amendment Act 2005.

24. What do you understand by Exclusive Marketing Right (EMR)?

TRIPS requires that member countries of the WTO not having provision in their laws for granting product patents in respect of drugs and agrochemical, must introduce Exclusive Marketing Rights (EMR) for such products, if the following criteria are satisfied:

1. A patent application covering the new drug or agrochemical should have been filed in any of the WTO member countries on or after 1January, 1995;

2. A patent on the product should have been obtained in any of the member countries (which provides for product patents in drugs and agrochemicals) after 1January, 1995;

3. Marketing approval for the product should have been obtained in any of the member countries;

4. A patent application covering the product should have been filed after 1 January, 1995 in the country where the EMR is sought;

5. The applicant should apply seeking an EMR by making use of the prescribed form and paying requisite fee.

EMR is only a right for exclusive marketing of the product and is quite different from a patent right. It was valid up to a maximum period of 5 years or until the time the product patent laws came into effect from January l, 2005.India would no more grant any EMR.

25. Does India have provision for grant of EMR?

Yes. The necessary amendment to the Patents Act, 1970 came into force on 26March, 1999. The provision was applicable with retrospective effect from 1January,

1995 this provision has been removed from the Act through h the amendment made in the Act in 2005

26. What is industrial property?

Industrial property includes:

(a) Patents (b) Utility models "(c) Industrial designs (d) Trademarks, service marks and trade names (e) Indication of source or appellations of origin (this is same as the geographical indications adopted in TRIPS). [Readers may note the use of the phrase 'Industrial Property' and how it is different from intellectual property].

27. What is the Paris Convention?

The Paris Convention is an international convention for promoting trade among the member countries, devised to facilitate protection of industrial property simultaneously in the member countries without any loss in the priority date. All the member countries provide national treatment to all the applications from the other member countries for protection of industrial property rights. The Convention was first signed in 1883. Since then, the Convention has been revised several times, in 1900 at Brussels, in 1911 at Washington, in 1925 at The Hague, in 1934 at London, in 1958 at Lisbon and in 1967 at Stockholm. The last amendment took place in 1979.1ndia became a member of the Paris Convention on December7, 1998.

28. What are the principal features of the Paris Convention?

The principal features of the Paris Convention have been listed below:-

- ☆ National treatment,
- ☆ Right of priority
- ☆ Independence of patents
- ☆ Parallel importation
- ☆ Protection against false indications and unfair competition

29. What is the meaning of national treatment under the Paris Convention?

This is a very important concept and is essential for successfully achieving the fundamental aim of the Paris Convention. The idea is to provide equal treatment to a patent applicant from other member countries, and not to differentiate between the nationals of your country and nationals of the other countries for the purpose of grant and protection of industrial property in your country. Imagine that a national of country X applies for grant of a patent in India. According to the Paris Convention, the Indian Patent Office shall apply the same norms and rules, to the applicant from X, as applicable to an Indian applicant, for granting a patent. Similarly the applicant from X shall have the same protection after grant and enjoy identical legal remedies against any infringement provided the conditions and formalities imposed upon Indians are complied with.

30. What do you understand by the right of priority and what is its significance?

The date from which patent right is deemed to start is the date of filing of a patent application. To obtain rights in other member countries, the application must be filed on the same day in other member countries if it is desired to have the rights started from the same day. However, there are practical. difficulties in synchronizing the activities. For facilitating simultaneous protection in member countries, the Convention provides that within 12 months of national filing, if patent applications are filed in those member countries, the patents, if granted in member countries, will be effective from the date of national filing. This right is known as the right of priority. In other words you maintain the priority or the same date of filing in all the member countries and no one else in; those countries can obtain the patent rights on a similar/identical invention from the same or a later date.

In case the applicant after a second look at the patent application finds that that parent contains more than one Invention or on his own accord wishes to divide the application he can claim the initial date of priority for subsequent patent applications. The applicant may also, on his own initiative, divide a patent application and preserve as the date of each divisional application the date of the initial application and the benefit of the right of priority, if any. Each country of the Convention shall have the right to determine the conditions under which such division shall be authorized. Priority may not be refused on the ground that certain elements of the invention for which priority is claimed do not appear among the claims formulated in the application in the country of origin, provided that the application documents as a whole specifically disclose such elements.

31. What is implied by 'importation' in relation to working of a patent under the convention?

Importation is considered as working of patent, provided that the patented product is manufactured in a member country and is imported into another member country which has also granted a patent on the same invention to the same applicant. Imagine that a.product X has been patented in two member countries A and B. The product X is then ! manufactured in country A and imported into the country B. This product X shall enjoy the same patent protection in the country B even though it has been manufactured in the country A. This would also be considered as if the patent has been worked in country B.

32. Is there a provision for compulsory license In the Paris Convention?

Yes, each member country shall have the right to provide for the grant of compulsory licenses to prevent the abuses resulting from the exclusive rights conferred by the patent. Compulsory licenses for failure to work or Insufficient. working of the invention may not be requested before the period of time of non-working or insufficient working has elapsed. This time limit is four years from the date of filing of the patent application or three years from the date of the grant. Such licenses will be a non-exclusive and non-transferable one.

33. Is there any relationship between the Paris Convention and the TRIPS Agreement?

It has been made mandatory for the member countries of the TRIPS Agreement to comply with the Article 1 to 12 and Article 19 of the Paris Convention.

34. What are the other advantages of joining the Paris Convention?

There are a number of international conventions and treaties, which are open only to the members of the Paris Convention. Some of these are

☆ Patent Cooperation Treaty (PCT)

☆ Budapest Treaty (for deposition of microorganisms)

☆ UPOV (for protection of new varieties of plants)

☆ Madrid Agreement (for repression of false or deceptive indications of source on goods)

☆ Madrid Protocol (concerning registration of marks)

☆ Hague Agreement (concerning deposits of industrial designs)

☆ Strasbourg Agreement (concerning the International Patent Classification)

35. What is the Budapest Treaty?

This is an international treaty governing the registration of deposits of microorganisms; cell lines *etc* in officially approved culture collections for the purpose of patent applications in any country that is a party to it. Because of the difficulties and, on occasion, of virtual impossibility of reproducing a microorganism from a description of it in a patent specification, it is essential to deposit a strain in a culture collection centre for \ testing and examination by others. The Treaty was signed in Budapest in 1973 and later on amended in 1980. India became a member of this 'Treaty, with effect from December 17, 2001.

36. Are there any differences in the filing of patent applications in respect of microbiological inventions and other inventions?

The basic difference is that an inventor is required to deposit the strain of a microorganism in a recognized depository which assigns a registration number to the deposited microorganism. This,n\11mber needs to be quoted in the patent application. Obviously a strain of microorganism is required to be deposited before filing a patent application. It may be observed that this mechanism obviates the need of describing a microorganism in the patent application. Further, samples of strains can be obtained from the depository for further working on the patent. There are many international depositories in many countries, which are recognized under the Budapest Treaty. Such a requirement of deposition of a physical element of an invention, in respect of inventions in other area, does not exist.

37. What is the system for protecting microbiological inventions and microorganisms in India?

The Indian Patent Act has included microorganisms discovered from nature in the list of items not patentable. However, genetically modified microorganisms will be patentable. The process for producing microorganisms or culture of microorganisms *etc.* will also be patentable. Many countries allow patenting of genetically modified microorganisms but a few also allow patenting of naturally occurring microorganisms if isolated from nature for the first time and if other conditions of patentability are satisfied.

38. What is the Patent Cooperation Treaty (PCT)?

The Patent Cooperation Treaty (PCT) is a multilateral treaty entered into force in 1978. Through PCT, an inventor of a member country (Contracting State) of PCT can simultaneously obtain priority for his/her invention in all the member countries, without having to file a separate application in those countries, by filing of PCT application, which is popularly known as International Application. India joined the PCT on December 7, 1998.

39. Is designation of countries in PCT applications required now?

No, it is no longer required to designate countries in a PCT application as was the practice until recently. All the PCT member countries are considered as designated countries.

40. Who coordinates the activities of PCT?

All activities related to PCT are coordinated by the World Intellectual Property, Organization (WIPO) situated in Geneva.

41. What is the need for PCT?

In order to protect your invention in other countries, you are required to file an independent patent application in each country of interest; in some cases, within a stipulated time to, obtain priority in these countries. This would entail a large investment, within a short time, to meet costs towards filing fees, translation, attorney charges *etc.* In addition you are making an assumption which, due to the short time available for making the decision on whether to file a patent application in a country or not, may not be well founded. Inventors of Contracting States of PCT on the other hand can simultaneously obtain priority for their inventions without haying to file separate application in the countries I of interest; thus saving the initial investments towards filing fees, translation *etc.* In addition the system provides much longer time for filing patent application member countries. The time available under Paris. Convention for securing priority in other countries is 12 months from the date of initial filing. Under the PCT, the time available could be a minimum of 20 months and maximum of 31 months. Further, an inventor is also benefited by the search report prepared under the PCT system to be sure that the claimed invention is novel. The inventor could also opt for preliminary examination before filing in other countries to be doubly sure about the patentability of the invention.

42. How are patent applications handled under PCT?

The.patent office or another office designated by each contracting state becomes a receiving office for receiving PCT patent applications. These applications are referred to International Searching Authorities (ISA), which usually are patent offices, appointed to carry out patent search on a global basis. In case the receiving office is also an ISA, a separate referral is not required. There is also a provision to get a patent application examined by International Preliminary Examining Authorities, which in most cases are ISAs.

43. What is the meaning of delayed processing of an application filed under the PCT system?

A search report on the patent application filed with a receiving office is received by the applicant/inventor 16 months after the priority date, the latter is nothing but the date of submitting the application in the receiving office (It has been observed that in some cases, the 16 months period is strictly not adhered to). The International Bureau of the WIPO publishes the application and the search report 18 months after the priority date. Within two months of the publication of the application *i.e.* by the 20th month, the applicant will have to formally apply to the patent offices of these countries for grant of patents by paying official fees and completing other formalities stipulated by these offices. However, it is now possible to enter national phase by 30th or 31st month in many countries even if no request is made for preliminary examination. In other words, the 20 month period has been virtually extended to 30 or 31 months. In case translated copies of the application are required, the same has to be furnished by the applicant. In spite of submitting the request for grant of patents in designated countries in the 20th or 31 st month after the priority date, the priority in these countries is the same as the date of filing the original.PCT application.

If applicant/inventor has requested for an examination report, the report is usually received by the applicant/inventor about 28 months after the priority date. Within two months of this and in some countries within three months, the applicant/inventor will have to formally apply for grant of patents in designated countries. The priority of the application is maintained in the designated countries.

44. What is the benefit of the delayed processing?

(a) By the end of the 20th to 31st month the applicant is in a better position to assess the quality of the invention being protected as a detailed search report or an examination report or both would be available to help making an assessment.

(b) Applicants can re-evaluate their decision about filing applications in all the member countries after a long gap of 20 to 31 months.

(c) If not satisfied, applicants may decide to select or drop a few countries from the list. This decision would also be influenced by the changing market conditions.

(d) Applicants can delay their investment in respect of the national phase or the regional phase applications by 20 to 31 months without sacrificing priority. As mentioned earlier, it is possible to file national phase applications in 30th or 31st months in many countries without opting for preliminary examination. The actual time limits for each country can be seen at http://www.wipo.int/pct/guide/en/for each country by clicking for the time limit for that country.

45. Which is the appropriate office in India in relation to PCT applications?

A PCT application can be filed in any of the Branch Offices of the Patent Office located at New Delhi, Chennai, Mumbai and Kolkata (Head Office). Any of these Offices shall futction as a receiving office, designated office and elected office for the purpose of international applications filed under the Treaty. An international application shall be filed in the Patent Office which would process the application in accordance with these rules and the provisions under the PCT.

46. What is the cost of filing a PCT application?

The cost of filing a PCT application by an Indian Applicant is as follows:

1.	Transmittal Fees	Rs.8,000/- for legal entity and Rs.2000/- for individuals
2.	International Filing Fee	US$ 1,086
3.	Fee per sheet over 30 sheets	US$12
4.	Search Fee	US$ 241 to US$ 2,059 depending on the International Searching Authority (ISA) selected by the applicant

All fees payable are reduced by 75 per cent for the applications filed by any individual as an applicant who resides in a PCT contracting state where the per capita national income is below US$3000.

47. Where do you pay the fees and in which currency?

All types of fees are payable at the receiving office and it is the responsibility of the receiving office to remit the search fees to the concerned office if the receiving office is not the search authority. Similarly, all other charges due to other agencies would be remitted by the receiving office. The fees are payable in the currency acceptable to the receiving office. As an Indian you can pay all the fees in Indian rupees.

48. What are the criteria to be used for naming inventors in an application for patent?

The naming of inventors is normally decided on the basis of the following criteria:

i. All persons who contribute towards development of patentable features of an invention should be named inventor(s).

ii. All persons, who have made intellectual contribution in achieving the final results of the research work leading to a patent, should be named inventor(s).

iii. A person who has not contributed intellectually in the development of an invention is not entitled to be included as an inventor.

iv. A person who provides ideas needed to produce the' germs of the invention' need not himself/herself carry out the experiments, constructs the apparatus with his/her own hands or make the drawings himself/herself. The person may take the help of others. Such persons who have helped in conducting the experiments, constructing apparatus or making the drawings or models without providing any intellectual inputs are not entitled to be named inventors.

Quite often difficulties are experienced in deciding the names of inventors. To avoid such a situation, it is essential that all the scientists engaged in research should keep factual, clear and accurate record of daily work done by them in the form of diary. The pages in the diary should be consecutively numbered and the entries made be signed both by the scientists and the concerned leader.

49. What is the nature of information needed while consulting a patent attorney?

As an inventor one should share the complete invention with a patent attorney in the same manner as a patient confides in a doctor; As the doctor may not be able to write a correct prescription without knowing the details of the disease/problem, a patent attorney may not be able to draft a good specification in the absence of details about the invention. Following points should be kept in mind while discussing with the attorney:

i. Provide complete details of the invention including failures, if any, on the way to the invention.

ii. Do not feel bad if attorney asks you questions like where did you get the idea from, or did you copy the idea from somewhere or are you keeping all inventors working with you on the inventorship or have you published the invention or disclosed it in a seminar/conference or have you displayed the invention in an exhibition? A patent document is a techno-legal document; hence precautions have to be taken right from the first step. You must provide/furnish correct information to the attorney and you may even show your laboratory note book/log book to the attorney. This will help the attorney/agent to explain the inventive step in a precise manner and draft a good specification and associated claims.

iii. Explain the central theme of the invention along with novelty, inventiveness and utility of the invention.

iv. Share all the prior art documents in your possession with the attorney.

v. If you have developed an improved version of a known product/process, admit it and be totally honest. This would help the attorney in drafting precise claims and avoid excessive claims, which might be struck down immediately or at a later date.

vi. A detailed description of the best way of putting the invention into practical use, results of your tests and trials, *etc.*, including all failures and defects should be shared with the attorney.

vii. Alternative ways of using the invention, and the substitutes or parts of it may be discussed.

viii. It may be worth drafting the patent wide enough to cover less satisfactory alternatives as well so as to prevent rivals from marketing a less satisfactory competing product which because of its defects might bring the whole genre of product into disrepute or which may be cheaper.

ix. Both after an initial search and during the course of the filing and grant of a patent application, it is important to respond quickly and accurately to queries which the patent attorney may have. In addition the client should also keep the patent attorney informed of any new developments in the field of invention carried by the patentee or some one else.

50. What is the distinction between patented inventions and know how?

Patent laws do not require that the information disclosed in the patent specification be sufficient for commercial exploitation of the invention. Thus, a patent will usually not disclose sufficient information for commercialization. Know how on the other hand, covers all information necessary to implement and commercialize the invention such as setting up a production plant, operating conditions, plant layout designs, details of the production methods, various designs and drawings *etc*. It is this know how which is traded while transferring technology. Know how is usually kept as a trade secret and not shared with public. Know how is not protected through patents as most of it may be non-patentable matter and one does not take patent on the remaining parts to avoid public disclosure. A know how developed around an existing patent and commercialized subsequently may be an infringement of the patent unless the patentee has agreed to commercialization on mutually agreed terms.

51. Is there any Act for protecting a new plant variety in India?

The Protection of Plant Varieties and Farmers' Rights Act 2001 was enacted in India to protect the new plant varieties. Rules for the same were notified in 2003. The Act has now come into force. The Protection of Plant Varieties and Farmers' Rights Authority has been set up and is responsible to administer the Act. The office of the Registrar has started receiving applications for registration of twelve notified crops *viz.* rice, lentil, maize, green gram, kidney bean, black gram, chickpea, pearl millet, pigeon pea, sorghum, field pea, bread wheat.

Under the TRIPS agreement it is obligatory on part of a Member to provide protection to new plant variety either through patent or an effective sui generis system or a combination of these two systems. India was therefore under an obligation to introduce a system for protecting new plant variety. India opted for sui generis system and enacted The Protection of Plant Varieties and Farmers' Rights Act 2001. However, in many countries such plants can be protected through Breeders' Rights, patents and UPOV Convention.

52. What are the objectives of Protection of Plant Varieties and Farmers' Rights Act in India?

The objectives of the Protection of Plant Varieties and Farmers' Rights Act are:

(i) to stimulate investments for research and development both in the public and the private sectors for the developments of new plant varieties by ensuring appropriate returns on such investments;

(ii) to facilitate the growth of the seed industry in the country through domestic and foreign investment which will ensure the availability of high quality seeds and planting material to Indian farmers; and

(iii) to recognize the role of farmers as cultivators and conservers and the contribution of traditional, rural and tribal communities to the country's agro biodiversity by rewarding them for their contribution through benefit sharing and protecting the traditional right of the farmers.

More importantly this act provides safeguards to farmers by giving farmers' rights while, providing for an effective system of protection of plant breeders' rights. The Act seeks to safeguard researchers' rights as well. It also contains provisions for safeguarding the larger public interest. The farmer's rights include his traditional rights to save, use, share or sell his farm produce of a variety protected under this Act provided. the sale is not for the purpose of reproduction under a commercial marketing arrangement.

53. What kind of varieties are registerable under the plant variety Act?

(i) A new variety if it conforms to the criteria of novelty, distinctiveness, uniformity and stability.

(ii) An extant variety if it conforms to criteria of distinctiveness, uniformity and stability.

54. How is an "Extant Variety" defined in PPVFR Act, 2001?

An "Extant Variety" means a variety, which is-

(i) notified under section 5 of the Seeds Act, 1966 (54 of 1966); or

(ii) a farmers' variety; or

(iii) a variety about which there is common knowledge; or

(iv) any other variety which is in the public domain.

55. What is the meaning of Farmers' Variety as per PPVFR Act, 2001

'Farmers' Variety" means a variety which:

(i) has been traditionally cultivated and evolved by the farmers in their fields; or

(ii) is a wild relative or land race of a variety about which the farmers possess the common knowledge;

where farmer means any person who (i) cultivates crops by cultivating the land himself; or (ii) cultivates crop by directly supervising the cultivation of land through any other, person; or (iii) conserves and preserves, severely or jointly, with any person any wild 1species or traditional.varieties or.adds value to such wild species or traditional varieties through selection and Identification of their useful properties.

56. What is the meaning of "Essentially Derived Variety" as per PPVFR Act, 2001?

"Essentially Derived Variety" is a variety which is predominantly derived from another variety (protected or otherwise) and conforms to the initial variety in all aspects except for the differences which result from the act of derivation, and yet is clearly distinguishable from such initial variety.

57. How novelty, distinctiveness, uniformity and stability have been defined in the Protection of Plant Varieties and Farmers' Rights Act?

A) **Novelty**: Plant variety is novel if at the date of filing of the application for registration for protection, the propagating or harvested material of such variety has not been sold or otherwise disposed of by or with the consent of breeder or his successor for the purpose of exploitation of such variety-

   (i) In India earlier than one year or

   (ii) outside India, in the case of trees or vines earlier than six years or in any other.

   case, earlier than four years,

   before the date of filing such application Provided that a trial of a new variety which has not been sold otherwise disposed of shall not affect the right to protection. Provided further that the fact that on the date of filing the application for registration, propagating or harvested material of such variety has become a matter of common knowledge other than through the aforesaid manner shall not affect the criteria of novelty for such variety.

B) **Distinctiveness**: New plant variety will be considered distinct if it is clearly distinguishable by at least one essential characteristic from any other variety whose existence is a matter of common knowledge in any country at the time of filing of the application.

58. What are farmers' rights?

The farmers' rights as defined in the Act are:

   (i) a farmer who has bred or developed a new variety shall be entitled for registration and other protection in like manner as a breeder of a variety under. this Act;

   (ii) the farmers' variety shall be entitled for registration if the application contains declaration as specified in clause (h) or sub-section (1) of section 18;

(iii) a farmer who is engaged in the conservation of genetic resources of land races and wild relatives of ' economic plants and their improvement through selection and preservation shall be entitled in the prescribed manner for recognition and reward from the Gene Fund. Provided that material so selected and preserved has been used as donors of genes in varieties registrable under this Act;

(iv) a farmer shall be deemed to be entitled, to save, use, sow, resow, exchange, share or sell his farm produce including seed of a variety protected under this Act in the same manner as' he was entitled before the coming into force of this Act.

Note: Branded seed means any seed put in a package or any other container and labeled in a manner which gives indication that such seed is of a variety protected under this Act. In addition to the above, where any propagating material of a variety registered under this Act has been sold to a farmer or. a group of farmers or any organization of farmers, the breeder of such variety shall disclose to the farmer or the group of farmers or the organization of farmers, as the case may be, the expected performance under given conditions, and if such propagating material fails to provide such performance under such given conditions, the farmer or the group of farmers or the organization of farmers, as the case may be, may claim compensation in the prescribed manner before the Authority and the Authority shall, after giving notice to the breeder of the variety and after providing him an opportunity to file opposition in the prescribed manner and after hearing the parties, direct the breeder of the variety to pay such compensation as it deems fit, to the farmer or the group of farmers or the organization of farmers, as the case may be.

59. Which plant varieties can not be protected under this Act?

A plant variety which is:

(i) not capable of identifying such variety: or

(ii) consists solely of figures; or

(iii) is liable to mislead or to cause confusion concerning the characteristics, value, identify of such variety, or the identity of breeder of such variety;

(iv) is likely to deceive the public or cause confusion in the public regarding the identity of such variety;

(v) is comprised of nay matter likely to hurt the religious sentiments respectively of (any class or section of the citizens of India;

(vi) is prohibited for use as a name or emblem for any of the purposes;

(vii) is comprised of solely or partly of geographical name.

60. What is the term of plant variety protection?

(i) in the case of trees and vines, eighteen years from the date of registration of the variety;

(ii) in the case of extant varieties, fifteen years from the date of the notification of that variety by the Central Government under section 5 of the Seeds Act, 1966.

(iii) in the other cases, fifteen years from the date of registration of the variety.

Initially the certificate of registration shall be valid for nine years in the case of trees and vines and six years in the case of other.crops and may be revived and renewed for the remaining period on payment of fees as may be fixed by the rules.

61. What is the cost of registering a plant variety?

Some important fees are given below:

Sl.No.	Action	Official Fee
1.	Conducting tests	Dependent on the nature and type of test subject to a maximum of Rs. 50,000 per entry
2.	Fees for. registration of essentially derived varieties	Individual-Rs. 5,000/- Educational -Rs. 7,000/- Commercial -Rs. 10,000/-
3.	Renewal fee per year	Individual- Rs. 5,000/- Educational -Rs. 7,000/- Commercial -Rs. 10,000/-
4.	Application for benefit sharing	Rs. 5,000/-

62. What is UPOV?

UPOV is an abbreviation of Union pour la Protection des Obtentious Vegetals (Union for protection of new varieties of plant). It is an international convention which.provides a common basis for the examination of plant varieties in different member States of UPOV for determining whether a plant variety merits protection under UPOV or not.

63. What does copyright cover?

(i) Literary, dramatic and musical work. Computer programs/software are covered within the definition of literary work;

(ii) Artistic work;

(iii) Cinematographic films which include sound track and video films;

(iv) Record-any disc, tape, perforated roll or other device.

64. What are the rights of a copyright holder (which when violated lead to infringement?

(a) In the case of literary, dramatic or musical work, not being a computer program:

    i. to reproduce the work in any material form including the storing of it in any medium by electronic means;

    ii. to issue copies of the work to the public not being copies already in circulation;

    iii. to perform the work in public, or communicate it to the public;

    iv. to make any cinematography film or sound recording in respect of the work;

    iv. to make any translation of the work; to make any adaptation of the work;

    v. to do, in relation to a translation or an adaptation of the Work, any of the acts specified in relation to the work in Sub-clauses (i) to (vi);

(b) In the case of **computer programme:**

    i. to do any acts specified in clause (a);

    ii to sell or give on hire, or offer for sale or hire any copy of the computer programme, regardless of whether such copy has been sold or given on hire on earlier occasions;

(c) In the case of an artistic work:

    i. to reproduce the work in any material form including depiction in three dimensions of a two dimensional work or in two dimensions of a three dimensional work;

    ii. to communicate the work to the public;

    iii. to issue copies of the work to the public not being copies already in circulation;

    iv. to include the work in any cinematography film;

    v. to make any adaptation of the work;, i

    vi. to do, in relation to a translation or an adaptation of the work, any of the acts specified in relation to the work in sub-clauses (i) to (vi);

(d) In the case of a cinematography film:

    i. to make a copy of the film including a photograph of. any image forming part thereof;

    ii. to sell or give on hire or offer for sale or hire, any copy of the film, regardless of whether such copy has been sold or given on hire on earlier occasions;

    iii. to communicate the film to the public;

(e) In the case of sound recording:

    i. to make any other sound recording embodying it;

    ii. to sell or give on hire or offer for sale or hire, any copy of the ~sound recording, regardless of whether such copy has been sold or given on hire on earlier occasions;

    iii. to communicate the sound recording to the public;

Explanation: For the purpose of this section, a copy which has been sold once shall be deemed to, be a copy already in circulation.

65. How is computer defined for the purpose of copyright?

Computer includes any electronic or similar device having information processing capabilities.

66. What is the definition of a computer program?

Computer program means a set of instructions expressed in words, codes, schemes or any other form, including a machine readable medium, capable of causing a computer to perform a particular task or achieve a particular result.

67. What is the term of a copyright?

a. If published within the life time of the author of a.literary work, the term is for the life time of the author plus 60 years.

b. For cinematography films, records, photographs, posthumous publications, anonymous publication, works of government and international agencies, the term is 60 years from the beginning of the calendar year following the year in which the work was published.

c. For broadcasting, the term is 25 years from the beginning of the calendar year following the year, in which the broadcast was made.

68. Is it necessary to deposit accompanying documents of the computer programme for which copyright is being sought?

Documentation which normally accompanies the program is regarded as separate work and for this reason if the same has to be registered, it must be separately registered and not combined with the computer program in a single application.

69. If an employee in a company develops a programme, would this employee own the copyright?

No. In the case of a program made in the course of author's employment under a contract of service or apprenticeship, the employer shall, in the absence of any agreement to the contrary, be the first owner of the copyright.

70. If an independent third party develops a programme for a company, who owns the copyright?

The copyright in works created by third parties on commission do not automatically belong to commissioning party. If the third party is an independent contractor, it is essential for the commissioning party to obtain the copyright through a written deed of assignment. It is a common misconception that the copyright automatically belongs to the commissioning party. Thus, it is only where the developer is an employee creating the work under a contract of service that the rights belong to the employer.

71. What is the rule for the transfer of copyright?

The owner of the copyright in an existing work or prospective owner of the copyright in a future work may assign to any person the copyright, either wholly or partially in the following manner.

i. for the entire world or for a specific country or territory; or

ii. for the full term of copyright or part thereof; or

iii. relating to all the rights comprising the copyright or only part of such rights.

72. Is there a possibility of divulging secrets through deposit of source code?

Once the copyright is registered, the work is open to public inspection. For this reason, it is advisable, only to file a small extract of the computer program rather than the full program itself. It is important, however, to know that the part of the computer program which is not being filed would remain the trade secret of the owner and can be subject matter of a protection against any person who wrongfully obtains and utilizes the said programme.

73. In order to further ensure that secrets are protected, is deposition of computer program in object code permissible?

Although the recent amendment (1994) in the Copyright Act enlarges the meaning of a computer program, it is still not very clear as to whether it includes both object code and source code. However, keeping in mind the proclaimed object of the amendment, presumably the benefit of the Copyright Act will be available to both. As per expert's opinion, it is easier to determine from source code whether the deposit represents copyrightable material. Deposit of object code may be possible, but registration presumably would be accepted pending on assurance that the code does represent copyrightable material. Procedures for these do not exist at present with the Copyright office.

74. In some of the programs, the screens could be the most commercially significant aspect. Is it necessary to register the program screen separately from the underlying code?

Generally, all copyrightable expressions embodied in a computer program, including screen displays, are protectable. However, unlike a computer program, which is a literary work, screen displays are artistic work and cannot therefore be registered in the same application as that covering the computer program. A separate application giving graphic representation of all copyrightable elements of the screen display is necessary.

75. What notice needs to he put on computer program copies to seek copyright protection?

When a work is published by authority of the copyright owner, a notice of copyright may be placed on publicly distributed copies. As per the Berne Convention for protection of literary and artistic works, to which India is a signatory, use of copyright notice is optional. It is, however, a good idea to incorporate a copyright notice.

76. What are the major provisions in the amended Copyright Act, 1999 with regards to computer programmes

The major provisions of Copyright Act are:

(i) he doing of any act necessary to obtain information essential for operating inter operability of an independently created computer program with other programs by a lawful possessor of a computer program provided that such information is not otherwise readily available;

(ii) the observation, study or test of functioning of the computer program in order to determine the ideas and principles which underline any elements of the program while performing such acts necessary for the functions for which the computer program was supplied;

(iii) the making of copies or adaptation of the computer program from a personally legally obtained copy for non-commercial personal use.

77. What is the cost of filing copyright application in India?

Some important fees are given below:

Sl.No.	Action	Official fee
1.	For an application for registration of copyright in a- (a) literary, dramatic, musical or artistic work (b) literary or artistic work which is used or iscapable of being used in relation to any goods	Rs. 50 per work Rs. 400 per work
2.	For an application for registration of copyright in a Cinematograph Film	Rs. 600 per work
3.	For an application for registration of copyright in a sound recording	Rs. 400 per work

78. What does the term 'Design' mean according to the Designs Act, 2000?

We see so many varieties and brands of the same product (*e.g.*, car, television, a piece of furniture, mineral water bottle *etc.*)in the market, which look quite different from each other. If the products have similar functional features or have comparable price tags, the eye appeal or visual design of a product determines the choice. Even if similarities are not close, a person may decide to go for a more expensive item because that item has a better look or color scheme suiting the person's taste and choice. What is being said is that the external design or color scheme or ornamentation of a product plays a key role, in determining the market acceptability of the product over other similar products. Therefore, if you have a good external design that gives you an advantage then you must have a system to protect the features. This protection is provided by the Designs Act, 2000 in India. "Design" means only the features of shape, configuration, pattern, ornament or composition of lines or colours applied to any article whether in two dimensional or three dimensional or in both forms, by any industrial process or means, whether manual, mechanical or chemical, separate or combined, which in the finished article appeal to and are judged solely by the eye; but does not include any mode or principle of construction or anything which is in substance a mere mechanical device. Functional aspects of design arc not covered in this Act.

79. What are the types of designs not registrable under the Act?

A design which:

(a) is not new or original; or

(b) has been disclosed to the public any where in India or in any other country by publication in tangible form or by use in any other way prior to the filing date, or where applicable, the priority date of the application for registration; or

(c) is not significantly distinguishable from known designs or combination of known designs; or

(d) comprises or contains scandalous or obscene matter, shall not be registered.

80. What is meant by an 'article' under the Designs Act, 2000?

Article means any article of manufacture and any substance, artificial, or partly artificial and partly natural; and includes any part of an article capable of being made and sold separately.

81. What is the object of registration of designs?

The Designs Act protects new or original designs so created to be applied or applicable to particular article to be manufactured by industrial process or means. Sometimes purchase of articles for use is influenced not only by their practical efficiency but also by their appearance. The object of design registration, therefore, is to see that the artisan. Creator or originator of a design is not deprived of his bonafide reward by others applying it to their goods.

82. What are the essential requirements for the registration of design under the Designs Act, 2000?

1. The design should be new or original, not previously published or used in any country before the date of application for registration. The novelty may reside in the application of a known shape or pattern to new subject matter. However, if the design for which application is made does not involve any real mental activity for conception, then registration may not be considered.

2. The design should relate to features of shape, configuration, pattern or ornamentation applied to an article. Thus, designs of industrial plans, layouts and installations are not registrable under the Act.

3. The design should be applied to any article by any industrial process. Normally, designs of artistic nature like painting, sculptures and the like which are not produced in bulk by any industrial process are excluded from registration under the Act. Paintings and sculptures are subject matter of copyright.

4. The features of the designs in the finished article should appeal to and are judged solely by the eye. This implies that the design must appear and should be visible on the finished article, for which it is meant. Thus,

any design in the inside arrangement of a box, money purse or almirah may not be considered for registration, as these are generally put in the market in the closed state.

5. Any mode or principle of construction or operation or any thing, which, in. substance is a mere mechanical device, would not qualify for registrable design. For instance, a key having its novelty only in the shape of its corrugation or bend at the portion intended to engage with levers inside the lock associated with, cannot be registered as a design under the Act. However, when any design suggests any mode or principle of construction or mechanical or other action of a mechanism, a suitable disclaimer in respect thereof is required to be inserted on its representation, provided there are other registrable features in the design.

6. The design should not include any trade mark or property mark or artistic work.

83. Can stamps, labels, tokens, cards, be considered an article for the purpose of registration of a design?

No. Because once the alleged design *i.e.*, ornamentation is removed only a piece of paper, metal or like material remains and the article referred ceases to exist. Article must have its existence independent of the design applied to it. So, the design as applied to an article should be integral with the article itself.

84. When does the applicant for registration of design get the registration certificate?

When an application for registration of a design is in order, it is accepted and registered, then a certificate of registration is issued to the applicant. However, a separate request should be made to the Controller for obtaining a certified copy of the certificate for legal proceedings with requisite fee.

85. What is a Register of Designs?

The Register of Designs is a document maintained by the Patent Office, Kolkata as a statutory requirement. It contains the design number, date of filing and reciprocity date (if any), name and address of proprietor and such other matters as would affect the validity of proprietorship of the design and it is open for public inspection on payment of prescribed fee and extract from register may also be obtained on request with the prescribed fee.

86. What is the duration of the registration of a design? Can it be extended?

The term of a registered design is 15 years. Initially the right is granted for a period of 10 years, which can be extended, by another 5 years by making an application and paying a fee of Rs. 2000/- to the Controller before the expiry of initial 10 years period. The proprietor of design may make application for such extension as soon as the design is registered.

87. Is it mandatory to make the article by industrial process or means before making an application for registration of design?

No, design means a conception or suggestion or idea of a shape or pattern, which can be applied to an article or intended to be applied by industrial process or means.

88. Why is it important for filing the application for registration of design at the earliest possible?

First to file rule is applicable for registrability of design. If two or more applications relating to an identical or a similar design are filed on different. dates, the first application will be considered for registration of design.

89. Can the same applicant make an application for the same design again, if the prior application has been abandoned?

Yes. the same applicant can apply again since no publication of the abandoned application is made by the Patent Office, provided the applicant does not publish the said design in the meanwhile.

90. Are the registered designs open for public inspection?

Yes, registered designs are open for public inspection only after publication in the official gazette on payment of prescribed fee of Rs.500/- on a request in Form -5. 1

91. Can an applicant file the application for registration of design himself?

The application for registration of design can be filed by the applicant himself or through a professional person (*i.e.* patent agent, legal practitioner). However, applicants not residents of India have to file applications through an agent residing in India.

92. Can appeal for cancellation of the registration of a design be made? What are the grounds for cancellation?

The registration of a design may be cancelled at any time after the registration of design on a petition for cancellation in Form 8 with a fee of Rs.1500/- to the Controller of Designs on the following grounds:

1. That the design has been previously registered in India;
2. That it has been published in India or elsewhere prior to the date of registration;
3. That the design is not new or original;
4. That the design is not registrable;
5. That it is not a design under Clause (d) of Section 2.

93. What is the penalty for the piracy of registered design?

If anyone contravenes the copyright in a design he is liable for every offence to pay a sum not exceeding Rs.25,000/- to the registered proprietor subject to a maximum of Rs.50,000/- recoverable as contract debt in respect of anyone design.

94. What is the cost of filing design application in India?

The fee for filing application for registration of design in India is Rs.1,000/-.

95. What does Hague Agreement on Industrial Designs signify?

The Hague Agreement Concerning the International Deposit of Industrial Designs first came into existence in 1925. The Agreement aims at providing a mechanism for securing protection of an industrial design in all the member countries by means of an international deposit. The international deposit could be in the form of the industrial product or drawing or photograph or any other graphic representation of the said design. The duration of protection was 15 years from the date of deposit; this term is divided in two periods namely, one period of five years and the other of ten years. This Agreement is now being implemented by the WIPO.

96. What are the changes made to the Hague Agreement?

A Diplomatic Conference was held in June and July 1999 to bring out some amendments in the Hague Agreement. The revised agreement will come into effect after. it has been ratified by six of the initial signatory nations to the Agreement. The idea is to provide a way through which a single design application can give rights to protection for that design in member countries. The international design application must designate countries where protection will be sought. The designated countries can refuse to award design rights, if the application does not meet the requirements of national laws. The Agreement does not lay down any particular standards for registrability of the design, leaving this to national laws. Once registered, the international registration will have the same effect as a national design registration in those designated countries that have not refused grant for national registration. Other main features of the revised agreement treaty are:

1. International design protection will be available to nationals of a contracting country, domiciled in a contracting country or have industrial or commercial establishment in a contracting state.
2. An international design application maybe filed either at the applicant's national office or directly with the International Bureau of WIPO.
3. Two-dimensional designs (textile designs) would be eligible for protection.
4. A formalities examination will be carried out by the.International Bureau and then the application will be published if it is found to satisfy the formalities. The publication will be made six months after the registration. This can be deferred to 30 months in some special cases.
5. The International Bureau will, after the registration, send a copy or the application to each of the designated countries. These countries have to inform the Bureau within six months if national requirements are not met. However, countries that examine design applications for novelty or where opposition system exists, this time is increased to 12 months.
6. Multiple designs may be included in the same application. It is however, required that all products to which such designs relate must be in the same class under the Locarno Classification.

It can be seen that there are some Similarities with the PCT system for patent applications. India is not yet a member of the Hague Agreement and hence, the above provisions or description may not be of immediate relevance to us. However, there is a strong need to monitor the developments in this area.

97. How is the Trademarks Act 1999 different from the Trade and Merchandising Marks Act 1958?

Enactment of the Trademarks Act 1999 is a big step forward from the Trade and Merchandise Marks Act 1958 and the Trademark Act 1940. The newly enacted Act has some features not present in the 1958 Act and these are:

1. Registration of service marks, collective marks and certification trademarks.
2. Increasing the period of registration and renewal from 7 years to 10 years.
3. Allowing filing of single application for registration in more than one class.
4. Enhanced punishment for offences related to trademarks.
5. Exhaustive definitions for terms frequently used.
6. Simplified procedure for registration of registered users and enlarged scope of permitted use. "
7. Constitution of an Appellate Board for speedy disposal of appeals and rectification applications which at present lie before High Court.

98. How is "Trademark" defined?

A trademark is a distinctive sign, which identifies certain goods or services as those produced or provided by a specific person or enterprise. A trademark may be once or a combination of words, letters and numerals. It may also consist of drawings, symbols, three-dimensional colours and combination of colours. It is used by traders/companies/firms *etc* to distinguish their goods and services from those of their competitors. A consumer associates some level of quality/price/prestige with the goods of a particular trademark. In other words the consumer uses the trademark for making a choice while buying a particular product. There are so many examples in our day to day life such as TATA, BATA, Liberty, Brooke Bond, Dabur, Baidyanath, Park Avenue, SAIL and so on. Trademarks do not protect the design- or the ideas behind the goods or services from imitation or duplication, but prevent other traders/company/firm from deceiving customers into believing that goods or services actually produced by them were produced by the trademark holder.

99. What are "Well-known Trademarks"?

Well-known trademarks in relation to any goods or services, means a mark which has become known to a substantial segment of the public which uses such goods or receives such services that the use of such mark in relation to other goods or services is likely to be taken as indicating a connection in the course of trade or rendering of services between those goods or services and a person using the mark in relation to the first-mentioned goods or services.

100. What is the meaning of ~'Service" in the Trademark Act 1999?

Service means service of any description which is made available to potential users and includes the provision of services in connection with business of any industrial or commercial matters such as banking, communication, education, financing, insurance, chit funds, real estate, transport, storage, material treatment, processing, supply of electrical or other energy, boarding, lodging, entertainment, amusement, construction, repair, conveying of news or information and advertising.

101. How are the terms "Certification Trademarks" and "Collective Marks" defined in the Act?

Certification trade mark means a mark capable of distinguishing the goods or services in connection with which it is used in the course of trade which are certified by the proprietor of the mark in respect of origin, material, mode of manufacture of goods or performance of services, quality and accuracy. Collective Mark means a trademark distinguishing the goods or services of an association of persons (not being a partnership within the meaning of the Indian Partnership Act, 1932).

102. What is the term of a registered trademark?

The initial registration of a trademark shall be for a period often years but may be renewed from time to time for an unlimited period by payment of the renewal tees.

103. What is the cost of filing trademarks application in India?

Some important fees are given below:

Sl.No.	Action	Official Fee
1.	Application to register a trade mark for a specification of goods or services included in one class	Rs. 2500/-
2.	Application to register a textile trademark consisting exclusively of numerals or letters or any combination thereof for specification of goods included in one item	Rs. 2500/-
3.	Application to register a collective mark for a specification of goods or services included in one class	Rs. 10,000/-
4.	Application to register a certification trade mark for specification of goods or services included in one class	Rs.10,000/-
5.	Renewal of registration	Rs.5;000/-
6.	Application for approval of Registrar	(i) For first mark, Rs.2500/- (ii) For every additional mark Rs.500/-
7.	Application under Geographical Indications of Goods -(Registration and Protection) Act, 1999 to refuse orinvalidate registration of a trade mark consisting of Geographical Indication *etc.*	Rs.3000/

104. What is the Madrid Agreement?

The Madrid Agreement was adopted on April 14, 1891 to facilitate protection of a trademark or service mark in several countries by means of a single international registration. As on July15, 1999, 54 countries are party to this Agreement mainly belonging to Europe, countries of Africa and four countries in the Far East namely, China, the Democratic People's Republic of Korea, Mongolia and Vietnam. The United Kingdom, the United States of America, most Latin American countries, Japan and India are not signatories to this agreement. The Agreement covers both trademarks and service marks.

105. What are the main features of the Madrid Agreement?

Main features of the Madrid Agreement are as follows:

1. An applicant must be a national of a member country. A person having his domicile or a real and effective industrial or commercial interest in such a country is also eligible. It may be noticed that this would be governed by the national laws of the country in question:

2. A mark to be resisted in member states should be first registered at the national level in the country of, origin of the applicant. The first registration is called 'basic registration.

3. The country having given the basic registration can only transmit there quest for international filing to the International Bureau of the World Intellectual Property Organization (WIPO) along with the list of the countries in which protection is being sought. There is no provision for directly filing a request under the Agreement.

4. It may be iterated that the country of origin has to be a member state. The role of the office of the country of origin is not only to send the application for international registration but also to certify that the mark which is the subject of, the international registration, is the same mark which is the subject of the basic registration.

5. For each application fees has to be paid for each designated country and WIPO. The fees paid for the designated countries, is called the 'complementary fee'.

6. The International Bureau notifies the international registration to the offices of the designated countries and publishes it in a monthly periodical called 'The WIPO Gazette of International Marks'.

7. If the basic registration is cancelled for some reasons, in the country of' origin, during the first five years, the International registration automatically stands cancelled in all the designated countries. This also gives an advantage to a person, to oppose the registration of a mark only in the country of origin and that person need not oppose it in all the designated countries. This possibility of challenging an international registration through a national registration is referred to as 'Central, Attack' feature of the Agreement.

106. What is the Madrid Protocol?

The Protocol relating to the Madrid Agreement concerning the International Registration, of Marks was adopted at Madrid on June27, 1989. The Protocol, which entered into force on December!, 1995, retains the basic features of the Madrid Agreement. As on July1S, 1999, 39 countries have acceded to the 'Protocol. The Protocol was formed to remove some of the features of the Madrid Agreement, which posed some obstacles to accession by several countries. These features are:

1. For an international registration, it is essential to first register a mark at the national level. The time required for obtaining a mark at the national level varies from country to country. Hence some parties do suffer.

2. Within one year, a designated member country has to examine and issue a notice of refusal by giving all the grounds for refusal. The period was considered short.

3. A uniform fee is paid for the designation of a member country. This was found to be inappropriate for countries with high level of national fees.

4. An international registration is linked to the basic registration during the initial five years and the former gets cancelled if latter is cancelled. The fact, that grounds under which a mark is cancelled in the country of origin need not necessarily exist in every other designated country, is overlooked.

5. The only working language of the Madrid Agreement is French. Innovations introduced by the Madrid Protocol are:

   a. An international application need not necessarily be based on a registration made by the Office of Origin but can also be based on an application filed with the Office of Origin. This makes it convenient for countries with full examination system where the national registration takes time. It also makes it possible to claim the right of priority of six months under the Paris Convention.

   b. A Contracting Party can receive the fee under the existing Madrid Treaty system through its share in the international fees collected for each designation made as in the Madrid Treaty. Alternatively, the member country can choose "Individual fee" system for each designation made, which should be an amount not more than the national fee for a ten-year registration. The "Individual fee" system makes an attractive proposition for countries with high level of national fees.

   c. It is possible to transform an international registration into national or regional application in the designated Contracting Parties, if the basic registration is cancelled for some reasons, as in the case of "Central Attack"'.

   d. An applicant may choose to base an international registration in any of the Contracting States with which he has connection through nationality, domicile or establishment.

107. What does Semiconductor Integrated Circuits Layout-Design Act, 2000 cover?

It provides protection for semiconductor IC layout designs. Layout design includes a layout of transistors and other circuitry elements and includes lead wires connecting such elements and expressed in any manner in a semiconductor IC. Semiconductor IC is a product having transistors and other circuitry elements, which are inseparably formed on a semiconductor material or an insulating material or inside the semiconductor material and designed to perform an electronic circuitry function.

108. What is not registrable as IC layout design?

An IC layout design is not registrable if it is:

1. Not original;
2. Commercially exploited anywhere in India or in a convention country:
3. Inherently not distinctive;
4. Inherently not capable of being distinguishable from any other registered layout design.

Note: Design not exploited commercially for more than 2 years from date of registration of application shall not be treated as commercially exploited for the purpose of this Act.

109. What is the term of an IC layout design protection?

The term is 10 years from the date of filing or from the date of first commercial exploitation anywhere in any country whichever is earlier.

110. What constitutes an infringement under the Act?

Reproducing, importing, selling and distributing the IC layout design for commercial purposes only constitutes infringement. A layout design created on the basis of scientific evaluation of a registered layout design shall not constitute any infringement.

111. What is the cost of registering IC layout design application in India?

Some important fees are given below:

No Action Official fee

1. On application to register a layout -design Rs.5000/-
2. On request for certificate of the Registrar Rs.1 000/-

112. What does the term 'Geographical Indications (GI)' stand for?

Geographical Indications are names associated with goods which identify such goods as agricultural goods, natural goods or manufactured goods as originating, or manufactured in the territory of a country, or a region or a locality in that territory) here a given quality, reputation or other characteristics of such goods is essentially attributable to its geographical origin. Some examples of GI are Darjeeling Tea, Pochampalli Saree, Chanderi Saree, Kanjeevaram Silk, Champagne Wine, Scotch Whisky.

113. Who can apply for GI's registration?

Any association of persons or producers or any organization or authority established by or under any law for the time being in force representing the interest of the producers of the concerned goods, who are desirous of registering geographical indication in relation to such goods can apply for GI's registration.

114. Who is a registered proprietor of a geographical indication?

Any association of persons or of producers or any organization or authority established by or under the law can be a registered proprietor. Their name should be entered in the Register of Geographical Indication as registered proprietor for the Geographical Indication applied fur.

115. Who can use the registered geographical indication?

An authorized user has the exclusive rights to the use of geographical indication in relation to goods in respect of which it is registered.

116. Who is an 'authorised user'?

'Authorised user' is any person claiming to be the producer of the goods in respect of which a geographical indication has been registered. For becoming an authorized user the person has to apply in writing to the Registrar in prescribed manner and by paying requisite fee.

117. Who can be considered a 'producer' for the purposes of this act?

'Producer' in relation to goods, means any person who:-

(a) Produces processes or packages agricultural goods

(b) Exploits natural goods

(c) Makes or manufactures handicraft or industrial goods.

118. Can all geographical indications be registered?

No, a GI cannot be registered, if

☆ its use is likely to deceive or cause confusion or be contrary to any law;

☆ it comprises or contains scandalous or obscene matter or any matter likely to hurt religious susceptibilities of any class or section of the citizens of India;

☆ it is a generic name;

☆ it has ceased to be protected in their country of origin or which have fallen into disuse in that country;

☆ it is falsely represented by persons claiming that goods originate in another territory, region or locality as the case may be.

119. What is the punishment in the Act for falsifying GI?

A sentence of imprisonment for a term between six months to three years and a fine between fifty thousand rupees and two lakh rupees is provided in the Act. The court may reduce the punishment under special circumstances.

120. What is the term of GI protection?

The registration of a GI shall be for a period of ten years but may be renewed from time to time for an unlimited period by payment of the renewal fees.

121. What is the cost of registering a GI in India?

Some important fees are given below:

Sl.No.	Action	Official Fee
1.	On application for the registration of a geographical indication for goods included in one class	Rs.5000/-
2.	On a single application for the registration of a geographical indication for goods in different classes	Rs.5000/-for each class
3.	On application for the registration of an authorized user of a registered geographical indication	Rs.500/-
4.	For renewal of an authorized user	Rs.1,000/-

Patent Facilitating Centre (PFC)

Technology Information, Forecasting and Assessment Council (TIFAC), Department of Science and Technology, Vishwakarma Bhavan, A – Wing, Saheed Jeet Singh Marg New Delhi – 110 016 e-mail: tifac@nda.vsnl.net.in

Annexure – II

Agri-clinics and Agri-business Centres (Revised Guidelines 2010)

(Excerpt from http://www.agriclinics.net/guidelines2010.pdf)

(Adopted from REVISED AGRI-CLINICS AND AGRI-BUSINESS CENTRES (ACABC) SCHEME – 2010, 2011 Department of Agriculture and Cooperation (DAC), Ministry of Agriculture, Government of India, New Delhi.**)**

Concept/Definition

Agri-clinics: Agri-clinics are envisaged to provide expert advice and services to farmers on various technologies including soil health, cropping practices, plant protection, crop insurance, post harvest technology and clinical services for animals, feed and fodder management, prices of various crops in the market *etc.* which would enhance productivity of crops/animals and ensure increased income to farmers.

Agri-business Centres: Agri-business Centres are commercial units of agriventures established by trained agriculture professionals. Such ventures may include maintenance and custom hiring of farm equipment, sale of inputs and other services in agriculture and allied areas, including post harvest management and market linkages for income generation and entrepreneurship development.

Objectives of ACABC

1. To supplement efforts of public extension by necessarily providing extension and other services to the farmers on payment basis or free of cost as per business model of agri-preneur, local needs and affordability of target group of farmers;
2. To support agricultural development; and

3. To create gainful self-employment opportunities to unemployed agricultural graduates, agricultural diploma holders, intermediate in agriculture and biological science graduates with PG in agri-related courses.

Eligibility Criteria for Candidates

1. The scheme is open to following categories of candidates:

   Graduates in agriculture and allied subjects from SAUs/Central Agricultural Universities/Universities recognized by ICAR/UGC. Degree in Agriculture and allied subjects offered by other agencies are also considered subject to approval of Department of Agriculture and Cooperation, Government of India on recommendation of the State Government.

2. Diploma (with at least 50 per cent marks)/Post Graduate Diploma holders in Agriculture and allied subjects from State Agricultural Universities, State Agriculture and Allied Departments and State Department of Technical Education. Diploma in Agriculture and allied subjects offered by other agencies are also considered subject to approval of Department of Agriculture and Cooperation, Government of India on recommendation of the State Government.

3. Biological Science Graduates with Post Graduation in Agriculture and allied subjects.

4. Degree courses recognized by UGC having more than 60 percent of the course content in Agriculture and allied subjects.

5. Diploma/Post-graduate Diploma courses with more than 60 percent of course content in Agriculture and allied subjects, after B.Sc. with Biological Sciences, from recognized colleges and universities.

6. Agriculture related courses at Intermediate (*i.e.* plus two) level, with at least 55 per cent marks.

The scheme covers full financial support for training and handholding, provision of loan and credit linked back ended composite subsidy as per the details given in the following section.

Training and Handholding

National Institute of Agricultural Extension Management (MANAGE) will be responsible for providing training to eligible candidates, through Nodal Training Institutes (**NTIs**) and motivating them for setting up of Agri-clinics and Agri-business centres.

Process of Selection of Candidates

Call for applications from eligible candidates: Advertisement in local newspapers will be issued under intimation to local employment exchanges separately by the NTI. The NTIs may also give publicity to the scheme through

Doordarshan, AIR, Private Television and Radio channels, ATMAs, Agricultural Universities, ICAR Institutions, KVKs, Banks, Agriculture Graduate Associations, Agri-Business Companies, Agriculture and allied departments, network of trained/ established Agri-preneurs, posters, leaflets *etc*. Besides, NTIs may adopt any innovative method for giving publicity about the scheme. Candidates can apply in the format prescribed to the NTIs personally or by post.

Online applications: In order to encourage submission of applications by candidates for training without personally visiting NTIs for completing enrolment formalities, MANAGE will develop a mechanism for submitting applications online, through its website:

http://www.agriclinics.net/onlineform.htm.

Selection of Candidates

(a) Scrutiny of applications of shortlisted candidates meeting the prescribed qualifications will be done jointly by an official of Nodal Training Institute (NTI) and any one official from ATMA/Block Technology Team/State Agriculture and allied departments. For this purpose, NTI will coordinate with Project Director, ATMA.

(b) Eligible candidates will be called for an interview. The Interview Committee will consist of representatives from NTI, MANAGE, KVK, NABARD, Bank, State Agriculture/allied departments, and an Agri Business Company in the area (if any). Besides NTI, presence of at least three officials is necessary to make the minimum quorum of the interview committee. NTI shall request the offices concerned in writing with proper notice.

(c) The candidates will be selected based on predetermined criteria and weightage

(d) To select the most eligible and genuine candidates for undergoing training under the Scheme, NTIs should ensure at least 60 applications are considered by the Screening Committee. Each batch shall not exceed 35 candidates.

Certificates to Successful Ventures

Certificates are being issued to trainees by MANAGE after successful completion of two month training. It has been decided that successfully established ventures will a get a certificate recognizing that their enterprise is set up under the Scheme. Such certificates will be issued to ventures after verification and recommendation by PD, ATMA, Programme Coordinator of KVK or Head of nearest Research Station of State Agriculture University/ICAR institute as per the following procedure.

On establishing the venture and running it successfully (including provision of extension services) for at least 6 months, the candidates will seek the certificate from MANAGE through respective NTI by submitting the verification report and recommendation to MANAGE as above.

Certificates consisting Candidate's ID No. and venture details will be issued by MANAGE indicating the date on which verification was conducted.

This certificate entitles the agri-preneurs to be recognized by the Development Departments in supplementing the efforts of public extension services.

Refresher and Sensitization Training

About 500 selected agri-preneurs will undergo refresher training every year. This training of about 3-5 days duration may be conducted in specialized Institutions like SAUs/ICAR Institutes/IIMs/IITs/CSIR Institutes/DST Institutes/reputed Private/Non-governmental Institutions.

NABARD will organize sensitization training/workshops to motivate the bankers across the country to provide credit to Agri-preneurs for establishing ventures. Budget of ' 25 lakh per year has been provided to NABARD for the purpose. NABARD will submit details of cost norms, numbers ofcourses, duration and officers to be trained. NABARD will necessarilyinform MANAGE and DAC for participation of their representativesin such sensitization programmes.

Linkage with Credit

Assistance under the scheme would be purely credit linked and subject to sanction of the project by banks based on economic viability and commercial considerations. The eligible financial institutions under the scheme are:

1. Commercial Banks
2. Regional Rural Banks
3. State Cooperative Banks
4. State Cooperative Agriculture and Rural Development Banks
5. Such other institutions eligible for refinance from NABARD.

Project Cost Ceiling

Ceiling of project cost for subsidy has been enhanced to ' 20 lakh for an individual project (25 lakh in case of extremely successful individual projects) and up to ' 100 lakh for a group project. (Established by a group comprising at least 5 trained persons under the scheme, out of which one could be from Management backgorund). The bank may, neverthless, subject to their own satisfaction, finance groups formed by 2 or more trained persons under the scheme. (Person with management background can only be included in groups of 5 or more) with in the TFO ceiling ' 20 lakh per trained person and overal ceiling of ' 100 lakh, whichever is less for the purpose of subsidy. However, the actual credit sanctioned by the bank for a venture established under the scheme could be higher depending on the financial viability and technical feasibility. Thus, for instance, if an individual is granted a loan for TFO of ' 35 lakh, subsidy shall be reckoned only on TFO of ' 20 lakh.

To encourage exceptionally successful individual agri-preneurs, the project cost limit for subsidy purposes may be extended by ' 5 lakh in addition to the generally applicable project cost limit of ' 20 lakh for calculating subsidy. This will serve as an incentive to an agri-preneur to expand his/her already established and successful venture. Format for considering such cases is given on the MANAGE website http://www.agriclinics.net/guidelines/Annexure-XVII.pdf. Such cases are to be approved by the Empowered Steering Committee using the criteria given at **AnnexureX** based on the recommendation of the Banker and ATMA official.

In order to provide extension services using web-enabled resources of DAC and other agencies, computer, printer and broadband connection (1 mbps at least) should be an integral part of TFO.

Term Loan:

The term loan would be composite in nature and participating bank(s) would extend bank loan as per the TFO, which would includes fixed capital cost and working capital for one operating cycle. Loan sanctioned will be the differential amount between TFO and margin money.

Composite Subsidy will be provided even for low capital investment cases has been introduced (for cases sanctioned by banks on or after 04.08.2010) as it is definitely possible that some agri-preneurs already have capital (*e.g.* a building) to start a venture, thereby requiring minimal capital investment. At least 10 per cent value of the Total Financial Outlay of the project should be in capital form.

The repayment schedule will be drawn on the total amount of the loan (including subsidy) in such a way that the subsidy amount is adjusted after liquidation of net bank loan (excluding subsidy).

Repayment period will depend on the nature of activity and will vary between 5 to 10 years. The repayment period may include a maximum grace period of 2 years (to be decided by the financing bank as per needs of individual projects).

Rate of interest on term loan shall be as per RBI guidelines and declared policy of the bank in this regard. Interest would be chargeable on borrower's accounts as per RBI/Bank's policy.

Margin Money

The stipulations on margin money shall be in accordance with the guidelines of Reserve Bank of India issued from time to time. In case of loans up to ' 5 lakh, no margin money is required as per present norms.

The margin money to be contributed by the general category entrepreneur will be as per prevailing norms. However, concessions would be made in respect of SCs/STs, women and beneficiaries of North- Eastern States, Hill areas. In such cases, a maximum of 50 per cent of the margin money prescribed by banks could be given by NABARD to meet the shortfall in borrower's contribution, if the bank is satisfied that the borrower is unable to meet the margin money requirements. Such assistance to banks by NABARD will be without any interest. The banks may, however, levy a service charge up to 2 per cent per annum from the borrowers.

Security

As most of the eligible activities pertain to agricultural input supply and services and the cost of investment will be less than ' 25 lakh in most cases, the security norms applicable to tiny industries as prescribed in RBI circular No.RPCD. PLNFS.BC.65/06.02.31/99-2000 dated 31.3.2000 would be made applicable to these units. Accordingly, up to a loan amount of ' 5 lakh, the loans can be secured against hypothecation of assets created and no further security would be necessary.

Time Limit for Completion of the Project

Time limit for completion of the project would be as envisaged under the project, subject to maximum of 6 months period from the date of disbursement of the first installment of loan by financial institution, which may be extended by a further period of 6 months, if reasons for such delay are considered justifiable by the financial institution concerned.

If the project is not completed within the stipulated period, benefit of subsidy shall not be available and advance subsidy placed with the participating bank, if any, will have Revised Agri-clinics and Agri-business Centres (ACABC) Scheme - 2010 11 to be refunded forthwith to NABARD.

Other Conditions

The participating banks will adhere to the norms of appraising the projects regarding technical feasibility and commercial/financial viability.

The participating banks should ensure insurance of the assets created under the project, wherever required statutorily. If beneficiary opts to buy insurance on these assets, even if insurance is not required statutorily, such expenditure will be an eligible component of the TFO.

A sign board displaying "Assisted under the Scheme of Agri-clinics and Agri-business Centres, Ministry of Agriculture, Government of India" will be exhibited at the unit.

Pre and post completion inspection of the project shall be undertaken by the participating bank to verify physical, financial and operational progress as and when required.

Refinance Assistance from NABARD

NABARD will provide refinance assistance to commercial banks, RRBs, SCBs, SCARDBs and other such eligible institutions at the rate of 100 per cent of the amount financed by the banks as term loan. Rate of interest on refinance will be as decided by NABARD from time to time.

Eligibility for Subsidy

In the Pre-revised Scheme, subsidy was admissible in respect of agriculture graduates trained under ACABC scheme on or after April 1, 2004, for fresh investments made after July 9, 2006. The candidates trained under the Scheme prior to April 1, 2004 who have made investments after July 9, 2006, were to be considered on a case by case basis. Subject to fulfillment of all other eligibility conditions under the scheme, a Committee headed by Director General (DG), National Institute of

Agricultural Extension Management (hereinafter referred to as MANAGE) used to take a decision in such matters. However, very few such cases have been received in MANAGE since it was set up, mainly due to ignorance among candidates and financing institutions and distance of MANAGE from their place of operation.

Therefore, it has been decided to remove this cut-off date of April 1, 2004 *qua* the training date and henceforth all candidates trained under the scheme shall be eligible for subsidy with respect to investments made after July 9, 2006 (Applicable for loan accounts sanctioned on or after this date). As agreed in the Review Meeting held on 09.11.2010, trained candidates whose knowledge acquired during the training might have become obsolete due to passage of time, may undergo a refresher course. Whether or not a candidate trained sufficiently long time ago needs his skills to be upgraded by refresher courses, shall be decided by the Selection Committee for candidates at the NTI level. However, need for such refresher trainings shall not be considered, if the training has been within 5 years before date of loan application. Funds for such skill upgradation courses can be pooled from various resources including MANAGE coordination charges, savings from funds allocated for refresher courses and budget earmarked for advertisement and publicity of ACABC Scheme with DOE.

The trained candidates could also undertake group projects. If the group consists of a total of five or more persons trained under the scheme, all except one of them would have to be from among the eligible categories mentioned under para 3 and the remaining person could be non-agriculture graduate with experience in business development and management.

Delivery of extension services shall be the main component of ACABC projects for availing of the benefit of subsidy under the Scheme.

Linking Subsidy Adjustment to Delivery of Extension Services

Adequate safeguards are put in place to ensure providing extension services necessarily by the agripreneurs to their clients. In particular, payment of back-ended subsidy will be linked to extension services provided by the agripreneurs.

For this purpose, any two out of a group of rank and status not below the following officials/persons would make a periodic/surprise inspection atleast on quarterly basis and submit its report to NABARD/financing bank/office of ATMA. Record of advisory services provided to farmers by the agri-preneurs may form the basis for providing the report.

1. Block Technology Manager/Subject Matter Specialists under ATMA
2. Chairman (or his nominee), ATMA Block Farmers' Advisory Committee
3. Panchayat Members
4. NABARD/Bank official
5. Block level officer of State Department of Agriculture and allied sectors.

The financing bank will consider the periodic or surprise visit reports mentioned above while releasing the back-ended composite subsidy to the agri-preneur under this Scheme. A consolidated report has to be finalized before the expiry of three years lock-in period. The financing bank has to send the report to the NABARD and ATMA for information. The report may highlight nature of enterprise initiated by agri-preneur, its coverage/reach, impact on income of agri-preneur, impact on farmers and agricultural development in the area.

States are encouraged to provide information on all government policies, programmes, schemes *etc* to agri-preneurs and also use their services in implementation of extension activities funded by the government.

Monitoring

ACABC Monitoring Cell will be strengthened at DOE and shall be responsible for proper Monitoring and Administration of the Scheme. The linkages of agri-preneurs with State Governments, Banks, Agro industries under GOI and other stakeholders will be reviewed by MANAGE and DAC at least twice a year. The DOE would employ contractual staff, organize review meetings/workshops/interfaces and carry out periodic monitoring of the Scheme and its impact evaluation study at the end of XI Plan. DOE along with MANAGE would also maintain an on-line monitoring system, telephone, fax, computer, *etc.* for the Scheme.

In order to carry-out the monitoring and evaluation of the Scheme on regular basis, the following financial support is provided to Directorate of Extension (DOE):

Besides the surprise and regular visits by teams of officers as per par above, monitoring of each project shall be done by the financing bank. Joint review will also be undertaken on a quarterly basis by Bank and NABARD.

NABARD may prescribe suitable formats for submission of various reports by participating banks, taking into account reporting requirements of the DAC and subject to guidance of DAC/Empowered Steering Committee.

The progress report of the scheme in the prescribed format shall be sent to the Department of Agriculture and Cooperation, Government of India by NABARD on monthly basis, with a copy to MANAGE.

The Scheme of ACABC and the progress there under shall be reviewed invariably in all District Level Consultative Committee and Stat Level Bankers Committee meetings with a view to sorting out problems/issues emerging in smooth implementation thereof.

Linkages with ATMA and Agro Industries

The Scheme should be linked with Agricultural Technology Management Agencies (ATMAs) constituted by States under the Scheme "Support to State Extension Programmes for Extension Reforms". The Extension Reforms Scheme mandates that minimum 10 per cent of resources on extension activities are to be utilized through the nongovernmental sector, which also includes Agri-preneurs. ATMAs would be encouraged to implement extension activities through Agripreneurs. The ventures could also be utilized to provide input services like plant saplings, seeds, and micronutrients under on-going flagship schemes implemented by the Department. Special emphasis will be made to review the progress of involvement of the agri-preneurs in ATMA activities on half-yearly basis by State

Nodal Officer, GoI and MANAGE.ATMAs shall also strive to achieve establishment of at least one agri-clinic every year in each Block depending on the availability of trained candidates under the Scheme. The Monthly Progress Report being submitted through EMS under ATMA scheme shall necessarily have a provision for reporting the progress achieved on this front.

In order to provide support to the Agri-preneurs for making their business more viable, Govt. of India Institutions such as National Seeds Corporation (NSC), State Farms Corporation of India (SFCI), Indian Farmers Fertilizers Cooperative Ltd (IFFCO), Krishak Bharati Cooperative Ltd. (KRIBHCO), Small Farmers Agri Business Consortium (SFAC), State Agro Industries Corporations *etc.* shall be advised to engage the Agri-preneurs as their authorized dealers on preferential basis without affecting their existing network. The linkages with these agro industries shall be reviewed on half-yearly basis.

INDICATIVE LIST OF AGRI VENTURES
UNDER ACABC SCHEME

☆ Extension consultancy services

☆ Soil and water quality cum inputs testing laboratories

☆ Crop protection services, including pest surveillance, diagnostic and control services (*with culture rooms,autoclaves, microscopes, ELISA Kits etc. for detection of plant pathogens including viruses, fungi, bacteria, nematodes, and insect pests*)

☆ Micro-propagation including plant tissue culture labs and hardening units

☆ Production, maintenance and custom hiring of agricultural implements and machinery including micro irrigation systems

☆ Seed production and processing units

☆ Vermiculture units

☆ Production of bio-fertilizers, biopesticides and other bio-control agents

☆ Apiaries (bee-keeping) and honey and bee products' processing units

☆ Agricultural insurance services

☆ Agri-tourism

☆ Agri journalism – film production, farm publications and exhibitions

☆ Poultry and fishery hatcheries

☆ Livestock health cover, veterinary dispensaries and services including frozen semen banks and liquid nitrogen supply and artificial insemination;

☆ Information technology kiosks

☆ Feed production, marketing and testing units

☆ Value addition centres

☆ Cool chain including cold storage units

☆ Postharvest management centres for sorting, grading, standardization, storage and packaging

☆ Metallic and non-metallic storage structures

☆ Horticulture clinic, nursery, landscaping, floriculture

☆ Sericulture

☆ Vegetable production and marketing

☆ Retail marketing outlets for processed agri-products

☆ Production and marketing of farm inputs and outputs

☆ Contract farming

☆ Crop production and demonstration

☆ Mushroom production

☆ Production, processing and marketing of medicinal and aromatic plants

☆ Production units like dairy, poultry, piggery, fisheries, sheep rearing, goat rearing, emu rearing, rabbit rearing *etc.*

Note: The above activities are indicative in nature. Any other activity in agriculture, horticulture, sericulture, animal husbandry, fisheries, allied sectors or combination of two or more of the above activities selected by the candidates, which, generate income to the agri preneur and render extension services to the farmers will also be eligible under the Scheme.

APPLICATION FORMAT FOR CANDIDATES TO ENROLL AS TRAINEES UNDER
ACABC SCHEME

1.	Name of the candidate	
	Name of Father/Spouse	
2.	Date of Birth	
3.	Sex (Male / Female)	
4.	Permanent Address	
	Village and Panchayat	
	Block and/or Tehsil	
	District	
	State	
	PIN code	
5.	Address for Correspondence	
6.	Educational Qualification	
	Name of the Degree / Diploma / Certificate / Course	
	Board/Institute/University where studied	
	University to which Affiliated	
	Marks / Grade obtained	
	Year of passing/completion	
7.	Contact details	
	Telephone/Mobile	
	Email	
8.	Experience	
8A	Family Background	
	Agriculture	
	Other than agriculture	
9.	Agri-Business interest	
	Nature of enterprise being planned to set up after the training	
	Experience in the enterprise being planned	
	Likely place of establishment of enterprise	
10.	Aptitude for extension work with brief details of extension work done and vision for future in serving farmers	

Date :   Signature :

CRITERIA FOR SELECTION OF CANDIDATES

S.No.	Parameter	Scoring pattern	
		Parameter	Score
1	Age	< 25 Yrs.	1
		25 – 50 Yrs	3
		> 50 Yrs.	2
2	Gender	Male	1
		Female	2
3	Family background	Agriculture	5
		Business	3
		Others	1
4	Qualification	Ph.D in Agriculture & allied disciplines	10
		Masters in Agriculture & allied disciplines	9
		Degree in Agriculture & allied disciplines with first class	8
		Degree in Agriculture & allied disciplines / PG Diploma in Agri with Degree in Biological Sciences	7
		Diploma in Agriculture with first class	6
		Diploma in Agriculture	5
		Degree Courses (> 60% content in agriculture) with first class	4
		Degree Courses (> 60% content in agriculture)	3
		Intermediate in Agri. with first class	2
		Intermediate in Agriculture	1
5	Experience in relevant area	> 5 years	5
		3-5 years	3
		1-3 years	1
		< 1 year	0
6	Techno-commercial viability of the proposed venture	High	4-5
		Medium	2-3
		Low	0-1
7	Knowledge of the chosen activity	High	4-5
		Medium	2-3
		Low	0-1
8	Risk bearing capacity	High	4-5
		Medium	2-3
		Low	0-1
9	Aptitude for Extension work	High	8-10
		Medium	4-7
		Low	0-3
		Total Score	50

Marks for item nos. 6-9 shall be decided during the interview. Minimum qualifying marks are 25.

CRITERIA FOR SUCCESSFUL ESTABLISHMENT OF VENTURES

S.No.	Parameters	Criteria	Maximum
1.	Age of the Project	2 Score for every year after 3 years	10
2.	Total Annual Turnover (for last 3 years)	Upto ₹ 25 lakh/year ₹ 25-50 lakh /year ₹ >50 lakh / year	5 10 15
3.	No. of farmers being served during the last 3 years	Upto 500 farmers/year 500-1000 farmers /year >1000 farmers/ year	5 10 15
4.	Linkages with Development Departments	1 Score per Department / Institution	5
5.	Nature of activities being carried out with the Departments	1 Score for each type of activity (Training, Demonstrations, Visits etc.)	5
6.	Impact Created (data of 30 farmers served by the Agri Clinic & Agri Business Centre to be enclosed with this format)		
	(i) Increase in production/ productivity of crops over previous year	Upto 10% increase 10-20% increase > 20% increase	5 10 15
	(ii) Increase in farmers income over previous year	Upto 10% increase 10-20% increase > 20% increase	5 10 15
7.	Status of Loan taken from Financial Institutions under the Scheme	No Overdue and prompt repayment as per repayment schedule	5
8.	Employment provided	> 6 persons throughout the year 6-10 persons throughout the year >10 persons throughout the year	5 10 15

FORMAT FOR MAINTAINING RECORDS OF ADVISORY SERVICES PROVIDED BY THE AGRI-PRENEUR

Part-I

Name of the Agri-Clinic & Agri-Business Centre:

Date	Name of the Farmer	Father's / Spouse Name	Village & Panchayat	Contact No.	Category (MF/SF/ others)	Advise Provided	Amount charged, if any (₹)	Signature / thumb impression of farmer if visited in person

Part-II

Format for Reporting Advisory Services provided by the Agri-preneur on Monthly basis

Name of the Agri-Clinic & Agri-Business Centre:

Period of Report:

Sl. No.	Farmer's Name, Address & contact numbers	Category (MF/SF/ others)	No. of Times interacted during the month	Details of Advisories provided	Amount charged (₹)

NABARD

Statement on Release of Subsidy under ACABC Scheme during the month ————

(i) Consolidated Statement:

Sl No.	State	No. of Projects (₹ lakhs)	Total TFO Sanctioned Sanctioned Amount (₹ lakhs)	Total Bank Loan Money (₹ lakhs)	Total Margin Amount (₹ lakhs)	Total Subsidy (₹ lakhs)

Appendix – III

Institutional Support for Entrepreneurship Development

Government Schemes and Incentives for Promotion of Eentrepreneurship

Micro and Small Enterprises Cluster Development Programme (MSE-CDP) - DC(MSME) launched MSE-CDP for holistic development of selected MSEs clusters through value chain and supply chain management on co-operative basis.

Scheme for Capacity Building - Scheme for capacity building, strengthening of database and advocacy by industry/enterprise associations, as envisaged in the promotional package for Micro and Small Enterprises (MSEs).

Credit Linked Capital Subsidy Scheme for Technology Upgradation - The Scheme was launched in October, 2000 and revised w.e.f. 29.09.2005. The revised scheme aims at facilitating Technology Upgradation of Micro and Small Enterprises by providing 15 per cent capital subsidy (12 per cent prior to 2005) on institutional finance availed by them for induction of well established and improved technology in approved sub-sectors/products. The admissible capital subsidy under the revised scheme is calculated with reference to purchase price of Plant and Machinery. Maximum limit of eligible loan for calculation of subsidy under the revised scheme is also been raised Rs. 40 lakhs to Rs. 100 lakh w.e.f. 29-09.2005.

Credit Guarantee Scheme - Collateral free loans upto a limit of Rs.50 lakhs - *for individual MSEs.*

ISO 9000/ISO 14001 Certification Reimbursement Scheme - Incentive Scheme of Reimbursement of expenses for acquiring Quality Management System (QMS)

ISO 9000 certification/environment management (EMS) ISO 14001 certification to the extent of 75 per cent or Rs.75,000/- whichever is lower.

MSME MDA - The scheme offers funding upto 75 per cent in respect of to and fro air fare for participation by MSME Entrepreneurs in overseas fairs/ trade delegations. The scheme also provide for funding for producing publicity material (upto 25 per cent of costs) Sector specific studies (upto Rs. 2 lakhs) and for contesting anti-dumping cases (50 per cent upto Rs. 1 lakh) - *for individual MSMEs and Associations.*

Scheme of National Award - The Micro, Small and Medium Enterprises (MSMEs) in India have seen a vast development in the last five decades. The MSMEs have registered tremendous growth as also progress in terms of quality production, exports, innovation, product development and import substitution, very much beyond the expected objectives of setting up MSMEs by the planners of industrial production base in the country.

Scheme of Micro Finance Programme - Creating self employment opportunities is one way of attacking poverty and solving the problems of unemployment. There are over 24 crore people below the poverty line in the country.

Scheme to Support 5 Selected University/Colleges to Run 1200 Entrepreneurship Clubs per Annum - A package for the promotion of Micro and Small Enterprises, based on the Circular No.2(6)/2006-MSME Policy dated the 7th November, 2006 has been approved by the Cabinet Committee for Economic Affairs(C.C.E.A.).

Ministry of Micro, Small and Medium Enterprises Nirman Bhavan, New Delhi-110 108 www.dcmsme.gov.in (MINISTRY OF SMALL SCALE INDUSTRIES)

Principal organisations of the Ministry are:

i) Small Industries Development Organisation (SIDO); and
ii) National Small Industries Corporation (NSIC)
iii) Three national-level Entrepreneurship Development Institutes at Hyderabad, Noida and Guwahati.
iv) National Commission for Enterprises for Unorganised Sector (NCEUS) headquarters in New Delhi.

Small Industries Development Organisation (SIDO): established in 1954

Major Activities

☆ Advising the Government on policy formulation for promotion and development and enhancement of competitiveness of small-scale industries.

☆ Facilitating credit flow to micro and small enterprises through support measures like the Credit Guarantee Fund Scheme.

| Classification | Investment Ceiling for Plant, Machinery or Equipments*@ | |
	Manufacturing Enterprises	Service Enterprises
Micro	Upto Rs.25 lakh ($50 thousand)	Upto Rs.10 lakh ($20 thousand)
Small	Above Rs.25 lakh ($50 thousand) & upto Rs.5 crore ($1 million)	Above Rs.10 lakh ($20 thousand) & upto Rs.2 crore ($0.40 million)

Quick Estimates of 4ᵗʰ Census (2006-07)	
• Number of MSMEs	26.1 million
• Number of Manufacturing Enterprises	7.3 million
• Number of Service Enterprises	18.8 million
• Number of Women Enterprises	2.1 million (8%)
• Number of Rural Enterprises	14.2 million (54.4%)
• Employment	59.7 million
• Per unit employment	6.24
• Per unit fixed investment	Rs.33.78 lakh
• Per unit original value of Plant & Machinery	Rs.9.66 lakh
• Per unit gross output	Rs.46.13 lakh
• Employment per one lakh fixed investment	0.19

☆ Providing techno-economic and managerial consultancy, common facility and extension services to small-scale units.

☆ Providing support measures for technology upgradation, (Credit Linked Capital Subsidy Scheme), modernisation of management, product/process quality improvement (including product testing and instrument/equipment calibration) and creation and upgradation of infrastructure facilities through individual schemes/programmes as well as to clusters under Small Industries Cluster Development Programme.

☆ Human resource development through training and skill upgradation.

☆ Providing economic information, including market analysis, services.

☆ Market (including export) promotion measures and facilitating linkage of small-scale industries as ancillaries to large and medium scale industries.

☆ Conducting periodical census/survey of the SSI and generating data/reports on various important parameters/indicators of growth and development of the sector.

☆ Maintaining close liaison with other Central Ministries, Planning Commission, State Governments, Financial Institutions and other organisations concerned with the development of small-scale industries.

☆ Maintaining a comprehensive web site for wide dissemination of all relevant information relating to the activities of SIDO.

National Small Industries Corporation Limited (NSIC) established in 1955

Major Activities

Promoting, aiding, and fostering the growth of small scale industries and industry related small scale services/business enterprises in the country, on a commercial footing.

The Corporation expanded its activities in the areas of Marketing, Technology, Financing and Information Support to small enterprises in the country.

The emerging competitive business environment and dismantling of the market barriers provide an added impetus for enlarging the activities of NSIC. In line with the current economic changes, the Corporation has introduced many new Schemes from time to time for the development of small enterprises.

The Corporation is now an ISO: 9001-2000 company, providing services of marketing assistance, technology upgradation and entrepreneurship development.

The scheme of assisting these enterprises in obtaining performance – cum – credit rating is the latest initiative of the Government being implemented by the NSIC with the assistance of noted rating agencies operating in the country.

National Institute of Small Industry Extension Training (NISIET), Hyderabad (1960)

Major Activities

To assist in promotion, development, and modernisation of small and medium enterprises (SMEs) in the country.

NISIET is involved in entrepreneurship development related training programmes, training research and consultancy, including the methodology of cluster development.

The National Institute for Entrepreneurship and small Business Development (NIESBUD)

Established in 1983 by the Ministry of Industry (now Ministry of Small Scale Industries). An apex body for coordinating and overseeing the activities of various institutions/agencies engaged in Entrepreneurship Development Particularly in the area of small industry and small business. The Institute which is registered as a society under Govt. of India Societies Act (XXI of 1860) started functioning from 6th july, 1983.

The major activities of the Institute, include,

1. Development of model syllabi for training of various target groups,
2. Designing effective training strategies, methodology, manuals and tools, facilitating and
3. supporting Central/State Governments and other agencies in executing programmes of entrepreneurship and small business development, *etc.*

Objectives

1. To evolve standardised materials and processes for selection, training, support and sustenance of entrepreneurs, potential and existing.
2. To help/support and affiliate institutions/organisations in carrying out training and other entrepreneurship development related activities.
3. To serve as an apex national level resource institute for accelerating the process of entrepreneurship development ensuring its impact across the country and among all strata of the society.
4. To provide vital information and support to trainers,promoters and entrepreneurs by organising research and documentation relevant to entrepreneurship development
5. To train trainers, promoters and consultants in various areas of entrepreneurship development.
6. To provide national/international forums for interaction and exchange of experiences helpful for policy formulation and modification at various levels.
7. To offer consultancy nationally/internationally for promotion of entrepreneurship and small business development.
8. To share internationally experience and expertise in entrepreneurship development.
9. To share experience and expertise in entrepreneurship development across National frontiers

The Initiatives of the Institute include:

☆ Evolving effective training strategies and methodology
☆ Standardising model syllabi for training various target groups
☆ Formulating scientific selection procedure
☆ Developing training aids, manuals and tools
☆ Facilitating and supporting Central/State/Other agencies in organising entrepreneurship development programmes
☆ Conducting training programmes for promoters, trainers and enterpreneurs.

Accreditation Programme for Entrepreneurial Motivation Trainers:

☆ Trainers' Training Programme for Enterprise Launching and Management

☆ Trainers/Promoters Programme for support organisations such as SISIs, DICs, Development Corporations *etc.*

☆ Small Business Promotion Programme

☆ Entrepreneurship Orientation Programme for HODs and Senior Executives.

Indian Institute of Entrepreneurship (IIE), Guwahati (1993)

Aim: Undertaking training, research and consultancy activities in the small industry sector focusing on entrepreneurship development as an autonomous national institute.

IIE is working towards strengthening the capacity in the field of entrepreneurship development, training, entrepreneurship education, research, consultancy, publication and sensitisation of environment for promotion of entrepreneurship, enterprise creation and self-employment in the North Eastern Region.

IIE, Guwahati has also taken initiatives for providing hand-holding services to the entrepreneurs in the North Eastern Region for which a Business Facilitation and Development Centre (BFDC) has been set up with financial assistance from the Ministry.

National Commission for Enterprises in the Unorganised Sector has been set up by the Government as an advisory body and a watchdog for the informal sector with a view to fulfilling the commitment in the National Common Minimum Programme of the Government.

The Commission will recommend measures considered necessary for bringing about improvement in the productivity of these enterprises, generation of large scale employment opportunities on a sustainable basis, particularly in the rural areas, enhancing the competitiveness of the sector in the emerging global environment, linkage of the sector with institutional framework in areas such as credit, raw material, infrastructure, technology upgradation, marketing and formulation of suitable arrangements for skill development

The Khadi and Village Industries Commission (KVIC)

The Khadi and Village Industries Commission (KVIC) is a statutory body established by an Act of Parliament (No. 61 of 1956, as amended by act no. 12 of 1987 and Act No.10 of 2006. In April 1957, it took over the work of former All India Khadi and Village Industries Board.

The world is moving towards natural and Eco-friendly Products. More and more people are becoming conscious of responsibility towards the environment and ecology and products like plastic, *etc.* are highly looked down upon and greater demand is being generated for bio-degradable and Eco-friendly products. The KVI Sector produces a Vide range of exclusive products which are having distinct characteristics. Some of the chief KVI products are as under.

PRODUCTION

Year	(Rs. In Crores)		
	Khadi	Village Industries	Total
2000-2001	431.57	6491.69	6923.26
2001-2002	411.00	7140.52	7551.52
2002-2003	443.07	8126.30	8569.37
2003-2004	453.50	9228.27	9681.77
2004-2005	461.54	10458.89	10920.43
2005-2006	468.30	11915.54	12383.84
2006-2007	491.52	13537.19	14028.71
2007-2008	543.39	16134.32	16677.71
2008-2009	585.25	16753.62	17338.87

SALES

Year	(Rs. In Crores)		
	Khadi	Village Industries	Total
2000-2001	570.55	7384.55	7955.10
2001-2002	518.25	8383.49	8901.74
2002-2003	577.63	9615.71	10193.34
2003-2004	587.04	10988.17	11575.21
2004-2005	617.84	12487.35	13105.19
2005-2006	628.69	14647.33	15276.02
2006-2007	663.19	16899.21	17562.40
2007-2008	724.39	20819.09	21543.48
2008-2009	799.60	21948.59	22748.19

EMPLOYMENT

Year	(Rs. In Crores)		
	Khadi	Village Industries	Total
2000-2001	9.56	50.51	60.07
2001-2002	8.48	54.16	62.64
2002-2003	8.58	57.87	66.45
2003-2004	8.61	62.58	71.19
2004-2005	8.64	68.14	76.78
2005-2006	8.68	74.09	82.77
2006-2007	8.84	80.08	88.92
2007-2008	9.16	90.11	99.27
2008-2009	9.50	94.41	103.91

RAJIV GANDHI UDYAMI MITRA YOJANA (A Scheme of "Promotion and Handholding of Micro and Small Enterprises")

Objective

The objectives of Rajiv Gandhi Udyami Mitra Yojana (RGUMY) are:

☆ To provide handholding support and assistance to the potential first generation entrepreneurs, who have already successfully completed or undergoing Entrepreneurship Development Training Programme (EDP)/Skill Development Training Programme (SDP)/Entrepreneurship cum Skill Development Training Programme (ESDP)/Vocation Training Programmes (VT), through the selected lead agencies *i.e.* 'Udyami Mitras', in the establishment and management of the new enterprise, in dealing with various procedural and legal hurdles and in completion of various formalities required for setting up and running of the enterprise.

To provide information, support, guidance and assistance to first generation entrepreneurs as well as other existing entrepreneurs through an 'Udyami Helpline' (a Call Centre for MSMEs), to guide them regarding various promotional schemes of the Government, procedural formalities required for setting up and running of the enterprise and help them in accessing Bank credit *etc.*

Role and Responsibilities of Udyami Mitras

The selected lead agencies *i.e.* Udyami Mitras would be expected to render assistance and handholding support for following services:

(i) Networking, coordinating and follow up with various Government departments/agencies/organizations and regulatory agencies on the one hand and with support agencies like Banks/financial institutions, District Industries Centers (DICs), technology providers, infrastructure providers on the other hand, to help the first generation entrepreneurs in setting up their enterprise. Udyami Mitras are expected to help the first generation entrepreneurs in:

   a) Identification of suitable project/product/enterprise and preparation of bankable project report for the same;

   b) Creation of the proprietorship firm/partnership firm/Company/Society/Self Help Group (SHG) *etc.*

   c) Filing of Memorandum (as prescribed under MSMED Act 2006);

   d) Accessing bank loans, admissible capital subsidy/assistance under various schemes of the Central/State Government and other agencies/organizations/financial institutions/Banks *etc.* by networking with respective agencies

   e) Assistance and support in establishment of work shed/office;

   f) Sanction of Power load/connection;

   g) Selection of appropriate technology and installation of plant and machinery/office equipment *etc.*

h) obtaining various registrations/licenses/clearances/No Objection Certificates (NOCs) *etc.* from the concerned regulatory agencies/ Government departments/local bodies/Municipal authorities *etc.*

i) Allotment of Income Tax Permanent Account Number (PAN) and Service Tax/Sales Tax/VAT registration *etc*;

j) Sanction of working capital loan from the banks;

k) Arranging tie up with raw material suppliers;

l) Preparation and implementation of marketing strategy for the product/service and market development; and

m) Establishing linkage with a mentor for providing guidance in future

n) Creation of web page and email identity;

(ii) Once the enterprise has been successfully set up, the Udyami Mitras would also monitor and follow up on the functioning of the enterprise for a further period of minimum 6 months and provide help in overcoming various managerial, financial and operational problems.

The National Science and Technology Entrepreneurship Development Board (NSTEDB)

☆ The National Science and Technology Entrepreneurship Development Board (NSTEDB), established in 1982 by the Government of India under the aegis of Department of Science and Technology.

☆ It is an institutional mechanism to help promote knowledge driven and technology intensive enterprises. The Board, having representations from socio-economic and scientific Ministries/Departments.

☆ Aims to convert "job-seekers" into "job-generators" through Science and Technology (S and T) interventions.

Objectives

☆ To promote and develop high-end entrepreneurship for S and T manpower as well as self-employment by utilising S and T infrastructure and by using S and T methods.

☆ To facilitate and conduct various informational services relating to promotion of entrepreneurship.

☆ To network agencies of the support system, academic institutions and Research and Development (R and D) organisations to foster entrepreneurship and self-employing using S and T with special focus on backward areas as well.

☆ To act as a policy advisory body with regard to entrepreneurship.

The Innovation and Entrepreneurship Development Centre (IEDC)

☆ The Innovation and Entrepreneurship Development Centre (IEDC) is being promoted in educational institutions to develop institutional mechanism to create entrepreneurial culture in S and T academic institutions and to foster techno-entrepreneurship for generation of wealth and employment by S and T persons.

☆ The IEDCs are established in academic institutions (science colleges, engineering colleges, universities, management institutes) having requisite expertise and infrastructure.

☆ The mission of the IEDCs is to "develop institutional mechanism to create entrepreneurial culture in academic institutions to foster growth of innovation and entrepreneurship amongst the faculty and students ".

Functions

☆ To organise Entrepreneurship Awareness Camps, Entrepreneurship Development Programmes, Faculty Development Programmes and Skill Development Programmes in the college/institution for the benefit of S and T persons.

☆ To initiate five innovative student projects each year for new innovative product development.

☆ To organize Business Plan Competitions every year.

☆ To guide and assist prospective entrepreneurs on various aspects such as preparing project reports, obtaining project approvals, loans and facilities from agencies of support system, information on technologies, *etc.*

☆ To arrange interaction with entrepreneurs and create a mentorship scheme for student entrepreneurs.

☆ To facilitate creation of entrepreneur's club in each college to foster culture of entrepreneurship amongst students

☆ To act as a Regional Information Centre on business opportunities, processes, technologies, market, *etc.* by creating and maintaining relevant data bases.

☆ IEDCs would also sensitise the management of the institutions regarding the importance of entrepreneurship and integrate their activities with the Host Institutions.

Science and Technology Entrepreneurship Development Scheme (STEDS)

☆ Science and Technology Entrepreneurship Development Scheme (STEDS) was launched in 1985 when NSTEDB prepared a scheme to map the available material resources in industrially backward regions and prepare a basket of technically feasible and economically viable project profiles of enterprises that could be promoted by local S and T entrepreneurs and thus tapping hitherto un-utilised/under-utilised resources.

☆ Initially the scheme was launched in 13 districts in consultation with the Planning Commission, and later replicated across the country

Objectives

☆ To identify possible resource based projects in the region covering both rural and urban areas.

☆ To identify S and T intervention (like technology selection, modification, alteration and dissemination) for exploiting the opportunities by prospective entrepreneurs.

☆ To improve working of existing enterprises through S and T intervention to upgrade the technology and modernisation of units.

☆ To organise regular enterprise awareness programmes and skill development programmes for creating suitable entrepreneurial environment in the district.

☆ To launch at least 200 micro-enterprises in the district during the four years of duration in which at least 50 technology-driven micro-enterprises are to be in technology specific areas as identified by the implementing agency earlier.

The Science and Technology Entrepreneurs Park (STEP)

☆ The Science Parks and similar initiatives help in creating an atmosphere for innovation and entrepreneurship; for active interaction between academic institutions and industries for sharing ideas, knowledge, experience and facilities for the development of new technologies and their rapid transfer to the end user.

☆ The Science and Technology Entrepreneurs Park (STEP) programme was initiated to provide a re-orientation in the approach to innovation and entrepreneurship involving education, training, research, finance, management and the government.

Objectives

☆ To forge a close linkage between universities, academic and R and D institutions on one hand and industry on the other.

☆ To promote entrepreneurship among Science and Technology persons, many of whom were otherwise seeking jobs soon after their graduation.

☆ To provide R and D support to the small-scale industry mostly through interaction with research institutions.

☆ To promote innovation based enterprises.

STEP Profile

☆ The department has so far catalysed 15 STEPs in different parts of the country, which have promoted nearly 788 units generating annual turnover of around Rs. 130 crores and employment for 5000 persons.

☆ More than 100 new products and technologies have been developed by the STEPs/STEP promoted entrepreneurs.

☆ In addition, over 11000 persons have been trained through various skill development programmes conducted by STEPs.

☆ STEPs are autonomous bodies registered as societies under the Societies Registration Act.

Appendix - IV

Practical Exercises on Entrepreneurship Development

Exercise – 1

Johari Window

Objectives

1. To understand the group dynamics.
2. To internalize the process of modifying the behavior of individual and group.
3. To understand the factors influencing in managing self and a group of people.

The Johari Window model was devised by American psychologists Joseph Luft and Harry Ingham in 1955, while researching group dynamics at the University of California Los Angeles. Luft and Ingham called their Johari Window model 'Johari' after combining their first names, Joe and Harry.

The Johari Window is a model for understanding and training self-awareness, personal development, improving communications, interpersonal relationships, group dynamics, team development and inter-group relationships.

The Johari Window model is a simple and useful tool for illustrating and improving self-awareness, and mutual understanding between individuals within a group. The Johari Window model can also be used to assess and improve a group's relationship with other groups.

The Johari Window model is also referred to as a 'disclosure/feedback model of self awareness', and by some people an 'information processing tool'. The Johari Window actually represents information - feelings, experience, views, attitudes, skills, intentions, motivation, *etc.* - within or about a person - in relation to their group, from four perspectives.

The Johari Window model can also be used to represent the same information for a group in relation to other groups. Johari Window terminology refers to 'self' and 'others': 'self' means oneself, ie, the person subject to the Johari Window analysis. 'Others' means other people in the person's group or team.

The four Johari Window perspectives are called 'regions' or 'areas' or 'quadrants'. Each of these regions contains and represents the information - feelings, motivation, *etc* - known about the person, in terms of whether the information is known or unknown by the person, and whether the information is known or unknown by others in the group.

Johari Window Four Regions

1. What is known by the person about him/herself and is also known by others - open area, open self, free area, free self, or 'the arena'
2. What is unknown by the person about him/herself but which others know - blind area, blind self, or 'blind spot'
3. What the person knows about him/herself that others do not know - hidden area, hidden self, avoided area, avoided self or 'facade'
4. What is unknown by the person about him/herself and is also unknown by others - unknown area or unknown self

Exercise – 2

The Entrepreneurial Orientation Inventory

Objectives

1. To understand the concept of entrepreneurial orientation.
2. Learning by undergoing the experiment and analyse the self in entrepreneurial situation.

Instructions

This inventory contains twenty pairs of statements. In each pair, you may agree with one statement more than the other. You have five points to distribute between the two statements in each pair, to indicate the extent to which you agree with each of the statements. You may distribute the five points hi any combination (0-5, 1-4, 2-3, 332, 4-1, 5-01. If you agree slightly more with statement "a" than with "b", then assign three points to ~a" and two points to "b". If you agree completely with "a" but do not agree at all with "b", assign five points to "a" and zero to "b".

You may not divide your points equally (ex: 2.5) between the two choices. You must choose one statement with which you agree more and then distribute the points.

1. a. How successful an entrepreneur one will be depends on a number of factors? One's capabilities may have very little to do with one's success.
   b. A capable entrepreneur can always shape his or her own destiny.
2. a. Entrepreneurs are born, not made.
   b. It is possible for people to learn to become more enterprising even if they do not start out that way.
3. a. Whether or not a salesperson will be able to sell his or her product depends on how effective the competitors are.
   b. No matter how good the competitors are, an effective salesperson always will be able to sell his or her product.
4. a. Capable entrepreneurs believe in planning, their activities in advance.
   b. There is no need for advance planning, because no matter how enterprising one is, there always will be chance factors that influence success.
5. a. Whether or not a person can become a successful entrepreneur depends on social and economic conditions.
   b. Real entrepreneurs can always be successful, irrespective of social and economic conditions.
6. a. Entrepreneurs fail because of their own lack of ability and perceptiveness.
   b. Entrepreneurs are bound to fail at least half the time because success or failure depends on a number of factors beyond their control.
7. a. Entrepreneurs are often victims of forces they can neither understand nor control.
   b. By taking an active part in economic, social and political affairs, entrepreneurs can control events that affect their business.
8. a. Whether or not you get a business loan depends on how fair the bank officer you deal with is.
   b. Whether or not you get a business loan depends on how good your project plan is.
9. a. When purchasing raw materials or any' other goods, it is wise to collect as much information as possible from various sources and then to make a final. Choice
   b. There is no point in collecting a lot of information; in the long run, the more you pay, the better the product is.
10. a. Whether or not you make profit in business depends on how lucky you are.

b. Whether or not you make a profit in business depend on how capable you are as an entrepreneur.

11. a. Some types of people can never be successful as entrepreneurs.

    b. It is possible to develop entrepreneurial ability in different types of people.

12. a. Whether or not you will be a successful entrepreneur depends on the social environment into which you were born.

    b. People can become successful entrepreneurs with effort and capability irrespective of the social strata from which they originated.

13. a. These days, people must depend at every point on the help, support, or mercy of others (governmental agencies, bureaucracies, banks, *etc.*)

    b. It is possible to generate one's own income without depending too much on the bureaucracy. What is required is a knack in dealing with people.

14. a. The market situation today is very unpredictable. Even perceptive entrepreneurs falter quite often.

    b. When an entrepreneur's prediction of the market situation is wrong, that person can blame 'only himself or herself for failing to read the market correctly.

15. a. With effort, people can determine their own destinies.

    b. There is 10 point in spending time planning or doing things to change one's destiny. What is going to happen will happen.

16. a. There are many events beyond the control of entrepreneurs.

    b. Entrepreneurs are the creators of their own experiences.

17. a. No matter how hard a person works, he or she will achieve only what is destined.

    b. The rewards one achieves d13pend solely on the effort one makes.

18. a. Organizational effectiveness can be achieved by employing competent' and effective people.

    b. No matter how competent the employees in a company are, if socioeconomic conditions are not good, the organization will have problems.

19. a. Leaving things to chance and letting time take care of them helps a person to relax and enjoy life.

    b. Working for things always turns out better than leaving things to chance.

20. a. The work of competent people always will be recognized.

    b. No matter how competent one is, it is almost impossible to get ahead in life without contacts.

The Entrepreneurial Orientation Inventory Scoring Sheet

Name...Date.....................................

Instructions

Transfer your point allocations from the inventory form onto this Scoring Sheet.

Internal Locus of Control	*External Locus of Control*
1. b	1. a..............
2. b..............	2. a..............
3. b..............	3. a..............
4. a	4. b..............
5. b..............	5. a..............
6. a..............	6. b..............
7. b..............	7. a..............
8. b..............	8. a..............
9. a..............	9. b..............
10. b..............	10. a..............
11. b..............	11. a..............
12. b..............	12. a..............
13. b..............	13. a..............
14. b..............	14. a..............
15. a..............	15. b..............
16. b..............	16. a..............
17. b..............	17. a..............
18. a..............	18. b..............
19. b..............	19. a..............
20. a..............	20. b..............
Total Internal..............	**Total Internal..............**

Determine the ratios of your internal/external locus of control scores by dividing the total internal score by the total external score. Record the amount here.......................................

Internal/external ratios above 3.0 indicate a high level of entrepreneurial internality; the chances are high that such individuals will initiate entrepreneurial activities. Ratios.below 1.0 indicates that the respondent has a more external. (less entrepreneurial) locus-of-control orientation. There is a need for this type of person to become more internal in order to be able to initiate and sustain entrepreneurial activities. Ratios above 1.b indicate possible entrepreneurs. The higher the ratio above 1.0, the more internal the respondent is.

*Test developed by T.V. Rao. Reproduced from the 1985 Annual, "Development Human Resources'. Leonard D Goodstein and J. William Pfeiffer Edition, San Diego, California University Associates, 1985.

Exercise – 3

Tower Building

Objectives

1. To understand the goal setting behavior in task performance,
2. To internalize the process of helping behavior and its consequences on performance,
3. To understand value orientation and its relation to entrepreneurial performance, and
4. To understand the factors in decision making.

Conditions

1. **The subject will be blind folded and will perform the task with his wrong hand.**
2. **The associate members will not be allowed to touch the subject or the wooden blocks. However they may do anything of their choice.**

Observations to be Recorded

Sl.No.	Observations	Remarks
I.	**Decision making**	
1.	Amount of time spent for decision making (in min)	
2.	Who tries maximum to influence	
3.	How is group taking decision	
4.	Agreement/disagreement during decision making process	
II.	**Help/encouragement**	
III.	**Confidence of direction**	
1.	What sort of guidance by whom	
2.	Who gives maxim mum guidance	
3.	What is the emotional reaction of subject receiving guidance	
4.	Level of anxiety in giving guidance	
IV.	**Confidence of worker**	
V.	**Planning and searching environment**	

Goal Setting

Sl.No.	Details	Remarks
1.	How did you choose a particular estimate	
2.	Did you consider the average figure given by me to decide your estimate	
3.	How could you under/overestimate your subject	
4.	What prompts you to say that you are above average	
5.	What prompted you to rate your subjects as being above average	
6.	What may happen if we under/overestimate a person	

Data Recording

3. How could you under/overestimate your subject

Group	I	II	III	IV
I Goal setting				
Associate member - I				
Associate member – II				
Subject				
II Consensus				
III Performance (target achieved)				

(Taken from MMP Akhouri, S P Mishra and Rita Sengupta (1999) Trainers Manual on Developing Entrepreneurial Motivation. National Institute for Entrepreneurship and Samll Business Development, New Delhi)

Exercise – 4

Treasure Hunting

Objectives

1. To comprehend the real situation as an entrepreneur,
2. To realize the risk taking behavior of an entrepreneur,
3. To internalize the process of task performance and to have the first hand experiences of problems faced by entrepreneur, and
4. To identify the factors in decision making and facing the new problems.

Procedure

1. All the students move out of the class.

2. Students will be divided into a group of 4 or 5 students and they will be blind folded.

3. Some valuable items will be kept in the class room and the blindfolded team will enter the class and try to find out the hidden items in the class room (in the entire process the team members will remain blindfolded).

4. The person who could get maximum treasure will be identified as winner.

The players will note down their observations

The non-players will note their observations

Conclusions of the class on the game

Exercise – 5

Achievement Motivation (Toss the Ring)

Objectives

1. To understand the target setting behavior in task performance.
2. To internalize the process of achievement motivation.
3. To understand and analyse the difficulties faced in undergoing the process.

Methodology

1. A peg is placed and three marks of different distances made.
2. There are three levels in this game a. the distance selected is fixed b. You can vary the distance c. you will deposit certain amount before playing.
3. Observations will be recorded for an individual payer in all the three rounds.
4. There may be certain students who may not play the game:
   A. Analysis of individual player based on his/her performance in each round.
   B. Analysis of successful players.
   C. Observations of then non-players.

(Taken from MMP Akhouri, S P Mishra and Rita Sengupta (1999) Trainers Manual on Developing Entrepreneurial Motivation. National Institute for Entrepreneurship and Samll Business Development, New Delhi)

Exercise – 6

Group Discussion

Introduction

The reason why should you go through a Group discussion is to get to know you as a person and gauge how well you will fit in to perform the specified job. The Group discussion tests how you function as a part of a team. As a leader, you will always be working in teams, as a member or as a leader. Therefore how you interact in a team becomes an important criterion for your selection to perform a specific task. leaders have to work in a team and get best results out of teamwork. Some of the personality traits the GD is trying to gauge may include Ability to work in a team, Communication skills, Reasoning ability, Leadership skills, Initiative, Assertiveness, Flexibility, Creativity and Ability to think on ones feet.

Methodology: Phases of Group Discussion

   i. Initiation/Introduction

   ii. Body of the group discussion

   iii. Summarisation/Conclusion

Summarisation Technique

Most Group Discussions do not really have conclusions. A conclusion is where the whole group decides in favour or against the topic.

But every Group Discussion is summarised. You can summarise what the group has discussed in the Group Discussion in a nutshell.

Keep the following points in mind while summarising a discussion:

☆ Avoid raising new points.

☆ Avoid stating only your viewpoint.

☆ Avoid dwelling only on one aspect of the Group Discussion.

☆ Keep it brief and concise.

☆ It must incorporate all the important points that came out during the Group Discussion.

If you are asked to summarise a Group Discussion, it means the Group Discussion has come to an end.

1. **A**lways be the initiator and concluder of the Group Discussion than being a participant.

2. But if you are participant always try to be the most key participant.

3. Put points firmly and always try to get others support too.

4. If you find that the discussion is going off-track then never loose an opportunity to bring it back to stream.

5. Try to keep latest information on the topic

6. Be very polite and try to keep cool.

7. Most important don't wait for your turn to speak when discussion is on. interrupt politely.

8. Last but not the least keep a tab on the time given for discussion. When you feel that the discussion is heating but the time is going to be over wrap up the discussion.

9. During conclusion, do end with the conclusion note that shows your leadership quality.

Best Impression Points

1. Initiation of discussion,
2. Always keeping/trying to keep discussion on track.
3. Conclusion on time.
4. Your capability to keep you cool and listen as well as putting your points.

Group Discussion basically a means of searching your team player, leadership, communication capability. (Taken from http://www.yuvajobs.com/gd-tips-freshers.asp)

Exercise – 7

Entrepreneurial Behaviour Index

Objective

1. To identify the entrepreneurial behavior of students.

Procedure

1. Read the statements carefully
2. Put √ mark at appropriate block where you feel you are like this

Entrepreneurship Index

Sl.No.	Statements	I am very much like this	I am some-what like this	I am not like this	I Cant Say
1.	I would like to do something which others have not done				
2.	Its my nature to take life easily				
3.	I want to work hard to become a well-known professional in our area				
4.	I want to do something which others can hardly do				
5.	When I fail to succeed I feel upset and give up the task				
6.	I feel my success depends upon my parents and my relatives				
7.	I feel my success depends upon my hard work				

Sl.No.	Statements	I am very much like this	I am some-what like this	I am not like this	I Cant Say
8.	I like to be very systematic in my work				
9.	I like to do my best in whatever work I undertake				
10.	When working in groups I desire to excel others in similar tasks				
11.	When working in groups I desire to take lead				
12.	My aim of life is to make long record of successful achievements				
13.	I believe that it is not possible for me to get sufficient power in my hands				
14.	I am not very serious about being a great man in my profession				
15.	I do not consider myself in doing better endeavor than others				
16.	Whatever I undertake I make advance plans				
17.	In whatever work I undertake I like to assume full responsibility for it				
18.	It is my nature to undertake tasks which require great skill				
19.	I am always keen to develop my knowledge and skill in farming				
20.	I feel best when I undertake difficult job				
21.	I aspire to be the centre of others attention				
22.	I like to do something which others regard as a manifestation of leadership				
23.	It gives me great satisfaction to supervise and direct others				
24.	I always try to perform my activities in my own way				
25.	I have a general tendency to continue a work till it is finished				

Sl.No.	Statements	I am very much like this	I am some- what like this	I am not like this	I Cant Say
26.	I have a general tendency to analyze others judgment critically				
27.	Before starting a difficult task do not plan its details so as to lead to success				
28.	I am sure that after a few years I will be earning a lot				
29.	I am sure that after a few years I will be a recognized authority in my profession				
30.	I am determined to work towards high goal				
31.	I enjoy long spell of continuous activity to solve difficult problem				
32.	I like those situations which are not competitive				

Scoring Procedure

Score of 3, 2, 1, 0 to be given Statements

The scoring Procedure is reversed for negative statements: Statements Number 2, 6, 13, 14, 15, 27, 32

Calculations

Total Score, Mean and Standard Deviation.

Entrepreneurship Index

Mean + SD = High

Mean ± SD = Medium

Mean – SD = Low

Interpretation of the scores obtained:

Exercise – 8

Understanding Entrepreneurship – Case Method

A real case of successful or unsuccessful entrepreneur will be circulated among the students a week before the practical session. Students will be asked to study the case and come to the class with observations. Instructor will begin the discussion with a brief introduction on the concept of entrepreneurship and will open the topic of the case for discussion.

Objectives

1. To understand Case study method of understanding Entrepreneurship.
2. To analyse the case topic wise and interpret the results.
3. To encourage participation of students in the analysis of the case.

Methodology

Among the total students in the class a group of two students will asked to act as rapporteaur for the activity. The course instructor will moderate the discussion. All the key elements emerged as a part of discussion will be noted by the rapporteaur.

Timings

1. Introduction to the case and concept: 15 minutes
2. Nomination of different role players: 05 minutes
3. Discussion on the case: 80 minutes
4. Summarisation of the discussion and concluding remarks on the case

Students will note the conclusion of the class and will note their observations in the practical record

Exercise – 9

Brain Storming

Introduction

Brainstorming is a procedure to generate a great number of thoughts for the solution to a problem. The method was first popularized in the late 1930s by Alex Osborn. The experiences reveal that individuals can enhance their creative output by exploiting the method of brainstorming.

Every individual suffers from a syndrome known as glossophobia (fear of public speaking). Hence for a dynamic brain storming and eliciting creativity of the group the participants should relax and have free mind while participating in brain storming. There should be free flow of ideas hence as thumb rule the focus should be on quantity. Because it is a means of enhancing divergent production, aiming to facilitate free flow of ideas through the maxim, quantity breeds quality. The supposition is that the greater the number of ideas generated, the greater the chance of producing a radical and effective solution. In the process fragmented ideas and just keywords are also welcome. It is the natural humans' tendency to embrace the first idea which comes to mind. If upon analyzing the problem further, we discover our solution is not sufficient to solve the problem, we try to modify that first solution rather than consider alternative solutions. However, the first solution to a problem is rarely the most creative. The first solution is usually the most commonplace one.

It is often emphasized that in group brainstorming, criticism should be put 'on hold'. Instead of immediately stating what might be wrong with an idea, the participants focus on extending or adding to it, reserving criticism for a later 'critical stage' of the process.

By suspending judgment, one creates a supportive atmosphere where participants feel free to generate unusual ideas.

Objectives

1. To analyze the communication ability of an individual.
2. To examine the innovative ideas the individual has and how he/she can impress upon the group.
3. To encourage participation of students in the participation in the process of brain storming.

Methodology

1. There will be three rounds in this activity.
2. In the initial round the group will be assigned with a topic, identification of moderator and an observer.
3. In the second round the group will begin with free flow of ideas on the given topic, observer will maintain the record of observations.
4. In the third round only the pertinent issues raised in the previous session will be culled out and focused discussion on the emerged issues. The observer will keep record of the discussions.
5. At the end the group will come to a conclusion and that will be opinion of the group:
   A. Analysis of individual participants based on his/her performance in each round.
   B. Analysis of successful participants.
   C. Observations of the group as a whole.

Exercise – 10

Broken Square

Objectives

1. To demonstrate the group dynamics in accomplishing task that require collaboration from different individuals.
2. To help participants to recognize their own tendencies to collaborate and compete.
3. To help participants develop their sensitivity to the needs of the group, make some sacrifice for the group effectiveness and joint problem solving.

Methodology

1. There will be five members in each group and one observer.
2. Each member will be provided with a cover containing four pieces for joining them as a square.
3. No member may speak during the process.
4. Time required approximately 45 minutes.
5. No member my ask another member for a piece or in any signal another person to give him/her a piece.

Experiences of each member:

Experiences of group as a whole:

(Taken from MMP Akhouri, S P Mishra and Rita Sengupta (1999) Trainers Manual on Developing Entrepreneurial Motivation. National Institute for Entrepreneurship and Samll Business Development, New Delhi).

Index

W

Wadhwani centre for entrepreneurship
development (WCED) 134

Weakness 249

White button mushroom 311

Women 309

Women entrepreneurs 89

World trade organisation 327

World trade organization (WTO)
standards 249

Y

Yogopluralist model 41

www.ingramcontent.com/pod-product-compliance
Lightning Source LLC
Chambersburg PA
CBHW050520190326
41458CB00005B/1605